T0296067

CAMBRIDGE AERONAUTICAL SERIES

General Editors

ERNEST F. RELF, c.b.e., f.r.s.
PROFESSOR W. A. MAIR

I

THE PRINCIPLES OF THE CONTROL AND STABILITY OF AIRCRAFT

THE PRINCIPLES OF THE
CONTROL AND STABILITY
OF AIRCRAFT

BY

W. J. DUNCAN, C.B.E. D.Sc., F.R.S.

*Mechan Professor of Aeronautics and Fluid Mechanics in the
University of Glasgow, Fellow of University College London
and formerly Professor of Aerodynamics in the
College of Aeronautics, Cranfield*

CAMBRIDGE
AT THE UNIVERSITY PRESS
1959

CAMBRIDGE
UNIVERSITY PRESS

University Printing House, Cambridge CB2 8BS, United Kingdom

Cambridge University Press is part of the University of Cambridge.

It furthers the University's mission by disseminating knowledge in the pursuit of education, learning and research at the highest international levels of excellence.

www.cambridge.org
Information on this title: www.cambridge.org/9781316509746

© Cambridge University Press 1959

First edition 1952
Reprinted 1959
First paperback edition 2015

A catalogue record for this publication is available from the British Library

ISBN 978-1-316-50974-6 Paperback

FOREWORD

The present volume is the first of a series of books on aeronautical subjects. The idea of producing such a series first originated in the College of Aeronautics at Cranfield, where the teaching is at post-graduate level and every endeavour is being made to take students up to the frontiers of existing knowledge in the various branches of aeronautics. This endeavour has of necessity led to the collection and systematization, by the staff, of a good deal of recent research, the results of which have not yet found their way into text-books, as well as to the recasting of older material to form, with the new, a comprehensive basis for the teaching. It was felt from the outset that there would emerge a wealth of material very suitable for conversion into text-book form and that the writing of such books would supply a much-felt need because of the rarity of modern works on the subjects. The work of establishing the College, which only began in 1946, has been very arduous, and not until now has it been possible to give serious consideration to the project, but it is hoped that in the future volumes written by members of the College staff will appear at not too infrequent intervals, and that authors elsewhere will be encouraged to contribute to the series. The resulting books should play an important part in augmenting and extending the range of available aeronautical literature.

This first volume of the series should fill a particularly large gap in existing writings. Its subject has usually been treated in one or two chapters of books on aerodynamics, but recent developments, and especially the effects of the compressibility of air and the deformation of aircraft structures, have introduced a complexity which demands much fuller treatment. The present book by such a well-known authority should do much to help those who wish to study modern concepts of stability and control, a subject which presents some of the most important problems in the design and operation of aircraft.

ERNEST F. RELF
(General Editor)

PREFACE

The stability and the control of aircraft are among the most important and intricate subjects in the realm of aerodynamics and they are closely related. Hence by logic and also by custom the two are treated together and often regarded as one. Accordingly this book is devoted to giving a systematic account of the stability and control of aircraft but the subject matter could be regarded as the broader one of the dynamics of aircraft, with, however, certain topics excluded. These include flutter, the dynamics of the propulsive machinery, and 'performance' which is the detailed study of the steady motion of aircraft with particular reference to speed, drag, power, rate of climb, altitude, range and endurance. It is now fully recognized that the frontier between aircraft stability and flutter is somewhat artificial, but, on account of the extent and complexity of the theory of flutter, it would be highly inconvenient to include it in a treatise on control and stability.

For several reasons it seemed opportune to produce a treatise on stability and control. In the first place, the text-book literature of the subject is extremely meagre. Thus, since the appearance in 1911 of G. H. Bryan's pioneer treatise *Stability in Aviation*, no book devoted solely to stability and control has been published, although Melvill Jones's contribution to Durand's *Aerodynamic Theory* (1935) is indeed a notable treatise on the subject, and Bairstow's *Applied Aerodynamics* (1920 and 1939) contains much valuable material. The only post-war account in a text-book is contained in *Airplane Performance, Stability and Control* by Perkins and Hage (1949). In the second place, the advances made, especially during the course of the World War of 1939–45, have transformed many aspects of stability and control and greatly extended the field of investigation, so that a new systematic exposition seems to be called for. Lastly, there is a need for the theory of stability and control to be set forth by someone familiar with the kindred theory of flutter—a qualification the writer happens to possess.

The constant aim of the writer has been to give a lucid and logical account of the principles of the subject and he has attempted to avoid the extremes of over-simplification and over-elaboration. The subject of the book is so vast that a really comprehensive, thorough, and detailed treatment would require many volumes and many authors. As one instance, the full discussion of the aerodynamic derivatives which appear in the theory would need the co-operation of several experts. It is opportune to say that elaborate discussions of aerodynamic derivatives have been avoided, partly because the theory is far from complete and is developing rapidly and more particularly because there is a lamentable want of reliable experimental determinations to check the theory. Subject to the avoidance of elaboration, the aim has been to make the book comprehensive and up-to-date. For example, the effects of distortion of the structure and of the compressibility of the air are allotted separate chapters and there is a chapter on the measurement of aerodynamic derivatives, while methods for conducting response calculations are given, with special emphasis on the use of 'impulsive admittances'. The writer believes he is justified in claiming to have avoided all difficult mathematics. Any reader having an elementary knowledge of algebra, trigonometry and the calculus should find no difficulty in following the mathematical arguments.

The author wishes to express his warm thanks to Principal E. F. Relf for contributing a Foreword to this book, but, more particularly, for his interest and encouragement. The author is very specially indebted to Professor A. D. Young for contributing the chapters on 'Stalling and the Spin' and on 'Flaps for Landing and Take-off'; Professor Young has also given valuable advice regarding other parts of the text. Helpful advice has also been received from Mr A. H. Yates, particularly on the more practical aspects of the subject. Mrs D. M. Stanton Jones and Mrs M. E. Pipes have contributed by their conscientious work in typing the text and mathematics, and valuable assistance in the preparation of diagrams has been given by Mr K. D. Ross and Mr S. Deards. The author would also wish to thank many friends, too numerous to mention individually, who have helped him in a great variety of ways.

Although the recent development of the subject matter of this book has been great and rapid, the author's sense of indebtedness to the early pioneers is vivid and their influence is manifest throughout the whole of the book. It is here fitting to mention in particular the immense influence and insight of Lanchester, Bryan and Bairstow. But it is impossible to write a book such as this without being strongly conscious of indebtedness to the great founders and constructors of the science of dynamics. Galileo, Newton, Euler, Daniel Bernoulli, Lagrange and Routh are among the true progenitors of aeronautical science.

Readers approaching the subject of control and stability for the first time are advised to begin by reading Chapters 1, 2, 7 and 10 in that order. The more advanced theoretical Chapters 3, 4, 5 and 6 then follow in logical order while the remaining chapters are largely independent and can be taken in any order convenient to the reader. Pains have been taken to provide all necessary cross-references in the text. The references to literature are given concurrently in footnotes, while more general references are given at the ends of some chapters.

W. J. DUNCAN

Cranfield,
July, 1950

Note to second impression

The issue of a second impression of this book has presented the opportunity to correct misprints and other errors. Some brief notes on additional topics have been inserted in Chapter 15.

W. J. D.

CONTENTS

AMERICAN AND BRITISH TERMS AND SYMBOLS

The only important difference in aerodynamic terminology between the U.S.A. and Britain is that the *stabilizer* or *horizontal stabilizer* of American usage is called the tailplane in Britain. Unfortunately there is a more serious divergence in the use of symbols but, even here, a large proportion are in common use. It is also true that there is not complete uniformity of usage in either country. Some corresponding symbols are listed below.

Symbol used in this book	U.S. equivalent	Definition
η	δ_e	Elevator angle
ξ	δ_a	Aileron angle
β	δ_t	Tab angle
Θ	γ	Flight path angle. (In this book γ is the downward glide angle.)
t	τ	Unit of aerodynamic time, $m/\rho S V$
A	I_X	Moment of inertia about OX-axis
B	I_Y	Moment of inertia about OY-axis
C	I_Z	Moment of inertia about OZ-axis
D	I_{YZ}	Product of inertia about OY- and OZ-axes
E	I_{ZX}	Product of inertia about OZ- and OX-axes
F	I_{XY}	Product of inertia about OX- and OY-axes
$\bar{u}$	u'	Dimensionless airspeed variable, u/V
$\bar{v}$	β	Dimensionless sideslip variable, v/V
$\bar{w}$	α	Dimensionless vertical velocity variable, w/V
a_1	C_{L_α}, a_w	Rate of change of lift coefficient with angle of attack
a_2	$C_{L_\delta}, \tau a_w$	Rate of change of lift coefficient with control surface angle, $\partial C_L/\partial \delta$
a_3	$C_{L_{\delta_t}}, \tau_t a_w$	Rate of change of lift coefficient with tab angle, $\partial C_L/\partial \delta_t$
b_0	$(C_h)_{\substack{\alpha=0\\\delta=0}}$	Hinge-moment coefficient for zero angle of attack and zero control surface angle
b_1	C_h	Rate of change of hinge-moment coefficient with angle of attack, $\partial C_h/\partial \alpha$
b_2	C_{h_δ}	Rate of change of hinge-moment coefficient with control surface angle, $\partial C_h/\partial \delta$

Symbol used in this book	U.S. equivalent	Definition
b_3	$C_{h_{\delta_t}}$	Rate of change of hinge-moment coefficient with tab angle, $\partial C_h/\partial \delta_t$
c_1	$C_{h_{t_\alpha}}$	Rate of change of tab hinge-moment coefficient with angle of attack, $\partial C_{h_t}/\partial \alpha$
c_2	$C_{h_{t_\delta}}$	Rate of change of tab hinge-moment coefficient with control surface angle, $\partial C_{h_t}/\partial \delta$
c_3	$C_{h_{t_{\delta_t}}}$	Rate of change of tab hinge-moment coefficient with tab angle, $\partial C_{h_t}/\partial \delta_t$
E	c_f/c	Flap-chord ratio
l_p	C_{l_p}	Non-dimensional form of derivative L_p
l_r	C_{l_r}	Non-dimensional form of derivative L_r
l_v	C_{l_β}	Non-dimensional form of derivative L_v
m_q	$\dfrac{\bar{c}}{2l_t}\mu_1 C_{m_{d\theta}}$	Non-dimensional form of derivative M_q
m_u	$\dfrac{\bar{c}}{2l_t}C_{m_u}$	Non-dimensional form of derivative M_u
m_w	$\dfrac{\bar{c}}{2l_t}C_{m_\alpha}$	Non-dimensional form of derivative M_w
$m_{\dot{w}}$	$\dfrac{\bar{c}}{2l_t}\mu_1 C_{m_{d\alpha}}$	Non-dimensional form of derivative $M_{\dot{w}}$
n_p	C_{n_p}	Non-dimensional form of derivative N_p
n_r	C_{n_r}	Non-dimensional form of derivative N_r
n_v	C_{n_β}	Non-dimensional form of derivative N_v
y_p	$\tfrac{1}{2}C_{y_p}$	Non-dimensional form of derivative Y_p
y_r	$\tfrac{1}{2}C_{y_r}$	Non-dimensional form of derivative Y_r
y_v	$\tfrac{1}{2}C_{y_\beta}$	Non-dimensional form of derivative Y_v
μ_1	$\dfrac{c}{\bar{l}_t}\mu$	$W/g\rho Sl_t$
μ_2	2μ	$W/g\rho Ss$

Note. The symbol μ has not the same meaning in the last two entries.

Chapter 1

INTRODUCTORY SURVEY

1·1 Aims and importance of the study of control and stability

An aircraft is a kind of vehicle and a vehicle is a kind of tool—a machine tool—for it is a mechanical appliance so designed and made that it enables men to do what they could not do with their unaided bodies. Vehicles are machine tools for transportation and each kind of vehicle is made to move and carry in a definite medium and subject to certain external constraints. For instance, a locomotive moves in the medium air and is constrained to follow a linear track while a ship moves in the twin media of water and air and is so constrained by the forces of gravity and buoyancy that its centre of gravity always lies on or near the surface separating water and air. The means for the control of vehicles vary greatly from type to type, and depend especially on such external constraints as may be present. Thus a motor car requires no control for vertical position since it is constrained to move on the surface of separation of earth and air and its track on this surface is controlled by varying the curvature of path in a suitable manner. An aircraft in flight shares with a submerged submarine the unusual condition of entire freedom from geometrical constraint and it can only be guided in its three-dimensional path by the indirect process of modifying the aerodynamic forces upon it.* Thus it is clear from the start that the control of aircraft will present problems of unusual complexity.

Another feature of aircraft in flight is that, on account of their numerous degrees of freedom and of the high speeds of flight, they are specially prone to instability unless appropriate precautions are taken, and the consequences of any but very mild instabilities will be disastrous. Hence the stability of aircraft is a subject well worthy of close study. But stability is only one aspect of the dynamical characteristics; the general

* For lighter-than-air craft we should add the forces of gravity and buoyancy as being under control.

1

dynamical behaviour profoundly affects the comfort of passengers and the suitability of the aircraft for a gun or bombing platform. Therefore we must extend our inquiry to a general examination of dynamical behaviour, including the response to control movements and to gusts.

1·2   Pilots—human and automatic

Most aircraft are controlled only by human pilots and hitherto nearly all have been designed for at least occasional human control. The capacities of the human pilot are therefore the basis for the design of controls; moreover, the question of stability can only be intelligently discussed in the light of the characteristics of the pilot.

From the present point of view the most important characteristics of pilots are:

(1) Muscular strength.

(2) Reaction time.

(3) Proneness to fatigue.

(4) Sensitiveness to acceleration.

(5) Sensitiveness to changes of the force reactions on the controls.

(6) Sensitiveness to changes of position, as of the limbs, or of visible objects such as the pointers of instruments.

(7) Adaptability.

These will be discussed separately.

It is obvious that the loads on the control stick and pedals required for the effective control of the aircraft in all operational conditions must be within the physical capacity of all fit pilots; moreover the sustained loads must not be great enough to cause undue fatigue. No very precise figures for the forces can be stated, but those given in Table 1·2, 1, which are due to M. B. Morgan, will serve as useful guides. It appears from experience that forces of even a few pounds, when continually sustained, cause discomfort and fatigue. Hence it is necessary that each control should be provided with a trimming device by which the control force can be balanced out for any setting of the control. Another important aspect of control forces is their harmony; it is well recognized that pilots strongly dislike a mixture of heavy and light controls. Again no very precise criteria can be laid down but most pilots would agree that controls are in satisfactory harmony when the maxima for the lateral stick force, fore-and-aft stick force and pedal force are in the proportion 1 : 2 : 4.

Reaction time may be defined as the time which elapses between the impact of a physical stimulus on some sense organ such as the eye or touch spot on the skin and the beginning of

TABLE 1·2, 1. MAXIMUM CONTROL FORCES EXERTED BY THE PILOT. ALL FORCES ARE IN POUNDS

Nature of force		Aileron (sideways or peripheral force)		Elevator (push or pull)		Rudder, push
		Stick	Wheel	Stick	Wheel	
Greatest effort of which pilot is capable in emergency for a very short time		90 (two hands)	120 (two hands)	180 (two hands)	220 (two hands)	400
Maximum force which it is permissible to demand of the pilot, even in an emergency, for a short time	Two hands	—	80	100	110	200
	One hand	50	50	70	70	
Maximum force which the pilot cares to exert for a short time	Two hands	—	30	—	40	60
	One hand	20	20	30	30	

Where blanks occur it is intended to convey that the use of two hands should not be demanded.

The maximum tolerable forces are reduced when the cockpit is cramped.

the responsive movement of hand or foot. Reaction time varies somewhat from individual to individual and depends on attention, practice, the nature of the task and of the stimulus and, most importantly, the physical state of the subject. It is reduced by attention and practice and increased by fatigue but particularly by oxygen starvation (anoxia). Visual reactions are appreciably slower than tactual, the additional lag being probably associated with the photochemical reaction in the retina. The normal reaction time to touch and to sound is about 0·12 second while it is about 0·17 second for a visual stimulus; these figures refer to determinations in the laboratory under favourable conditions.

Reaction time is important in relation to the steadiness of flight which the pilot is able to maintain, particularly in bumpy air, and it influences the amount of instability of the aircraft which can be tolerated. It is recognized that an aircraft may behave quite satisfactorily in the hands of an experienced pilot when it is subject to one or more instabilities, provided that these are not too severe. One important variable is the ratio of the time required to double the amplitude of the disturbance to the reaction time of the pilot, although the magnitude of the disturbance is also of great importance. When this ratio is of the order of 50 or greater the instability will probably not be dangerous. The foregoing criterion is intended to apply only to aperiodic movements and to periodic movements of low frequency. When the frequency is greater than about $\frac{1}{2}$ cycle per second even the slightest instability will probably be uncontrollable.

Fatigue of the pilot may be of two extreme types, although in practice the effect will often be mixed. The first is physical fatigue caused by the prolonged application of control forces which are too great for comfort and the second is mental fatigue caused by the effort of attention needed to overcome the departures of the aircraft from the desired attitude, etc. Fatigue of the first kind can be minimized by designing the controls so that excessive control forces are avoided; that of the second kind will not be unduly severe provided the controls are not too sensitive and that the aircraft itself has the right dynamical characteristics.

The pilot's sensitiveness to acceleration must be considered from two aspects. First, there is the maximum acceleration which he can withstand without 'blackout' or unconsciousness, for it is clearly without profit to design the aircraft for accelerations which would cause these disabilities. A pilot is most sensitive to acceleration directed along the major axis of his body, while maximum accelerations of the aircraft in the fore-and-aft and lateral directions are much smaller than in the 'normal' direction. Hence large normal accelerations are better tolerated when the pilot is prone than when he sits vertically; the maximum tolerable acceleration can also be increased by the use of 'anti-g' suits. In the design of large aircraft the accelerations which can occur at parts far distant from the centre of gravity during angular motions must be considered (see § 3·3). Angular accelerations unaccompanied by linear accelerations, such as may occur in the vicinity of the c.g. in the rapid initiation of a roll, do not appear to be physiologically important. Second, the pilot's sensitiveness to *small* accelerations may be of value to him as an aid to maintaining steady flight.

The pilot's sensitiveness to changes of control force and of control position are of importance in helping him to maintain steady flight, but the sensitiveness to changes of force seems to be the more important.

An adaptable pilot is one who quickly becomes accustomed to and masters a new aircraft or a new system of control. All human beings, fortunately, have a tendency to like what they have become accustomed to, even when the familiar thing is very faulty, but this is accompanied by a more or less pronounced and irrational dislike of change. The bearing of this on the design of controls is that almost any innovation will be disliked at first by a large proportion of pilots, but this dislike will be removed by education and experience if the innovation is intrinsically good.

The most important fact about automatic pilots is that the aircraft and its automatic pilot form a *single dynamical system* and the mechanism and gearing of the pilot must be so adapted to the aircraft that the whole system has satisfactory characteristics. One obvious requirement is that the system shall be definitely stable in all circumstances. It is, however, incorrect to suppose that the automatic pilot which makes a given aircraft the most stable is therefore the best. The fact is that stability is concerned with the *ultimate* consequence of a disturbance and stability, in the technical sense of the word, is high when the motion induced by a disturbance ultimately decays rapidly. But this condition may be compatible with a large amplification of the disturbance in the *early* stages of the induced motion, which is clearly most undesirable. Hence an automatic pilot must be designed to give the greatest steadiness or *stabilization* at all stages of the induced motion. The optimum arrangement will be one for which the *stability* is good, but usually not at an absolute maximum.

We cannot enter here into the technicalities and theory of automatic control (see §§ 15·2 and 15·3), but the importance of avoiding lag in the mechanism must be emphasized. Lag in an automatic pilot corresponds with reaction time in a human pilot and is similarly detrimental; special devices have been invented for its effective neutralization.

1·3   The dynamical basis for the study of stability and control

All aircraft are deformable—that is, the relative positions of *all* their parts vary somewhat with the loads applied in flight. However, it has been customary to ignore the deformations of the main structure in discussing the theory of stability and control and only to take account of them in special inquiries concerning such matters as flutter and reversal of control. This simplification, though never exact, is often justified and may be accepted as necessary in the first approach to the subject, since the theory is vastly complicated when the deformations of the structure are allowed for. Hence we begin by assuming that the main aircraft structure and each individual control surface is a rigid body, but we are forced to abandon this assumption when the speed of flight is high in relation to the stiffnesses of the structure.

The dynamically dominant feature of the whole problem is that the aircraft moves in the gravitational field of the earth. The value of the apparent acceleration due to gravity varies slightly with latitude and with altitude, but the variations are negligible. Exceptionally, the variation of g with altitude would become appreciable for trans-atmospheric craft.

1·4   The aerodynamic basis

Most of the current theory of control and stability is based on very elementary aerodynamic theory helped by the results of ad hoc aerodynamic experiments which are not yet fully explained by fundamental aerodynamic theory. In many instances the aerodynamic data do not go beyond the ordinary non-dimensional coefficients C_L, C_D, etc., appropriate to the aerofoils concerned. Such coefficients are never strictly applicable to motions in which the angles of incidence of the various surfaces vary with time, but usually the error involved here is small. However, more serious errors arise where the aerodynamic forces or moments deviate importantly from being linear functions of incidence or of control angle, as is assumed in all elementary treatments of aircraft dynamics. It is notorious that the graphs of hinge moment against control angle for certain controls, e.g. Frise ailerons, are far from straight; a curve with a point of inflexion in the neutral region is quite usual.

The non-dimensional coefficients such as C_L for an aerofoil or body of given shape set at a given attitude are functions of the non-dimensional parameters R and $\mathbf{M}$. Here R is the Reynolds number given by

$$R = \frac{Vl}{\nu}, \qquad (1\cdot4, 1)$$

with V = velocity of flight,

l = a typical linear dimension such as mean wing chord

and ν = kinematic viscosity of the air

$$= \mu/\rho, \qquad (1\cdot4, 2)$$

where μ = viscosity of the air

and ρ = air density.

The Mach number $\mathbf{M}$ is the ratio of the speed of flight to the speed of sound in the undisturbed air; thus

$$\mathbf{M} = \frac{V}{a}, \qquad (1\cdot4, 3)$$

where a is the velocity of sound in the surrounding atmosphere. The Mach number is a very convenient parameter for indicating the importance of the compressibility of the air in the circumstances of the flight considered. When $\mathbf{M}$ is less than about 0·3 the influence of compressibility is usually negligible and for thin aerofoils at small incidences the effect is slight for Mach numbers up to about 0·7.

The Reynolds number serves to indicate in a broad way the relative importance of the forces of fluid friction and of inertia within the fluid; the lower the value of R the more relatively important is the fluid friction. Variation of a non-dimensional aerodynamic coefficient, such as C_L for a given aerofoil at a given angle of incidence, with R is usually called *scale effect*. Even for values of R in excess of several millions, scale effect may not be negligible. This may be associated with chordwise shift of the region of transition from laminar to turbulent motion in the boundary layer as R varies and to surface roughnesses penetrating beyond the laminar sub-layer at high values of R. An apparent scale effect, especially in wind tunnel tests, may be associated with increasing turbulence in the air stream as the speed rises. In as much as transition region and surface roughness may vary largely from aircraft to aircraft of the same type, and with time for a given aircraft, it must be expected that the full-scale aerodynamic coefficients will exhibit corresponding variations.

An interesting interpretation of the Reynolds number for motion in a gas is provided by the following formula

$$R = k\mathrm{M}\left(\frac{l}{\lambda}\right), \qquad (1\cdot 4, 4)$$

where k is a numerical constant which depends slightly on the nature of the gas and differs little from unity, M and l have their former meanings, and λ is the mean free path of the molecules of the gas. We see, therefore, that in place of taking M and R as fundamental parameters we may use M and (l/λ).* This has the advantage of being applicable when the conception of viscosity no longer has meaning, as when the mean free path is comparable with the linear dimensions of the body. In such circumstances it might be appropriate to substitute for M the ratio V/C, where C is the root mean square velocity of the molecules of the gas, for C and a are proportional for a given gas.

1·5   Methods of control

We have already remarked in § 1·1 that the path of an aircraft is controlled by varying the aerodynamic forces upon it. This can be done in two main ways:

(a) By varying the propulsive force, provided by an airscrew or jet, in magnitude or direction.

(b) By changing the relative positions of parts of the aircraft.

Method (a) is used chiefly in controlling the inclination to the horizontal of a rectilinear flight path but it has already been used for directional control in helicopters. In method (b) a change of configuration is so arranged that a local change of aerodynamic force occurs, giving rise to a moment about the centre of gravity of the aircraft. The aircraft then rotates and,

* λ/l is known as the Knudsen number.

in general, alters its attitude to the flight path with the consequence that the aerodynamic forces are further modified. The local change of configuration may be secured by distortion of the structure, as with wing warping which was used for lateral control on some early aeroplanes, but usually at the present day by rotating a more or less rigid part of the structure, called a control surface, about hinges fixed in the main structure. Sometimes the relative movement is compounded of rotation and sliding, as in some landing flaps. Pure sliding might also be used.

The foregoing by no means exhausts the possibilities of control. For example, control could certainly be achieved by localized suction suitably arranged, but no such arrangement has yet been used. Spoilers, which reduce wing lift locally by spoiling the 'circulation', have been tried but not widely adopted. In emergencies helpful control forces can be produced by launching drogues or parachutes, e.g. anti-spin parachutes.

1·6 Some needs and difficulties

An outstanding difficulty in the design of non-power operated controls arises from the manner in which the hinge moments on flap controls vary with speed of flight. Apart from scale effect and the influence of compressibility, the hinge moment on a given rigid flap when deflected to a given angle varies as the square of the speed. Thus if the greatest and least speeds are say in the ratio 3 : 1 the hinge moments and control forces will be in the ratio 9 : 1. This would imply that the control was either unduly light at low speeds or unduly heavy at high speeds, although the disparity may be mitigated by the circumstance that somewhat smaller control movements usually suffice at the higher speeds. The designer is thus faced with the problem of 'defeating the V^2 law'. Again, on large aircraft the control flaps are also large and the hinge moments correspondingly great. Hence the pilot would be unable to operate them except by use of an excessively low gear ratio between control column or pedal and control flap; since some slowing of control movement may be tolerated on large aircraft on account of their slowness of response some reduction in the gear ratio may be acceptable.

The difficulties just mentioned are met or at least mitigated by providing the control flaps with some kind of aerodynamic balance which reduces the hinge moments. The primitive and most obvious method of balancing is to set back the hinge from the nose of the flap but this is limited by the imperative need to avoid overbalance in *all* circumstances. Much ingenuity has been shown in the design of balancing arrangements and one of the most useful and effective is undoubtedly the spring tab. This has the most valuable feature that it gives the greatest effect in lightening the control just when it is most needed, i.e. it succeeds, in some measure, in overcoming the usual tendency towards excessive heaviness at high speeds without incurring excessive lightness at low speeds. It is worthy of remark here that the hinge moments on flap controls are affected by the responsive movement of the aircraft and it is possible to arrange matters so that this response effect is helpful.

1·7 Power-operated controls

One radical method of overcoming the difficulties outlined in the last section is to provide servo motors to operate the control surfaces. Here the force exerted by the pilot can be as small as desired in all circumstances. The paramount requirement of power operated controls is absolute reliability and this will usually require duplication of the servo motors and of the source of power. Obviously the main objections to the use of power are the extra weight and complication, but it is probable that it will be used increasingly in the future for large aircraft and for small aircraft which fly very fast. Two outstanding questions concern:

(*a*) The need to provide a manual control for use in emergency.

(*b*) The need to provide the pilot with 'feel' by feeding back to his control lever a small fraction of the full control load.

Neither of these questions is yet settled, but it appears probable that both of these needs will disappear as technique develops and pilots become accustomed to the new system.

1·8 The place of theory

It is beyond dispute that the observed behaviour of aircraft is so complex and puzzling that, without a well developed theory, the subject could not be treated intelligently. Theory has at least three useful functions:

(*a*) It provides a rational background for the analysis of actual occurrences.

(*b*) It provides a rational basis for the planning of experiments and tests, thus securing economy of effort.

(*c*) It helps the designer to design intelligently.

Theory, however, is never complete, final or exact. Like design and construction it is continually developing and adapting itself to circumstances. We close with two quotations:

'Nothing is so practical as a really good theory'

(L. Boltzmann);

'A scientific theory is a policy rather than a creed'

(J. J. Thomson).

Chapter 2

ELEMENTARY MECHANICS OF FLIGHT

2·1 Introductory remarks

In this chapter we consider in a preliminary way various topics of aircraft dynamics which can be treated by very elementary methods. Most of the problems are simplified by neglecting certain factors; frequently the simplification consists in omitting one or more of the degrees of freedom. But in all cases the treatment adequately brings out the main features of the motion, although it may not be accurate in detail.

Throughout this chapter and elsewhere, unless the contrary is expressly stated, the atmosphere in which flight occurs is supposed to be uniform and at rest, apart from the disturbances caused by the aircraft itself. Only heavier-than-air craft are considered and the force of buoyancy is neglected.

2·2 Steady rectilinear flight

We begin by considering the steady rectilinear motion of an aircraft. Since there is no acceleration in such motion, the external forces acting must be in equilibrium. At present we consider only the forces in the plane of symmetry of the aircraft; this plane is assumed to be vertical and the path of the centre of gravity (C.G.) lies in the plane.

Let V = true speed of flight relative to the undisturbed air,

$V_e = V\sqrt{(\sigma)}$ = equivalent air speed (E.A.S.),

W = total weight of the loaded aircraft

  $= mg$.

m = total mass of the aircraft,

g = acceleration in free fall under gravity,

S = total wing area,

w = wing loading

  $= W/S$.

L = lift force,

D = drag force,

T = propulsive thrust,

M = total pitching moment of the aerodynamic forces about the c.g. of the loaded aircraft with tailplane and elevators in their standard settings,

M_t = trimming moment in pitch caused by displacement of tailplane and elevators from their standard settings. In the subsequent argument we shall suppose the tailplane to be fixed,

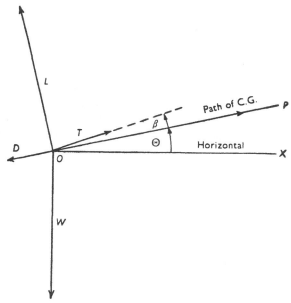

Fig. 2·2, 1. Forces in steady rectilinear flight.

α = angle of incidence
  = upward inclination of datum wing chord to direction of flight,

η = angular setting of elevators, measured from standard setting,

ρ = air density,

ρ_0 = standard air density,

$\sigma = \rho/\rho_0$ = relative air density,

Θ = upward inclination of flight path to the horizontal,

β = upward inclination of the propulsive thrust to the flight path,

C_L = non-dimensional lift coefficient

$$= \frac{L}{\frac{1}{2}\rho V^2 S}.$$

C_D = non-dimensional drag coefficient

$$= \frac{D}{\frac{1}{2}\rho V^2 S}.$$

C_m = non-dimensional pitching moment coefficient

$$= \frac{M}{\frac{1}{2}\rho V^2 S \bar{c}}.$$

$\bar{c}$ = mean wing chord,

$$C_{mt} = \frac{M_t}{\frac{1}{2}\rho V^2 S \bar{c}}.$$

We shall assume here for simplicity that in a glide C_L, C_D and C_m are functions of the angle of incidence α only although in reality they depend on the Reynolds and Mach numbers (see § 1·4). Similarly we shall assume that C_{mt} is a function only of the elevator setting η.

By definition the lift L is perpendicular to the flight path upwards* and the drag D is along the flight path backwards (see Fig. 2·2, 1). Hence the equation of equilibrium of the forces in the direction of motion is

$$D + W \sin \Theta = T \cos \beta, \tag{2·2, 1}$$

and the equation of equilibrium of the forces at right angles to the flight path is

$$L - W \cos \Theta = - T \sin \beta. \tag{2·2, 2}$$

Lastly, the equation of equilibrium of the pitching moments is

$$M + M_t = 0, \tag{2·2, 3}$$

where we assume that the line of thrust passes through the c.g.

We shall begin by considering steady motion in a glide, so T is zero. Equation (2·2, 3) is equivalent to

$$C_m + C_{mt} = 0, \tag{2·2, 4}$$

and, in accordance with our assumptions, C_m is a function only of α while C_{mt} is a function only of η. Hence we see that α is a

* The positive sense of L is downwards in inverted flight.

function only of η, so C_L and C_D are likewise functions only of η. Equation (2·2, 2) becomes

$$L = W \cos \Theta$$

$$= \tfrac{1}{2}\rho V^2 S C_L.$$

Hence

$$V = \sqrt{\left(\frac{2w \cos \Theta}{\rho C_L}\right)} \qquad (2·2, 5)$$

$$= \sqrt{\left(\frac{2w}{\rho_0}\right)} \times \sqrt{\left(\frac{\cos \Theta}{\sigma C_L}\right)}, \qquad (2·2, 6)$$

which can also be written

$$V_e = \sqrt{\left(\frac{2w}{\rho_0}\right)} \times \sqrt{\left(\frac{\cos \Theta}{C_L}\right)}. \qquad (2·2, 7)$$

Equation (2·2, 1) becomes

$$D = - W \sin \Theta,$$

so

$$\frac{D}{L} = - \tan \Theta.$$

This shows that Θ is negative and the glide angle is

$$\gamma = - \Theta = \tan^{-1}\!\left(\frac{D}{L}\right) = \tan^{-1}\!\left(\frac{C_D}{C_L}\right). \qquad (2·2, 8)$$

Usually the glide angle is small enough for unity to be substituted for $\cos \Theta$ in (2·2, 7) with negligible error. Hence we have for a given aircraft

$$V_e \sqrt{(C_L)} = \text{constant}. \qquad (2·2, 9)$$

We have seen that C_L and C_D are functions of η and it now follows that both the glide angle and the E.A.S. are determined by the elevator setting.

Conditions are not quite so simple when the propulsive thrust is not zero for the slipstream in general gives rise to a change in the pitching moment and to an increment of lift coefficient. Hence the angle of incidence becomes a function of η and of the non-dimensional thrust coefficient

$$T_c = \frac{T}{\rho V^2 D^2}, \qquad (2·2, 10)$$

where D is the diameter of the airscrew. However, except in the case of helicopters, $T \sin \beta$ is negligible and then equation

(2·2, 7) remains valid. Hence we now find that the E.A.S. is a function of η and of T_c, but usually η remains the dominating factor. It follows from equation (2·2, 1) that

$$\sin \Theta = \frac{T \cos \beta - D}{W},\qquad (2\cdot2, 11)$$

so the upward inclination of the flight path increases with the thrust. For steady horizontal flight the drag is

$$D = T \cos \beta$$
$$= T \qquad (2\cdot2, 12)$$

with sufficient accuracy when, as usual, β is small.

We now revert to the condition of longitudinal trim expressed by the equation (2·2, 3). It is sometimes convenient to take moments about a fixed datum point in the aircraft and to bring in explicitly the gravity moment due to the carried load.

Let M' = aerodynamic moment about the datum point with standard settings of tailplane and elevators,

M'_t = additional aerodynamic moment about the datum point brought into action by displacement of the tailplane and elevators from their standard settings,

M'_l = moment about the datum point of the weight of the carried loads and of the aircraft itself,

then the condition of trim is

$$M' + M'_t + M'_l = 0. \qquad (2\cdot2, 13)$$

It is necessary that the tailplane and elevators be so proportioned and arranged that their maximum correcting moment is always adequate to give balance when M'_l is at the numerical maxima required by the duty of the aircraft, with a satisfactory margin for manoeuvring and safety.

2·3   Rectilinear flight in a non-uniform atmosphere

The density of the earth's atmosphere is not constant but decreases with altitude. The atmosphere can, as a working approximation, be assumed to be horizontally stratified, as in the standard atmosphere, so the density is a function of altitude alone. In such an atmosphere horizontal steady flight at constant attitude and speed is still possible, but inclined steady flight is not possible in the same conditions. Thus, as altitude and σ change, equation (2·2, 6) cannot continue to be satisfied with both V and C_L (or attitude) unchanged. We shall confine attention here to the case where C_L is kept constant. This implies that the E.A.S. is constant, but the true speed varies. There is therefore an acceleration in the direction of motion and an additional term must be brought into equation (2·2, 1). It obviously becomes

$$D + W \sin \Theta + m \frac{dV}{dt} = T, \qquad (2\cdot3, 1)$$

where m is the mass of the aircraft. But

$$\sigma V^2 = V_e^2 = \text{constant},$$

and if h is the altitude

$$2\sigma V \frac{dV}{dh} + V^2 \frac{d\sigma}{dh} = 0.$$

Also
$$\frac{dh}{dt} = V \sin \Theta.$$

Hence
$$\frac{dV}{dt} = -\frac{V^2}{2\sigma} \frac{d\sigma}{dh} \sin \Theta, \qquad (2\cdot3, 2)$$

and equation (2·3, 1) becomes

$$D + W \sin \Theta = T + \frac{mV^2}{2\sigma} \frac{d\sigma}{dh} \sin \Theta. \qquad (2\cdot3, 3)$$

The additional term on the right is called the *kinetic energy correction* to the thrust. Since $\dfrac{d\sigma}{dh}$ is negative there is effectively a thrust deduction in climbing and a thrust augmentation in diving. Clearly, these effects are most pronounced at high speeds and in steep paths. Equation (2·3, 3) can be rewritten neatly in the form

$$D + W \left[1 + \frac{V_e^2}{2g} \frac{d}{dh} \left(\frac{1}{\sigma} \right) \right] \sin \Theta = T. \qquad (2\cdot3, 4)$$

The effect is therefore the same as if the weight were increased in the ratio

$$\left[1 + \frac{V_e^2}{2g} \frac{d}{dh} \left(\frac{1}{\sigma} \right) \right] \quad : \quad 1.$$

For example, with $V_e = 300$ m.p.h., the ratio varies between 1·09 at sea level and 1·6 at 40,000 ft. in the standard atmosphere.

In a similar manner it can be shown that for rectilinear flight at constant Mach number $\mathbf{M}$ the acceleration can be allowed for by altering the weight in the ratio

$$\left(1 + \frac{\gamma R}{2g}\frac{d\mathsf{T}}{dh}\mathbf{M}^2\right) : \quad 1$$

where γ is the ratio of the specific heats of air, R is the gas constant for unit mass of air and T is the absolute temperature. In the standard troposphere the temperature lapse rate is 1·98° C. per 1000 ft. Hence the ratio is

$$(1 - 0\cdot133\,\mathbf{M}^2) : \quad 1.$$

In the stratosphere the ratio is unity since T is constant. The reversal of sign of the correction in the troposphere as compared with the case of flight at constant E.A.S. is of course associated with the fall in the speed of sound with increasing altitude. It is to be remembered that in all cases the true weight must be used in calculating the lift coefficient and the induced drag.

2·4   Horizontal circling flight

It is possible for an aircraft to make a 'flat turn', i.e. to change its horizontal direction of flight while keeping the wing

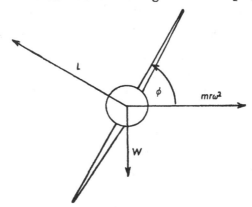

Fig. 2·4, 1. Aircraft making correctly banked steady turn.

span horizontal, but with usual proportions the turning is very slow and accompanied by much sideslip, which may be objectionable. The efficient way of turning consists in first rolling the aircraft so that the lift force on the wings has a component in the direction in which it is desired to turn. In a correctly banked steady turn the lift, weight and centrifugal force are in equilibrium and there is no sideslip (see Fig. 2·4, 1).

Let ϕ be the angle of bank, i.e. the angle between the wing span and the horizon. Then we obtain by resolution of forces in the vertical and horizontal directions

$$W = L \cos \phi, \tag{2·4, 1}$$

$$mr\omega^2 = L \sin \phi, \tag{2·4, 2}$$

where r is the radius of the path of the C.G. and ω is the angular velocity in the circular path. Hence

$$\tan \phi = \frac{mr\omega^2}{W} = \frac{r\omega^2}{g} = \frac{V^2}{rg} \tag{2·4, 3}$$

and
$$L^2 = W^2 + m^2 r^2 \omega^4$$

$$= W^2 \left(1 + \frac{V^4}{r^2 g^2} \right). \tag{2·4, 4}$$

The apparent value of gravity as measured by an accelerometer carried at the C.G. of the aircraft is

$$\frac{L}{W} g = g \sec \phi. \tag{2·4, 5}$$

In a turn the velocity is not constant over the wing span and this gives rise to a rolling moment which must be corrected by the controls. For a given apparent g this variation is less the higher the speed since the radius of turn is proportional to V^2 [see equation (2·4, 3)].

2·5 Incidence changes associated with changes of velocity: aerodynamic derivatives

We consider an aircraft or other body, regarded as rigid, moving without rotation in an atmosphere at rest. We take an origin O and perpendicular axes OX, OZ fixed in the body and suppose that the plane OXZ moves in its own plane, so the velocity of O is always in the plane. Suppose first that the velocity has the components U and W in the directions OX, OZ respectively, as shown in Fig. 2·5, 1. The resultant velocity is V inclined at the angle α to OX, where

$$V^2 = U^2 + W^2 \tag{2·5, 1}$$

and
$$\tan \alpha = \frac{W}{U}. \tag{2·5, 2}$$

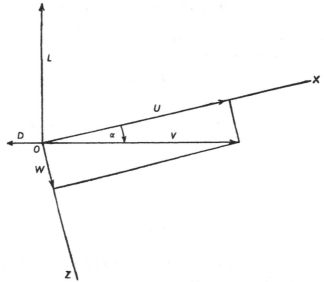

Fig. 2·5, 1. Velocity components and angle of incidence of non-rotating body.

Since the velocity of the air relative to the body is V reversed, the angle of incidence referred to OX as 'chord line' is α. Next, let U and W become $(U+u)$ and $(W+w)$ respectively, while V and α become V' and α' respectively. Then we have the accurate relations

$$
\begin{aligned}
\sin (\alpha' - \alpha) &= \sin \Delta\alpha \\
&= \sin \alpha' \cos \alpha - \cos \alpha' \sin \alpha \\
&= \frac{(W+w)\,U - (U+u)\,W}{VV'} \\
&= \frac{wU - uW}{VV'}, \tag{2·5, 3}
\end{aligned}
$$

$$
\begin{aligned}
V'^2 &= (U+u)^2 + (W+w)^2 \\
&= V^2 + 2(uU + wW) + (u^2 + w^2). \tag{2·5, 4}
\end{aligned}
$$

When u and w are small quantities of the first order and second order quantities are neglected, the formulae for the increments become

$$\Delta\alpha = \frac{w\cos\alpha - u\sin\alpha}{V}, \qquad (2\cdot5,\,5)$$

$$\Delta V^2 = 2V(u\cos\alpha + w\sin\alpha). \qquad (2\cdot5,\,6)$$

In the special case where α is zero, OX coincides with the original direction of the resultant velocity and the axes are then rather misleadingly called 'wind axes'. The formulae reduce to

$$\Delta\alpha = \frac{w}{V}, \qquad (2\cdot5,\,7)$$

$$\Delta V^2 = 2uV. \qquad (2\cdot5,\,8)$$

Even when α is not strictly zero, but small, the last equations are usually sufficiently accurate.

Now let F be some force which depends on the velocity and angle of incidence and given by

$$F = \tfrac{1}{2}\rho V^2 S C_F, \qquad (2\cdot5,\,9)$$

where we shall at first suppose C_F to be a function of α only and take ρ and S to be constant. Then, to the first order of small quantities, the increment of the force is given by

$$\frac{\Delta F}{\tfrac{1}{2}\rho S} = V^2\Delta C_F + C_F\Delta V^2$$

$$= V^2\frac{dC_F}{d\alpha}\Delta\alpha + C_F\Delta V^2,$$

and on substitution from $(2\cdot5,\,5)$ and $(2\cdot5,\,6)$ we obtain

$$\frac{\Delta F}{\rho VS} = u\left(C_F\cos\alpha - \tfrac{1}{2}\frac{dC_F}{d\alpha}\sin\alpha\right) + w\left(C_F\sin\alpha + \tfrac{1}{2}\frac{dC_F}{d\alpha}\cos\alpha\right).$$

$$(2\cdot5,\,10)$$

The increment of a moment can be obtained in exactly the same way.

The most important general conclusion to be drawn from the foregoing investigation is that the increment in the force F is a linear and homogeneous function in the velocity increments u, w. Clearly the coefficients of u and w are the partial differential coefficients of F with respect to these variables. In aerodynamics it has become usual to adopt a notation for these partial differential coefficients exemplified by

$$\left.\begin{aligned} \frac{\partial F}{\partial u} &\equiv F_u, \\ \frac{\partial F}{\partial w} &\equiv F_w, \end{aligned}\right\} \qquad (2\!\cdot\!5,\,11)$$

and they are called *aerodynamic derivatives*. It follows from (2·5, 10) that

$$F_u = \rho V S\!\left(C_F \cos\alpha - \tfrac{1}{2}\frac{dC_F}{d\alpha}\sin\alpha\right) \qquad (2\!\cdot\!5,\,12)$$

and

$$F_w = \rho V S\!\left(C_F \sin\alpha + \tfrac{1}{2}\frac{dC_F}{d\alpha}\cos\alpha\right). \qquad (2\!\cdot\!5,\,13)$$

These particular derivatives are typical of *force-velocity derivatives* and it is important to note that they are proportional to the *first power* of the velocity of flight V. The same is clearly true of *moment-velocity derivatives* for the foregoing analysis holds good when the force F is replaced by a moment M and a length factor c is appended to S in equation (2·5, 9) and onward. Aerodynamic derivatives of other kinds are considered later (see § 3·6).

Hitherto we have not particularized the force F but we may remark that in discussing the motion of aircraft we are usually concerned with the components of force along axes fixed in the aircraft, whereas the results of aerodynamic measurements are usually given as forces (lift, drag, cross-wind force) referred to axes fixed with respect to the relative air stream. This complicates the expressions for the derivatives, as shown below. We shall assume at present that C_F can be expressed in terms of the lift and drag coefficients by the equation

$$C_F = C_L \cos(\alpha + A) + C_D \cos(\alpha + B), \qquad (2\!\cdot\!5,\,14)$$

where A and B are constants. For example, the component of force in the direction OX is (see Fig. 2·5, 1)

$$\left.\begin{aligned} X &= L\sin\alpha - D\cos\alpha \\ &= L\cos\!\left(\alpha - \frac{\pi}{2}\right) + D\cos(\alpha + \pi), \end{aligned}\right\} \qquad (2\!\cdot\!5,\,15)$$

while the component of force in the direction OZ is

$$\left.\begin{aligned}Z &= -L\cos\alpha - D\sin\alpha \\ &= L\cos(\alpha+\pi) + D\cos\left(\alpha+\frac{\pi}{2}\right).\end{aligned}\right\}\qquad (2\cdot5,16)$$

The values of the angles A and B are accordingly as shown in Table 2·5, 1.

TABLE 2·5, 1. FIXED ANGLES A AND B FOR FORCE COMPONENTS IN TERMS OF LIFT AND DRAG

Force coefficient	Angle	
	A	B
C_x	$-\dfrac{\pi}{2}$	π
C_z	π	$\dfrac{\pi}{2}$

By equation (2·5, 14)

$$\frac{dC_F}{d\alpha} = \frac{dC_L}{d\alpha}\cos(\alpha+A) + \frac{dC_D}{d\alpha}\cos(\alpha+B)$$

$$- C_L\sin(\alpha+A) - C_D\sin(\alpha+B).$$

Hence (2·5, 12) and (2·5, 13) yield

$$\frac{F_u}{\rho VS} = \cos\alpha\left[C_L\cos(\alpha+A) + C_D\cos(\alpha+B)\right]$$

$$- \tfrac{1}{2}\sin\alpha\left[\frac{dC_L}{d\alpha}\cos(\alpha+A) + \frac{dC_D}{d\alpha}\cos(\alpha+B)\right.$$

$$\left. - C_L\sin(\alpha+A) - C_D\sin(\alpha+B)\right]. \quad (2\cdot5,17)$$

$$\frac{F_w}{\rho VS} = \sin\alpha\left[C_L\cos(\alpha+A) + C_D\cos(\alpha+B)\right]$$

$$+ \tfrac{1}{2}\cos\alpha\left[\frac{dC_L}{d\alpha}\cos(\alpha+A) + \frac{dC_D}{d\alpha}\cos(\alpha+B)\right.$$

$$\left. - C_L\sin(\alpha+A) - C_D\sin(\alpha+B)\right]. \quad (2\cdot5,18)$$

When we substitute for A and B from Table 2·5, 1 and reduce we obtain the following results:

$$\frac{-X_u}{\rho VS} = C_D - \tfrac{1}{2}\left(C_L + \frac{dC_D}{d\alpha}\right)\cos\alpha\sin\alpha + \tfrac{1}{2}\left(-C_D + \frac{dC_L}{d\alpha}\right)\sin^2\alpha,$$

$$(2·5, 19)$$

$$\frac{-X_w}{\rho VS} = -C_L + \tfrac{1}{2}\left(C_L + \frac{dC_D}{d\alpha}\right)\cos^2\alpha - \tfrac{1}{2}\left(-C_D + \frac{dC_L}{d\alpha}\right)\cos\alpha\sin\alpha,$$

$$(2·5, 20)$$

$$\frac{-Z_u}{\rho VS} = C_L - \tfrac{1}{2}\left(C_L + \frac{dC_D}{d\alpha}\right)\sin^2\alpha - \tfrac{1}{2}\left(-C_D + \frac{dC_L}{d\alpha}\right)\cos\alpha\sin\alpha,$$

$$(2·5, 21)$$

$$\frac{-Z_w}{\rho VS} = C_D + \tfrac{1}{2}\left(C_L + \frac{dC_D}{d\alpha}\right)\cos\alpha\sin\alpha + \tfrac{1}{2}\left(-C_D + \frac{dC_L}{d\alpha}\right)\cos^2\alpha.$$

$$(2·5, 22)$$

It follows that

$$-\frac{X_u + Z_w}{\rho VS} = \frac{3}{2}C_D + \tfrac{1}{2}\frac{dC_L}{d\alpha} \qquad (2·5, 23)$$

and

$$\frac{X_w - Z_u}{\rho VS} = \frac{3}{2}C_L - \tfrac{1}{2}\frac{dC_D}{d\alpha}, \qquad (2·5, 24)$$

so both these expressions are independent of the choice of the reference axes in the body and are thus *invariants*.

When the force coefficient C_F is not independent of the speed we must write

$$\Delta C_F = \frac{\partial C_F}{\partial\alpha}\Delta\alpha + \frac{\partial C_F}{\partial V}\Delta V, \qquad (2·5, 25)$$

where, by (2·5, 6),

$$\Delta V = u\cos\alpha + w\sin\alpha. \qquad (2·5, 26)$$

Additional terms are accordingly introduced in equations (2·5, 10), (2·5, 12) and (2·5, 13). Suppose that C_F depends on V solely through the Mach number

$$\mathbf{M} = \frac{V}{a},$$

where the velocity of sound a is a constant. Then

$$V\frac{\partial C_F}{\partial V} = \mathbf{M}\frac{\partial C_F}{\partial\mathbf{M}} \qquad (2·5, 27)$$

and the expressions for the derivatives become

$$\frac{F_u}{\rho VS} = C_F \cos\alpha - \tfrac{1}{2}\frac{\partial C_F}{\partial\alpha}\sin\alpha + \tfrac{1}{2}M\frac{\partial C_F}{\partial M}\cos\alpha, \quad (2\cdot5,\,28)$$

$$\frac{F_w}{\rho VS} = C_F \sin\alpha + \tfrac{1}{2}\frac{\partial C_F}{\partial\alpha}\cos\alpha + \tfrac{1}{2}M\frac{\partial C_F}{\partial M}\sin\alpha. \quad (2\cdot5,\,29)$$

At small and moderate values of M, $\dfrac{\partial C_F}{\partial M}$ is usually small and the additions to the non-dimensional forms of the derivatives to allow for the influence of the compressibility of the air are still smaller.

The aerodynamic pitching moment derivatives are of particular importance in relation to stability. Now the pitching moment M about the origin is related to the corresponding non-dimensional coefficient C_M [cp. equation (2·5, 9)] by the equation

$$M = \tfrac{1}{2}\rho V^2 Sl C_M, \quad (2\cdot5,\,30)$$

where l is a length which may be identified with the mean chord $\bar{c}$ of an aerofoil or the 'tail arm'. When C_M is a function of incidence only we deduce by the method used in obtaining equations (2·5, 12) and (2·5, 13) that

$$\frac{M_u}{\rho VSl} = C_M \cos\alpha - \frac{1}{2}\frac{dC_M}{d\alpha}\sin\alpha, \quad (2\cdot5,\,31)$$

$$\frac{M_w}{\rho VSl} = C_M \sin\alpha + \frac{1}{2}\frac{dC_M}{d\alpha}\cos\alpha, \quad (2\cdot5,\,32)$$

while the additions to the expressions on the right-hand sides of these equations to allow for compressibility are $\tfrac{1}{2}M\dfrac{\partial C_M}{\partial M}\cos\alpha$ and $\tfrac{1}{2}M\dfrac{\partial C_M}{\partial M}\sin\alpha$ respectively.

For 'wind axes' ($\alpha = 0$) we get from (2·5, 31) and (2·5, 32) respectively

$$\frac{M_u}{\rho VSl} = C_M, \quad (2\cdot5,\,33)$$

$$\frac{M_w}{\rho VSl} = \frac{1}{2}\frac{dC_M}{d\alpha}. \quad (2\cdot5,\,34)$$

When the body is 'trimmed' for the datum incidence, C_M is zero and we deduce the important result that in these circumstances

$$M_u = 0. \qquad (2\cdot5, 35)$$

If, however, C_M depends on Mach number or the body considered is not rigid, this equation will cease to be valid.

It is important to recognize that in the foregoing calculations the increments and derivatives correspond strictly to steady conditions before and after the changes in the velocity components. Derivatives calculated in this way are in fact applicable when the changes of velocity are sufficiently slow. The precise conditions in which such *quasi-static derivatives* can legitimately be used are considered in § 3·5.

2.6   The rapid longitudinal oscillation

When an aircraft is flying steadily in a straight line the resultant pitching moment about the c.g. is zero. The aircraft is said to be trimmed and the angle of incidence is, for a given condition of loading, determined by the elevator setting (see § 2·2). Suppose now that the aircraft is thrown out of trim, as by a gust of short duration or a purposeful movement of the elevator which is then brought back to its original setting. If the aeroplane is thoroughly stable the angle of incidence will quickly but not instantaneously become steady at its original value. There is a movement of the aircraft which Melvill Jones has called the rapid incidence adjustment. Strictly, this movement has normal and longitudinal translational components, but the important component is in pitch. In the elementary and approximate theory now advanced it will be assumed that the rapid incidence adjustment is a motion in pitch alone.

We shall assume that during the incidence adjustment the elevator is fixed and that a certain point P of the aircraft continues to move in a straight line with the speed of flight V. This point will be near the c.g. but may not coincide with it. Accordingly the motion has just one freedom, namely, angular pitch θ. The equation of motion can with sufficient accuracy be written

$$\mathscr{A}\frac{d^2\theta}{dt^2} + \mathscr{B}\frac{d\theta}{dt} + \mathscr{C}\theta = 0, \qquad (2\cdot6, 1)$$

where $\mathscr{A}$ = moment of inertia of the aircraft about P with an added allowance for the 'virtual inertia' of the air,

$\mathscr{B}$ = aerodynamic damping coefficient, which is proportional to ρV

and $\mathscr{C}$ = aerodynamic stiffness coefficient, which is proportional to ρV^2.

Both the coefficients $\mathscr{B}$ and $\mathscr{C}$ are mainly attributable to the tailplane. To calculate these we must find the total change of incidence of the tailplane.

First, the general change of attitude θ would give an equal change in the effective incidence of the tailplane were it not that increase of the incidence of the main lifting surface brings about an increase ϵ in the angle of downwash at the tail. The downwash can be taken to be proportional to θ,* and the net increase of tail incidence associated with θ is

$$\theta\left(1 - \frac{d\epsilon}{d\theta}\right).$$

But the angular velocity $\dfrac{d\theta}{dt}$ also gives rise to an incidence change at the tail. Let l' be the distance from P to the aerodynamic centre of the tailplane. Then the linear velocity of this centre due to the angular velocity of the aircraft is $l'\dfrac{d\theta}{dt}$, and this will be approximately at right angles to the airflow at the tail. Hence there is an incidence change at the tail given approximately by

$$\frac{l'}{V}\frac{d\theta}{dt}$$

since the angular velocity is supposed to be small (see §2·5). Thus the whole incidence change of the tailplane is approximately

$$\Delta\alpha' = \frac{l'}{V}\frac{d\theta}{dt} + \left(1 - \frac{d\epsilon}{d\theta}\right)\theta. \qquad (2\cdot6, 2)$$

The resulting change of tail lift is

$$\Delta L' = \tfrac{1}{2}k\rho S' V^2 \frac{dC_{L'}}{d\alpha'}\Delta\alpha',$$

---

* There is an approximation here since the downwash is not instantaneously established at the tail.

where S' is the area of the tailplane and the factor k allows for the tailplane being somewhat affected by the wake from the wings and body. Accordingly the pitching moment due to the incidence changes at the tail is

$$\Delta M' = -l'\Delta L'$$

$$= -\tfrac{1}{2}k\rho S'l'\frac{dC_{L'}}{d\alpha'}\left[Vl'\frac{d\theta}{dt} + V^2\left(1 - \frac{d\epsilon}{d\theta}\right)\theta\right]. \quad (2\cdot 6,\ 3)$$

Hence the contribution of the tail to the damping coefficient $\mathscr{B}$ is

$$\mathscr{B}' = \tfrac{1}{2}\rho Vkl'^2 S'\frac{dC_{L'}}{d\alpha'} \quad\quad\quad (2\cdot 6,\ 4)$$

and the contribution to $\mathscr{C}$ is

$$\mathscr{C}' = \tfrac{1}{2}\rho V^2 kl' S'\left(1 - \frac{d\epsilon}{d\theta}\right)\frac{dC_{L'}}{d\alpha'}. \quad (2\cdot 6,\ 5)$$

The part contributed by the wings and body to $\mathscr{B}$ is relatively unimportant for a conventional aeroplane; their contribution to $\mathscr{C}$ is commonly negative.

We now return to the dynamical equation $(2\cdot 6,\ 1)$. It is usual for the inequality

$$4\mathscr{A}\mathscr{C} > \mathscr{B}^2 \quad\quad\quad (2\cdot 6,\ 6)$$

to be satisfied and accordingly the solution can be written

$$\theta = \Theta e^{-\mu t}\sin(\omega t + \eta), \quad\quad (2\cdot 6,\ 7)$$

where Θ and η are arbitrary constants of integration, while

$$\mu = \frac{\mathscr{B}}{2\mathscr{A}} \quad\quad\quad (2\cdot 6,\ 8)$$

and

$$\omega = +\sqrt{\left(\frac{\mathscr{C}}{\mathscr{A}} - \frac{\mathscr{B}^2}{4\mathscr{A}^2}\right)}. \quad (2\cdot 6,\ 9)$$

Since $\mathscr{A}$ is independent of V, it follows* from $(2\cdot 6,\ 4)$ and $(2\cdot 6,\ 5)$ that μ and ω are both proportional to V. Hence for a normal stable aircraft the rapid incidence adjustment takes the form of a damped oscillation whose damping coefficient and frequency are both approximately proportional to the speed of flight.

* The contributions of other parts of the aircraft to $\mathscr{B}$ and $\mathscr{C}$ depend on speed in the same manner as $\mathscr{B}'$ and $\mathscr{C}'$.

Equation (2·6, 5) shows that the larger is the downwash factor $\dfrac{d\epsilon}{d\theta}$ the smaller is the restoring moment coefficient $\mathscr{C}$. This implies reduced static stability (see Chapter 10). The downwash factor for propeller-driven aircraft is substantially greater with power on than in a glide and the effect of the slipstream is specially marked for twin-engined aircraft. It appears that the slipstream in the region of the tail has the shape of a wide but rather shallow band, and within it the downwash is particularly large. The loss of stability with power on may amount to as much as 0·15 on the 'c.g. margin'. (See § 5·6.)

2·7 The phugoid motion

When a symmetrical aircraft has its attitude in space or angle of incidence altered by use of the controls or by a gust in such a manner that after the disturbing action has ceased the velocity of the c.g. is in the vertical plane of symmetry and the only angular movement is in pitch, it is said to have sustained a symmetrical disturbance. The first result of such a disturbance is the rapid incidence adjustment already considered in § 2·6, but when the incidence adjustment is complete the motion may still be unsteady. The remaining motion is of the kind called by Lanchester a *phugoid*. We shall now give a much simplified theory of the phugoid motion.

In a phugoid motion the angular pitching movement of the aircraft is at all times very slow and this implies that the aerodynamic pitching moment is nearly zero. Hence the angle of incidence is at all times very nearly that corresponding to zero pitching moment. With fixed elevator and for flight at a low or moderate Mach number this implies nearly constant incidence. We shall assume that the incidence, and with it the lift coefficient, is strictly constant. We shall also neglect the drag for the present since this has a minor influence on the motion.*

Since we have assumed the net drag to be zero, the total mechanical energy of the aircraft is constant by the Principle of the Conservation of Energy.† Also the rotary kinetic energy

---

* Or we may suppose that the propulsive thrust is adjusted exactly to balance the drag at all instants. The drag is important in relation to the stability of the motion but has only a small effect within a single period.

† The lift, being perpendicular to the velocity, does no work.

is negligible since the pitching velocity is very small. Hence if z is the height above a fixed datum we have

$$m(gz + \tfrac{1}{2}V^2) = \text{const.} \tag{2·7, 1}$$

Since the lift coefficient is constant we have

$$L = kV^2, \tag{2·7, 2}$$

where

$$k = \tfrac{1}{2}\rho S C_L. \tag{2·7, 3}$$

At first we shall suppose the path to deviate only slightly from the horizontal and we can accordingly equate the vertical aerodynamic force to the lift. Hence the equation for the vertical acceleration is

$$m\frac{d^2z}{dt^2} = kV^2 - mg. \tag{2·7, 4}$$

Suppose now that V_0 is the velocity of flight for equilibrium. Then by the last equation

$$kV_0^2 = mg \tag{2·7, 5}$$

and it can be rewritten

$$m\frac{d^2z}{dt^2} = k(V^2 - V_0^2). \tag{2·7, 6}$$

Define z_0 by the equation [see (2·7, 1)]

$$gz_0 + \tfrac{1}{2}V_0^2 = gz + \tfrac{1}{2}V^2, \tag{2·7, 7}$$

so z_0 is the height for steady flight in equilibrium with the total energy unchanged. By (2·7, 7) equation (2·7, 6) becomes

$$m\frac{d^2z}{dt^2} = 2kg(z_0 - z)$$

or by (2·7, 5)

$$\frac{d^2z'}{dt^2} + \frac{2g^2}{V_0^2}z' = 0, \tag{2·7, 8}$$

where

$$z' = z - z_0. \tag{2·7, 9}$$

Equation (2·7, 8) represents a simple harmonic motion with periodic time

$$\left.\begin{aligned} T &= \pi\sqrt{(2)}\left(\frac{V_0}{g}\right) \\ &= 4\cdot44\left(\frac{V_0}{g}\right). \end{aligned}\right\} \tag{2·7, 10}$$

It now readily follows that V_0 is the mean speed of flight in a complete period and that the wavelength of the undulatory path is*

$$\lambda = V_0 T = 4.44\left(\frac{V_0^2}{g}\right). \qquad (2.7, 11)$$

The amplitude of the oscillation in height depends on the initial conditions and can be investigated as follows. Suppose that at time $t = 0$ the path of the c.g. is inclined upward at the angle Θ relative to the equilibrium path and that the true speed is $V_0 + u$, so the deviations from the state of equilibrium are Θ and u, which are taken as small quantities of the first order. The general solution of (2.7, 8) is

$$z' = A \cos \omega t + B \sin \omega t, \qquad (2.7, 12)$$

where
$$\omega = \frac{g\sqrt{(2)}}{V_0}. \qquad (2.7, 13)$$

At time $t = 0$ we obtain

$$\frac{dz'}{dt} = \omega B = V_0 \Theta,$$

when small quantities of the second order are neglected. Hence

$$B = \frac{V_0 \Theta}{\omega}. \qquad (2.7, 14)$$

By the equation of energy

$$(V_0 + u)^2 = V_0^2 - 2gz'$$

or, when the second order term is neglected,

$$V_0 u = -gz' = -gA, \quad \text{when} \quad t = 0.$$

Therefore
$$A = -\frac{V_0 u}{g}, \qquad (2.7, 15)$$

and if h is the amplitude in height in the phugoid oscillation, then

$$h = \sqrt{(A^2 + B^2)}$$

$$= \frac{V_0}{g}\sqrt{(u^2 + \tfrac{1}{2}V_0^2 \Theta^2)}, \qquad (2.7, 16)$$

* The Froude number based on V_0 and λ is thus $\dfrac{V_0^2}{\lambda g} = 0.225$, a constant.

on substitution from (2·7, 13), (2·7, 14) and (2·7, 15). It is noteworthy that the amplitude of the phugoid can only be zero when both u and Θ vanish.

With the same simplifying assumptions as before we can investigate the general phugoid motion as follows. The equation of energy (2·7, 1) requires the velocity to vanish at a certain height and if h be measured *downwards* from this datum level we can write

$$V^2 = 2gh. \tag{2·7, 17}$$

Also the net force on the aircraft along the normal to its path is

$$F = L - mg \cos \Theta.$$

But, if R be the radius of curvature of the path, the normal acceleration is

$$\frac{V^2}{R} = \frac{F}{m} = \frac{kV^2}{m} - g \cos \Theta.$$

Hence

$$\frac{1}{R} = \frac{k}{m} - \frac{\cos \Theta}{2h}$$

by (2·7, 17). Let h_0 be the value of h for steady horizontal flight with the same total energy. Then the last equation gives, since the radius of curvature is now infinite,

$$\frac{k}{m} = \frac{1}{2h_0}$$

and it can therefore be rewritten

$$\frac{1}{R} = \frac{1}{2h_0} - \frac{\cos \Theta}{2h}. \tag{2·7, 18}$$

This equation defines geometrically the flight path which can indeed be constructed approximately from short circular arcs by its application.

Equation (2·7, 18) can be integrated once exactly for, if ds is an element of arc of the path, the curvature is

$$\frac{1}{R} = \frac{d\Theta}{ds} = -\frac{\sin \Theta \, d\Theta}{dh},$$

since

$$dh = -ds \sin \Theta,$$

or

$$\frac{1}{R} = \frac{d \cos \Theta}{dh}.$$

Hence we get $$\frac{d\cos\Theta}{dh}+\frac{\cos\Theta}{2h}=\frac{1}{2h_0}$$

or $$\frac{d}{dh}\left(h^{\frac{1}{2}}\cos\Theta\right)=\frac{h^{\frac{1}{2}}}{2h_0}.$$

Therefore $$\cos\Theta=\frac{1}{3}\left(\frac{h}{h_0}\right)+A\sqrt{\left(\frac{h_0}{h}\right)},\qquad(2\cdot7,19)$$

where A is a constant of integration; this is Lanchester's equation for the path. Let Θ_0 correspond to the level h_0. Then the last equation yields

$$A=\cos\Theta_0-\tfrac{1}{3}.\qquad(2\cdot7,20)$$

Let H stand for the function of h on the right-hand side of $(2\cdot7,19)$ and let x be the abscissa. Then, since

$$\tan\Theta=-\frac{dh}{dx},$$

we derive $$x=\int\frac{H\,dh}{\sqrt{(1-H^2)}},\qquad(2\cdot7,21)$$

but this does not, in general, lead to a simple expression for x. In the special case where A is zero, or $\cos\Theta_0$ is 1/3, $(2\cdot7,19)$ shows immediately that the equation to the path is

$$x^2+h^2=9h_0^2.\qquad(2\cdot7,22)$$

Since the velocity vanishes when h is zero, only the lower semicircle is relevant. The complete phugoid should be regarded as an indefinitely extended set of semicircles of radius $3h_0$ touching at the ends of their horizontal diameters. That a semicircle is a possible phugoid path can be seen very easily by considering a simple pendulum with the bob released from rest at the level of the fulcrum. It follows from elementary dynamics that the tension in the shank is then proportional to the square of the velocity and the tension can therefore be replaced by a lift force proportional to the square of the velocity.

The value $\arccos(\tfrac{1}{3})$ for Θ_0 is critical. When Θ_0 is smaller than this and positive, A is positive and equation $(2\cdot7,19)$ shows that $\cos\Theta$ is always positive, for h and h_0 are necessarily positive. Hence the path never becomes vertical or looped; it

3

is in fact undular and tends to the sinusoidal form already investigated when Θ_0 is small. In general the curvature at the peaks of the path is numerically greater than at the troughs.

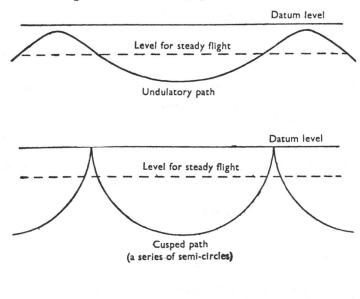

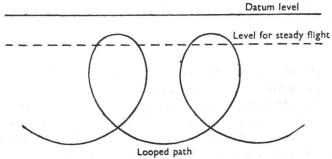

Fig. 2·7, 1. The three types of idealized phugoid path.

But when Θ_0 is positive and exceeds the critical angle, A is negative and (2·7, 9) can be satisfied with $\cos \Theta$ equal to -1 and h positive. Hence the paths are looped when Θ_0 is positive and exceeds the critical angle.* The three types of phugoid path are shown in Fig. 2·7, 1.

* Changing the sign of Θ_0 of course does not alter the nature of the path.

2·8   Pure rolling

In order to rotate the aircraft about the longitudinal axis it is necessary to apply a rolling moment. This is usually effected by a device which transfers a fraction of the lift on one wing to the opposite wing. The Wright brothers did this by 'wing warping', i.e. deliberately applied twist of the wings themselves, but at present separately hinged parts of the wings, called ailerons, are usually adopted. The ailerons on opposite wings are geared together so that when one moves up the other moves down.

We shall examine the theory of rolling subject to certain simplifying assumptions. Thus we shall suppose that the only effect of applying the ailerons is to cause a rotation ϕ about the longitudinal axis OX, the uniform motion in all other respects continuing. This treatment is incomplete since the ailerons when deflected from their neutral setting usually give rise to a yawing moment as well as a rolling moment. Also, as soon as the wing span has rolled over from the horizontal, the *forces* on the aircraft cease to be in equilibrium, for weight and lift no longer balance. The result is the development of sideslip with lateral and vertical deviations of the flight path. However, perfect ailerons would give no yawing moment while the perturbations of the general motion are slight in the early stages of the roll and in rapid continued rolling.

We shall suppose that when the ailerons are deflected from their neutral setting through the angle ξ the rolling moment is

$$L = l\rho V^2\xi, \qquad (2\cdot8, 1)$$

where l is a constant having the physical dimensions of a volume. As soon as the wing has acquired an angular rolling velocity p there is brought into play an aerodynamic moment which *opposes* the rotation and is proportional to p. This can be regarded as caused by an effective increase of wing incidence on the downgoing wing with a corresponding decrease on the upgoing wing. These incidence changes are inversely proportional to the speed of flight and we shall accordingly assume that the resisting moment is given by

$$L' = k\rho Vp, \qquad (2\cdot8, 2)$$

where k is a constant having the physical dimensions of (length)[4]. Then, if the moment of inertia of the aircraft about

OX is A, the equation of rolling moments is

$$A\frac{dp}{dt}+L'=L$$

or $$A\frac{dp}{dt}+k\rho Vp=l\rho V^2\xi. \tag{2·8, 3}$$

Subject to the simplifications introduced, this equation enables us to answer all questions about the response of the aircraft to aileron movement.*

In a first application of the equation, let us suppose that at time $t = 0$ the angular velocity is zero while the aileron deflection is steady. Then it is easy to verify† that the initial condition and equation (2·8, 3) are both satisfied by

$$p = \frac{lV\xi}{k}(1-e^{-\lambda t}), \tag{2·8, 4}$$

where $$\lambda = \frac{k\rho V}{A}. \tag{2·8, 5}$$

When t is large the exponential term becomes negligible and there is a steady rate of roll given by

$$p = \frac{lV\xi}{k}. \tag{2·8, 6}$$

This is proportional to the aileron deflection and to the speed of flight. For a flexible aircraft the rate of roll is only proportional to the speed at low and moderate speeds (see § 2·12). Numerical calculations for typical aircraft show that the steady rate of rotation is very closely attained in a fraction of a second.

Next suppose that ξ varies linearly with time, say

$$\xi = \xi_1 t, \tag{2·8, 7}$$

while p is again zero when $t = 0$. The solution is now easily shown to be

$$p = \frac{lV\xi_1 t}{k} - \frac{Al\xi_1}{\rho k^2}(1-e^{-\lambda t}), \tag{2·8, 8}$$

---

* In terms of the standard non-dimensional derivative coefficients
$$l = \tfrac{1}{2}bSl_\xi$$
and $k = -\tfrac{1}{4}b^2Sl_p$.

† The general solution of (2·8, 3) is given in (2·8, 10).

where λ is given by (2·8, 5). When t is large* we obtain

$$p = \frac{lV}{k}\left(\xi - \frac{A\xi_1}{k\rho V}\right), \qquad (2\cdot 8, 9)$$

so there is effectively a small constant deduction from the aileron angle. This may be called the angle of apparent lag (see Fig. 2·8, 1).

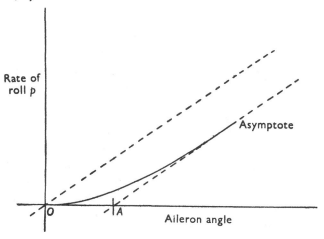

Fig. 2·8, 1. Response in roll to linearly applied aileron. OA is the apparent angle of lag.

When the aileron angle varies with time in accordance with a function $\xi(t)$ the solution of (2·8, 3) can be written, for the condition $p = 0$ when $t = 0$,

$$p = \frac{l\rho V^2}{A}e^{-\lambda t}\int_0^t e^{\lambda t}\xi(t)\,dt, \qquad (2\cdot 8, 10)$$

where λ is given by (2·8, 5). Equations (2·8, 4) and (2·8, 8) are special cases of this.

2·9   The influence of sideslip

When the velocity vector for the c.g. of an aircraft does not lie in the plane of symmetry of the aircraft there is a component of the velocity perpendicular to the plane. This is known as the velocity of sideslip. We shall now consider in an elementary

---

* The exponential term is negligible in practice when $\lambda t > 2$.

way the influence of sideslip when all three components of the angular velocity are zero.

Suppose for definiteness that the sideslip is towards the right, as shown in Fig. 2·9, 1. Then the velocity of the air stream relative to the aircraft is backward and towards the left, so the plane of symmetry is inclined to the wind at an angle

$$\beta = \frac{v}{V}, \qquad (2·9, 1)$$

where v is the velocity of sideslip (supposed small in relation to V). The fin, rudder (which will be supposed fixed in the neutral position) and sides

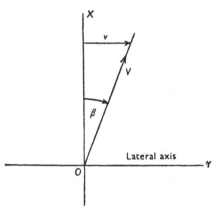

Fig. 2·9, 1.   Lateral velocity and angle of sideslip.

of the body are then surfaces making an angle of incidence with the wind. Consequently there is a lateral force Y perpendicular to the plane of symmetry and, in general, also a yawing moment N tending to rotate the aircraft about an axis perpendicular to the plane of the wings.

The first matter of importance is the sense of the moment N. If in the case considered this tends to turn the aircraft in the clockwise sense when seen from above, the response of the aircraft will be an angular movement in the clockwise direction and the angle β of sideslip will be reduced. Consequently the yawing moment will also be reduced and a new position of equilibrium can be reached for which the sideslip is zero. The aircraft then has positive 'weathercock stability'. This can be secured by making the fin and rudder sufficiently large and the whole discussion closely follows that given for the case of pitch in § 2·6 with angle of sideslip replacing angle of pitch and angular velocity in yaw replacing angular velocity in pitch. However, the induced angle of sidewash at the fin is much less than the induced angle of downwash at the tailplane under corresponding conditions.

The second important effect of sideslip is the rolling moment which is present when the wings have a dihedral angle Γ. In Fig. 2·9, 2 a pair of monoplane wings is supposed viewed from

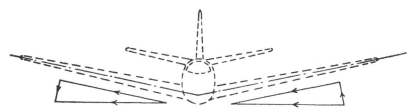

Fig. 2·9, 2. Monoplane with dihedral angle. Resolution of velocity of sideslip on starboard and port wings.

the rear and the dihedral angle is exaggerated for the sake of clearness. The velocity v of sideslip can be resolved into a component $v \cos \Gamma$ in the plane of the starboard wing and a component $v \sin \Gamma$ perpendicular to it. For a wing of infinite span a velocity parallel to the span has no influence on the aero-dynamic forces and for a finite wing the effect is slight. Hence, as a first approximation, we may neglect the component $v \cos \Gamma$. However, the component $v \sin \Gamma$ gives an increase of incidence

$$\Delta \alpha = \frac{v \sin \Gamma}{V} \qquad (2·9, 2)$$

(see § 2·5). The lift on the starboard wing is thus increased and there will be an equal or nearly equal decrease of lift on the port wing. Thus the net effect is a rolling moment tending to raise the starboard and depress the port wing. The presence of the fuselage complicates the phenomena but leaves the effect of the sideslip broadly the same.

Suppose now that an aircraft with positive weathercock stability and positive dihedral angle (as in Fig. 2·9, 2) is accidentally rolled so that the starboard wing is slightly depressed. The weight and lift force are no longer coincident in direction and there is an unbalanced component of force tending to cause sideslip to starboard. Hence sideslip develops in this sense and brings into play moments tending to raise the starboard wing and to rotate the aircraft in yaw in a clockwise sense when viewed from above. The full consequences of the disturbance can only be worked out by aid of the complete dynamical equations, but the foregoing argument shows how a positive dihedral angle tends to give stability in roll. However, the thorough treatment shows that an increase of dihedral angle is not necessarily beneficial.

2·10   Preliminary ideas on stability

The ideas of *static stability* and of *stiffness* are closely related. The dictionary definition of *stiff* is *not easily bent* and this can be at once extended to *not easily deflected*. Admiral Smyth's *Sailor's Word-Book* gives the definition: '*Stiff*. Stable or steady; the opposite to *crank*; a quality by which a ship stands up to her canvas, and carries enough sail without heeling over too much.' Now these same ideas apply as well to aircraft as to ships or springs, but the *deflection* may be a change of attitude or of velocity or, more usually, of both together, and the disturbing cause of the deflection may be the operation of a control, an air bump, etc. We shall now consider the manner in which the aircraft responds to displacement of the elevator, and for simplicity we shall confine attention to the ultimate result of this displacement, which will be supposed steady. Thus, after the elevator has been moved to a new setting and steady flight has been resumed, the lift will be the same as before.*

The essential quantities to be recorded in a test of elevator control are the control setting, the speed, the all-up weight and the c.g. position. From the weight and the e.a.s. the lift coefficient is calculated and a graph is drawn showing the relation between C_L and the elevator angle η for a given fixed position of the c.g. Such a graph is called a *trim curve* and it is usual to plot the trim curves for a number of c.g. positions in the same diagram (see Fig. 2·10, 1). Now it is clear that the slope of the trim curves must be negative for positive stability when elevator deflections are regarded as positive when the trailing edge moves downward. For, a positive elevator deflection increases the tail load, giving a nose-down pitching moment on the aircraft. If it is stable its angle of incidence will change *in the same sense* as the applied moment, i.e. the incidence will be reduced and C_L likewise. Moreover, the more stable or stiff

* The flight paths considered are so nearly horizontal that cos Θ can always be taken as equal to 1.

the aircraft is the less will be the change in incidence or C_L for a given movement of the elevator, and the greater will be the negative slope of the trim curve. Now, for a rigid aircraft flying at speeds below those where the compressibility of the air begins to tell, and at incidences not too close to the stalling angle, the trim curves are straight or very nearly so. Hence the slope is characteristic of the c.g. position and there is a particular c.g. position for which the slope is zero. (See the graph marked c.g. neutral in the figure.) This is the c.g. position for

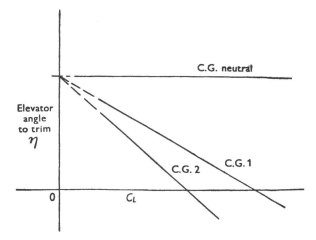

Fig. 2·10, 1. Trim curves showing the dependence of stick-fixed longitudinal stability on c.g. position.

neutral stability, *stick-fixed*, or the stick-fixed neutral point. The reason for adding the epithet *stick-fixed* will be made clear shortly.

The statement made above that the angle of incidence of a stable aircraft changes in the same sense as the applied pitching moment requires examination and leads us to the fundamental conception of static stability. For simplicity, consider first a pendulum hanging under gravity and an inverted pendulum (see Fig. 2·10, 2).

When the bob of the hanging pendulum is deflected, say to the right, it tends to return to its position of equilibrium and a force towards the right must be applied in order to hold it. Thus the tendency is for a small displacement of the bob to be annulled, and positive work must be done on the bob to pull it aside. The hanging pendulum is therefore stable and has positive stiffness for angular displacements about the fulcrum. All this is reversed for the inverted pendulum; a small displacement from the position of equilibrium tends to grow and the force required to hold the bob when displaced to the right is directed towards the left. The inverted pendulum is therefore unstable and it is capable of doing positive work when allowed to deflect.

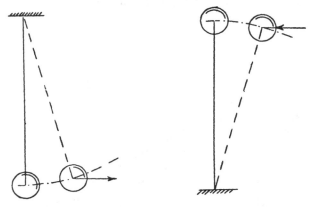

Hanging pendulum (stable)   Inverted pendulum (unstable)

Fig. 2·10, 2.

If the hanging pendulum were provided with a jockey weight which could be slid along the shank and beyond the fulcrum, a position for this weight could be found which rendered the pendulum neutrally stable. Here the effective stiffness is zero and a finite deflection can be produced by a vanishingly small force. Returning now to the aircraft we see that it is stable when a nose-down applied pitching moment brings about a nose-down angular displacement in pitch. Detailed calculation shows that this is accompanied by a decrease in the angle of incidence and an increase in the speed of flight.

The trim curves so far described enable us to find the stability characteristics of the aircraft with fixed elevator, usually known as the stick-fixed case. Although the elevators are moved between tests, they are kept fixed while the aircraft is settling down to its new condition of flight. The movement of the elevator is the most convenient means of applying a pitching moment, but the result is the same as if the elevator were kept fixed at all times and pitching moments applied independently.*

Now suppose that the elevators are left quite free to take up the angular setting for which their total hinge moment is zero, and let the elevator trimmer be adjusted until the desired speed of flight is attained. In general the stability of the aircraft in this condition will differ from that at the same speed with fixed elevator; the stability will be greater or less according to the manner in which the elevator responds to changes of incidence of the tailplane.† The stability with free elevator is usually

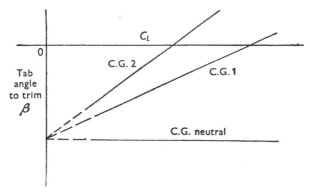

Fig. 2·10, 3. Trim curves showing the dependence of stick-free longitudinal stability on c.g. position.

called the *stick-free* stability. Static stick-free stability is explored with the help of trim curves in which the angle β at which the elevator trimmer is set to the elevator is plotted against the lift coefficient in the corresponding steady state (see Fig. 2·10, 3). The principle of the test is that a moment is applied to the free elevator by the displacement of the tab, and the nature of the response of the aircraft as a whole is related to the stability in the same general manner as in the stick-fixed case.

When the c.g. is at the stick-fixed neutral point an exceedingly small *movement* of the elevator suffices to bring about a large (theoretically infinite) change in the trimmed lift coefficient. The distance of the c.g. forward of this stick-fixed neutral

* This is not quite accurate since deflection of the elevator with constant wing incidence alters the lift coefficient of the aircraft slightly and the drag coefficient also.

† On this, see § 10·4.

point is proportional to the stick movement per unit of lift coefficient. On the other hand, when the c.g. is at the stick-free neutral point, the application of an exceedingly small *stick force* suffices to bring about a large change in the trimmed lift coefficient. The distance of the c.g. forward of the stick-free neutral point is proportional to the stick force per unit of lift coefficient. Since most pilots are more sensitive to changes of stick force than to changes of stick position, it may be concluded that it is usually of greater importance to secure the right degree of stick-free rather than of stick-fixed longitudinal stability. *Both* stabilities should normally be positive.

The general concept of stability is related to the *ultimate* result of the transitory application of a small disturbance. If the deviations of the motion ultimately die away the aircraft is stable; otherwise it is unstable. In accordance with this definition, it is possible for the deviations to grow larger for a time without violating the condition for stability and we recognize that the aircraft which is steadiest in service may not have the highest measures of stability. The consideration of the whole history of the deviations caused by disturbances falls under the heading of *response* and the means for minimizing unwanted responses at all stages of the motion is called *stabilization*. The study of stabilization is of particular importance for aircraft with automatic controls.

Instabilities are of two main types, which are described as *divergent* and *oscillatory* respectively. A *divergence* is a motion in which the deviations from the initial state of steady motion are proportional to a real growing exponential function of the time, i.e. $exp\,(\mu t)$, with μ real and positive; the corresponding damped or stable motion, for which μ is real and negative, is called a *subsidence*. For an oscillatory instability the deviations are proportional to $exp\,(\mu t)\sin\,(\omega t + \epsilon)$, with μ real and positive, and the corresponding stable motion is a damped oscillation. The definitions as given apply strictly only so long as the deviations are very small. More generally, we may say that a divergence is an instability in which the deviations grow and are unidirectional, while an oscillatory instability is one in which the deviations change sign with more or less regular periodicity while the amplitudes of the excursions increase with time. It is possible for an instability to begin as a divergence and change into an oscillation and vice versa. A simple example of divergent instability is provided by torsional divergence of a wing (see § 2·11) while unstable phugoid oscillations and wing flutter are examples of oscillatory instability. Stability, response and stabilization can only be investigated thoroughly by means of the complete dynamical equations of the disturbed motion.

2·11   Torsional divergence of a wing

Consider first a rigid aerofoil, untapered and untwisted, mounted on a fixed spindle parallel to its span, and quite free to rotate in pitch about the spindle except for the restraint of a spring giving a restoring moment proportional to the angle of pitch measured from a certain datum. When this is placed in a uniform stream whose direction is perpendicular to the spindle, the effective stiffness of the spring will be modified by the addition to it of an aerodynamic stiffness which is proportional to ρV^2. The sign and magnitude of this aerodynamic stiffness depend on the chordwise position of the axis of the spindle, which we shall for simplicity assume to intersect the chord line. It vanishes when the axis is at the aerodynamic centre, which lies near the quarter-chord point, for the aerodynamic pitching moment about the aerodynamic centre is independent of the angle of incidence so long as this is small. For an axis forward of the aerodynamic centre the aerodynamic stiffness is positive, while for an axis aft of this point it is negative. Suppose now that the axis is situated at a fraction h of the chord c aft of the aerodynamic centre. Then the aerodynamic pitching stiffness will be given by

$$m_a = -h\rho c V^2 S k, \qquad (2\cdot11, 1)$$

where k is a non-dimensional coefficient which, according to the elementary theory of aerofoils, can be identified with $\frac{1}{2}\frac{dC_L}{d\alpha}$. Let m_θ be the pitching stiffness due to the elasticity of the spring. Then the total pitching stiffness is

$$m = m_\theta + m_a$$
$$= m_\theta - h\rho c V^2 S k. \qquad (2\cdot11, 2)$$

The total stiffness vanishes and the stability becomes neutral at the critical divergence speed V_d given by

$$V_d = \sqrt{\left(\frac{m_\theta}{h\rho c S k}\right)}. \qquad (2\cdot11, 3)$$

For speeds less than V_d the stiffness is positive and the aerofoil stable, but for speeds greater than V_d the stiffness is negative and the aerofoil unstable. The equation shows that the divergence speed can be raised by increasing the elastic stiffness or by reducing h. If we determine the effective stiffness of an aerofoil mounted as described, by measuring the angular deflection

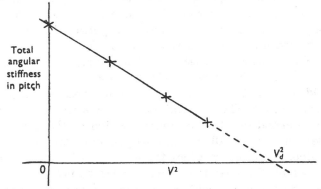

Fig. 2·11, 1. Determination of critical divergence speed from plot of stiffness against square of speed.

caused by the application of a known pitching moment, at a number of air speeds and plot the stiffness against V^2 we shall obtain a straight line, as shown in Fig. 2·11, 1. The line cuts the axis of abscissae at the point corresponding to the divergence speed, which can thus be determined by observations made only at lower speeds (see also § 12·3).

Next let us consider an elastic cantilever wing, provided with rigid support at the root, and having its quarter-chord axis perpendicular to the airstream. This behaves in the same general manner as the pivoted aerofoil with the pivot axis at the flexural axis of the wing, for the twist is determined by the total effective moment about the flexural axis.* The divergence

* We assume here for simplicity that the flexural centres at all sections lie on a straight axis, and that a normal load applied anywhere on this axis produces no pitch or incidence change anywhere.

speed can be found from a diagram similar to Fig. 2·11, 1, but the stiffness must be measured at the same fore-and-aft wing section throughout.

2·12   Reversal of control

When a rigid flap is hinged at or near its leading edge on a rigid fixed aerofoil towards its rear and the angle of incidence is distinctly less than the stalling angle, a deflection of the flap in the trailing edge down sense gives an increase both of the lift force on the wing and of the rolling moment about an axis near the wing root.* The deflection of the flap also brings into play a twisting moment tending to depress the leading edge and raise the trailing edge of the wing, i.e. tending to reduce the angle of incidence. When the wing is flexible the angle of incidence is in fact reduced by an amount proportional to the applied twisting moment. Hence there is a reduction of the additional lift and rolling moment caused by the flap deflection, and the flap loses some of its effectiveness. Moreover, the fractional loss of effectiveness increases with speed, for the twisting moment corresponding to a given flap setting is proportional to ρV^2 and the wing twist increases accordingly. Hence there will be a critical speed V_r at which the flap is totally ineffective, while for yet higher speeds the rolling action will be reversed. It is worth remarking that in the Flettner rudder and in any control with pure aerodynamic servo operation by a tab, the normal operation is the reversed one since the rudder or other principal control surface has no stiffness for angular movement about its hinge. This is the case where the reversal speed V_r is zero.

Reversal of control can easily be shown by a wind tunnel experiment in which a flexible wing provided with an aileron, which can be set at any required angle, is mounted on a balance which permits the rolling moment to be measured while the angle of incidence at the root is kept constant.† The rolling moment for each aileron setting is plotted against V or V^2 and it is found that all the curves pass through a common point at

* With the usual sign convention for rolling moment this would imply that we were considering a port wing and aileron.

† In a simpler version of the experiment the wing is rigidly fixed at the root and the flexural displacement of the flexural centre at the wing tip is observed. This is very nearly proportional to the rolling moment.

a certain speed. This is the critical reversal speed, for here the rolling moment is independent of the angular setting of the aileron (see Fig. 2·12, 1).

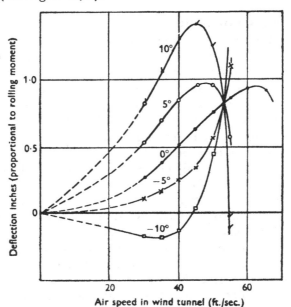

Fig. 2·12, 1. Determination of critical reversal speed.

An elementary theory of reversal can easily be worked out for the simple case of a rigid rectangular aerofoil mounted on a spindle at a distance hc aft of the aerodynamic centre (cp. § 2·11), provided with a full span rigid flap and having spring constraint against rotation in pitch. Let θ and ξ be the angle of pitch and flap setting respectively. Then the increment of lift coefficient above the value when ξ and θ are both zero is

$$\Delta C_L = a_1 \theta + a_2 \xi, \qquad (2\cdot12, 1)$$

where
$$a_1 = \frac{\partial C_L}{\partial \theta} \left.\right\}$$

and
$$a_2 = \frac{\partial C_L}{\partial \xi}. \qquad (2\cdot12, 2)$$

Equation (2·12, 1) implies that the load produced by the flap deflection alone is
$$\tfrac{1}{2} a_2 \rho V^2 S \xi.$$

Let jc be the distance of the centre of pressure of the aero-dynamic load caused by operation of the flap aft of the aero-dynamic centre, so the distance aft of the pitching axis is $(j-h)c$. Hence the pitching moment due to the flap deflection alone is

$$-\tfrac{1}{2}a_2(j-h)\rho V^2 cS\xi,$$

where the moment is reckoned positive in the trailing edge down sense. The pitching moment due to the displacement in pitch is

$$+\tfrac{1}{2}a_1 h\rho V^2 cS\theta.$$

When the total pitching moment is equated to the elastic restoring moment we obtain

$$-\tfrac{1}{2}\{a_2(j-h)\,\xi - a_1 h\theta\}\rho V^2 cS = m_\theta\theta.$$

Hence
$$\theta = \frac{-(j-h)\,a_2\,\xi}{\left(\dfrac{2m_\theta}{\rho V^2 cS}\right)-ha_1}. \qquad (2\cdot12,\,3)$$

It follows from $(2\cdot12,\,1)$ that the relative effectiveness of the flap is

$$\eta = \frac{\Delta C_L}{a_2\xi}, \qquad (2\cdot12,\,4)$$

for the denominator is what ΔC_L would be if θ were zero (rigid constraint in pitch). By $(2\cdot12,\,3)$ the last equation becomes on reduction

$$\eta = \frac{\left(\dfrac{2m_\theta}{\rho V^2 cS}\right)-ja_1}{\left(\dfrac{2m_\theta}{\rho V^2 cS}\right)-ha_1}. \qquad (2\cdot12,\,5)$$

The effectiveness vanishes at the critical reversal speed V_r and therefore*

$$V_r = \sqrt{\left(\frac{2m_\theta}{ja_1\rho cS}\right)}. \qquad (2\cdot12,\,6)$$

* If we take the aerodynamic centre to be at the quarter-chord point in accordance with the theory of thin aerofoils we have

$$k_{cp} = 0\cdot25 + j = 0\cdot25 + \frac{m}{a_2} \text{ by equation } (7\cdot2,\,20).$$

Hence
$$j = \frac{m}{a_2}$$

and
$$V_r = \sqrt{\left(\frac{a_2}{ma_1}\right)}\sqrt{\left(\frac{m_\theta}{\tfrac{1}{2}\rho cS}\right)}.$$

This is a much used form of equation $(2\cdot12,\,6)$.

4

It appears that the reversal speed is proportional to the square root of the elastic stiffness and independent of the position of the pitching axis. The latter conclusion may be surprising, but it can be proved independently as follows. At the critical speed the lift force caused by operation of the flap is zero and the load reduces to a couple. The twist brought about by the couple is independent of the situation of the pitching axis and the conclusion follows at once. Another conclusion to be drawn from (2·12, 6) is that a high reversal speed can be attained by making j small, but it is not easy to design for this.

Equation (2·12, 6) gives

$$\frac{2m_\theta}{\rho V^2 cS} = ja_1\left(\frac{V_r}{V}\right)^2$$

and equation (2·12, 5) can accordingly be written

$$\eta = \frac{1-\left(\frac{V}{V_r}\right)^2}{1-\left(\frac{h}{j}\right)\left(\frac{V}{V_r}\right)^2}. \qquad (2·12, 7)$$

Thus, although V_r is independent of h, the effectiveness of the flap at any lower speed is increased by moving the pitching axis aft. Unless h is negative, which is a most unusual circumstance, the effectiveness below the reversal speed will be underestimated by use of the simple formula

$$\eta = 1-\left(\frac{V}{V_r}\right)^2.$$

When the axis of rotation (or the flexural axis in the case of an elastic wing with fixed root) coincides with the centre of pressure of the aerodynamic load, operation of the flap does not twist the wing and there can be no reversal of control. This is the case where h and j are equal and equation (2·12, 7) confirms that the relative effectiveness of the control is unity. However, in this condition the divergence speed of the wing (see § 2·11) is equal to the invariant reversal speed given by (2·12, 6). When the axis of rotation lies still further aft divergence occurs at a speed lower than the invariant reversal speed (see Fig. 2·12, 2).*

* Figs. 2·12, 1 and 2·12, 2 are taken, with slight modifications, from *Reports and Memoranda, Aeronautical Research Committee, London. (R. & M.)* 1499 (1932).

Most of the conclusions we have reached remain true when the rigid aerofoil with elastic constraint in pitch is replaced by an elastic wing with fixed root and when the flap does not cover the whole span. The formula for the reversal speed remains valid with only a numerical correcting factor, but m_θ is then measured at some convenient reference section, say the section at mid-span of the flap.

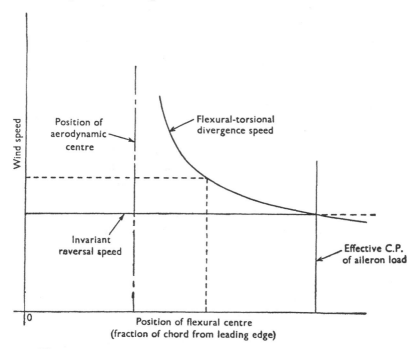

Fig. 2·12, 2. Dependence of divergence and reversal speeds on position of flexural centre.

2·13 Dives and pull-outs

Suppose that an aeroplane is flying horizontally and steadily at a certain speed, that the elevator is suddenly moved so as to give a nose-down pitching moment, and held in the new position while the throttle control is kept fixed. The state of equilibrium corresponding to the new setting of the elevator will, for a stable aeroplane, be one with a downward path and an increased flight speed. Thus the aeroplane is now trimmed for a dive. However, the speed and inclination of the path do not change instantly to the new equilibrium values and there is, in effect, a phugoid motion, possibly of large amplitude, of the kind considered in § 2·7, but the oscillation takes place about an inclined dive path. When the aircraft is quite stable the phugoid oscillation will be damped and the final state will be a steady rectilinear dive; the pilot can hasten the attainment of the steady state by appropriate movement of the elevator. The phenomena are more complicated when the atmospheric density varies with height, for then there is no steady speed for equilibrium, and when the Mach number of the flight is high enough for the effects of compressibility to be appreciable. In the real atmosphere, where air density increases as the aircraft descends, there is a tendency towards pulling out of the dive or flattening of the path. The influence of compressibility is complicated, but as a first approximation it can be assumed that flight with fixed elevator occurs at such a varying angle of incidence that the aerodynamic pitching moment is constantly zero. Distortion of the structure is another complicating factor which, however, cannot be discussed here.

In order to get out of the dive the pilot must pull back on the stick so as to bring a nose-up pitching moment into play. To simplify the discussion we may suppose that the pilot moves the stick in such a manner that the apparent acceleration normal to the flight path is constant and that the speed of flight remains constant during the manœuvre. Since, except in the initiation of the pull-out from a very steep dive, gravity is nearly normal to the path, we may simplify the matter still further and regard the motion in the pull-out as occurring at constant speed in a vertical circle, so that the resultant acceleration is radial and constant. Now, as emphasized by Gates, the longitudinal stability of the aircraft is not the same for this kind of flight as for trimmed horizontal flight. The behaviour of the aircraft depends vitally on the position of its c.g. relative to a *manœuvre point* which plays the part of the neutral point for this kind of motion. Thus, broadly speaking, the behaviour in the pull-out will be satisfactory when the c.g. is forward of and not too near the manœuvre point. Just as there are 'stick-fixed' and 'stick-free' neutral points (see § 2·10), so there are 'stick-fixed' and 'stick-free' manœuvre points. The 'stick force per g' of the normal acceleration is proportional, for a given aircraft, to the distance of the c.g. forward of the stick-free manœuvre point. In order that an 'aerobatic' aeroplane shall be safe and pleasant to fly it is essential that the stick force per g shall be neither too small nor too large. If it is too small the aircraft will be tricky to fly and it will be fatally easy for an inexperienced pilot to pull the wings off when attempting a pull-out. When it is too large, the aircraft will be lacking in manœuvrability and the pilot will have to use excessive force when pulling out. Usually the manœuvre point lies slightly aft of the corresponding neutral point (see further § 10·7).

2·14 Rudimentary theory of gust effects

It is important to be able to estimate the influence of gusts on the motion and loading of an aircraft, more particularly for large civil aircraft where the gust loading may be the most severe to be provided for and where the nature of the response to a gust largely affects the comfort of passengers. This is one of the most difficult problems in aeronautics, firstly because knowledge of gusts themselves is scanty (although rapidly increasing) and, secondly, because the effects of a gust are exceedingly complicated. Some of the complicating factors are as follows:

(1) The gust itself is complex. In general, it will have both horizontal and vertical components which will vary as the aircraft flies through the gust. Moreover, these components may vary over the region occupied by the aircraft at any given instant, so giving rise, for example, to unsymmetrical loadings.

(2) The response of the aircraft, even if rigid, influences the loading on its parts. The vertical and pitching motions induced by the gust are of particular importance.

(3) The elasticity of the structure still further complicates the question.

(4) Lift does not instantly adjust itself to incidence change (Wagner effect).

In order to make any headway with the problem it is necessary to make simplifying assumptions and, as our aim is merely to bring certain salient conclusions to light, we shall greatly simplify the problem as follows:

(*a*) It will be assumed that the gust is uniform in the region occupied by the aircraft at any given instant. This is equivalent to assuming that the whole effect of the gust depends on the values of its components at some particular point of the aircraft.

(*b*) We shall neglect any lateral component of the gust and assume that it meets the aircraft symmetrically.

(*c*) The aircraft is rigid.

(*d*) The aircraft maintains a fixed attitude in space.

(*e*) The initial flight is rectilinear, horizontal and trimmed.

(*f*) The Wagner effect is neglected.

Only horizontal and vertical gusts will be considered, the latter first.

In the discussion of the influence of vertical gusts we shall make the additional assumption that the horizontal speed of flight remains constant. Let

$G(t)$ = upward velocity of gust at time t,

$R(t)$ = response of the aircraft, upward velocity at time t,

V = constant horizontal true airspeed.

Then the vertical relative velocity is

$$G(t) - R(t)$$

and, if this is small in relation to V, the corresponding increment of the angle of incidence (see § 2·5) is

$$\Delta\alpha = \frac{G(t) - R(t)}{V}. \qquad (2\cdot14, 1)$$

Hence, if we neglect the increment of the resultant relative velocity, the increase of lift is

$$\Delta L = \tfrac{1}{2}\rho V^2 S \frac{dC_L}{d\alpha}\Delta\alpha$$

$$= \tfrac{1}{2}\rho V S \frac{dC_L}{d\alpha}[G(t) - R(t)]. \qquad (2\cdot14, 2)$$

If m is the mass of the aircraft the equation for the vertical acceleration is

$$m\frac{dR(t)}{dt} = \Delta L,$$

which can be reduced to

$$k\frac{dR(t)}{dt} + R(t) = G(t), \qquad (2\cdot14, 3)$$

where
$$k = \frac{2m}{\rho V S \dfrac{dC_L}{d\alpha}} \qquad (2\cdot14,\ 4)$$

$$= \frac{2t}{\dfrac{dC_L}{d\alpha}} \qquad (2\cdot14,\ 5)$$

and
$$\hat{t} = \frac{m}{\rho V S} \qquad (2\cdot14,\ 6)$$

is 'the unit of aerodynamic time' (see further § 2·15). The initial condition is

$$R(0) = 0 \qquad (2\cdot14,\ 7)$$

and accordingly the solution of (2·14, 3) is

$$R(t) = \frac{e^{-t/k}}{k} \int_0^t G(t)\, e^{t/k}\, dt. \qquad (2\cdot14,\ 8)$$

Whenever $G(t)$ is known we can now calculate $R(t)$ and the vertical acceleration. The initial acceleration is given by (2·14, 3) as

$$\left(\frac{dR(t)}{dt}\right)_{t=0} = \frac{G(0)}{k}, \qquad (2\cdot14,\ 9)$$

which is zero except for a sharp-edged gust.

For the sharp-edged gust of uniform velocity $G(t)$ is constant, say G, and equation (2·14, 8) yields

$$R(t) = G(1 - e^{-t/k}), \qquad (2\cdot14,\ 10)$$

$$\frac{dR(t)}{dt} = \frac{G}{k} e^{-t/k}. \qquad (2\cdot14,\ 11)$$

Thus the upward velocity of response tends asymptotically to the velocity of the gust while the acceleration tends to zero. The greatest acceleration occurs at time zero and its value, measured in g, is

$$\frac{1}{g}\left(\frac{dR(t)}{dt}\right)_{t=0} = \frac{G}{kg} = \frac{\rho V G S \dfrac{dC_L}{d\alpha}}{2mg} = \frac{\rho V G \dfrac{dC_L}{d\alpha}}{2w}, \qquad (2\cdot14,\ 12)$$

where
$$w = \frac{mg}{S} \qquad (2\cdot14,\ 13)$$

is the wing loading. This result tells us many important things about the acceleration produced by the gust:

(1) It is proportional, for a given aircraft, to the velocity of flight (true air speed) as well as to the gust velocity.

(2) It is proportional to the air density.

(3) It is proportional to the lift curve slope.

(4) It is inversely proportional to the wing loading.

A particularly simple expression for the acceleration is obtained from (2·14, 12) on dividing numerator and denominator by ρV:

$$\frac{1}{g}\left(\frac{dR(t)}{dt}\right)_{t=0} = \left(\frac{G}{V}\right)\frac{\left(\frac{dC_L}{d\alpha}\right)}{C_L}. \tag{2·14, 14}$$

This form emphasizes that the acceleration is expressed non-dimensionally and permits easy comparison of different aircraft.

Suppose, next, we take a linearly graded vertical gust for which

$$G(t) = jt, \tag{2·14, 15}$$

where j is a constant. We find from (2·14, 8) that

$$R(t) = G(t) - jk(1 - e^{-t/k}). \tag{2·14, 16}$$

When t is large

$$R(t) = G(t) - jk, \tag{2·14, 17}$$

so the response lags behind the gust velocity by a constant amount and the acceleration is constant and equal to j; it easily follows from (2·14, 15) that this is the maximum value of the acceleration. It is of great interest that this acceleration is independent of the characteristics of the aircraft. However, for a steady gust distributed in space according to the equation

$$G = lx \tag{2·14, 18}$$

we derive, for an aircraft flying at speed V, the relation

$$G(t) = jt = Vlt. \tag{2·14, 19}$$

Hence the greatest acceleration is Vl, again proportional to the true speed of flight. If the gust lasted such a short time that the asymptotic value of the acceleration was not approached, these simple conclusions would cease to be valid.

The horizontal gust is less simple to discuss because it is no longer permissible to assume that the horizontal speed of flight is constant. At its simplest, the problem essentially involves the two degrees of freedom of the c.g. in the plane of symmetry, and we shall not attempt the mathematical formulation here. However, we can give the following qualitative account. When a sharp-edged horizontal gust strikes the aircraft the immediate effect is to give a vertical acceleration proportional to the gust velocity, when this is small in relation to the speed of flight, and a much smaller horizontal acceleration caused by the increment in drag. The effect of the drag is ultimately to restore the original relative air speed and at this stage the vertical velocity and acceleration are nil. Accordingly the total effect of the gust is to add its velocity vectorially to the ground speed and to displace the flight path vertically. The displacement is upward when the gust velocity is towards the aircraft. It is easy to calculate the initial vertical acceleration when the gust is sharp edged; since the relative air speed is then a maximum, this is the greatest acceleration. Let the gust velocity be U towards the aircraft. Then the increment in lift is

$$\Delta L = \tfrac{1}{2}\rho S C_L [(V+U)^2 - V^2]$$
$$= \tfrac{1}{2}\rho S C_L (2UV + U^2).$$

Hence

$$\frac{\text{vertical acceleration}}{g} = \frac{\rho S C_L (2UV + U^2)}{2mg}$$

$$= \frac{\rho C_L (UV + \tfrac{1}{2}U^2)}{w}, \qquad (2\!\cdot\!14, 20)$$

$$= \frac{\rho V U C_L}{w}, \qquad (2\!\cdot\!14, 21)$$

when U/V is small. By comparison with $(2\!\cdot\!14, 12)$ we find that

$$\frac{\text{acceleration for sharp vertical gust}}{\text{acceleration for sharp horizontal gust}} = \left(\frac{G}{U}\right)\frac{\left(\dfrac{dC_L}{d\alpha}\right)}{2C_L}. \quad (2\!\cdot\!14, 22)$$

When G and U are of comparable magnitude the vertical gust will be more important since, at any rate in cruising flight,

$$\left(\frac{dC_L}{d\alpha}\right) > 2C_L.$$

Finally, we must emphasize that several of the factors neglected in this rudimentary theory have important effects, especially as regards the loading of the aircraft structure. From this point of view assumptions (c), (d) and (f) are particularly important.

2.15   The natural time scale in aircraft dynamics

Suppose we have two geometrically and dynamically similar aircraft performing similar free motions and let us inquire how the times required for corresponding manœuvres depend on the linear dimensions and other relevant factors. To be precise, let us define corresponding manœuvres as such that the percentage changes of corresponding components of velocity are equal. It is clear that mass is a relevant factor since acceleration varies inversely as the mass; air density and the velocity of flight are plainly relevant since the aerodynamic forces depend on them. Let us apply dimensional analysis to find how time t depends on the following factors:

l a typical linear dimension,

V a typical velocity (say the velocity of undisturbed flight),

ρ air density,

m mass of aircraft.

Assume that

$$t = \text{const. } l^a V^b \rho^c m^d, \qquad (2 \cdot 15, 1)$$

so T has the same dimensions as

$$L^a (LT^{-1})^b (ML^{-3})^c M^d.$$

Equate indices:

$T. \quad 1 = -b \quad \text{or} \quad b = -1.$

$M. \quad 0 = c + d \quad \text{or} \quad c = -d.$

$L. \quad 0 = a + b - 3c \quad \text{or} \quad a = 3c - b = 1 - 3d.$

Hence
$$t = \text{const. } \frac{l}{V} \left(\frac{m}{\rho l^3} \right)^d. \qquad (2 \cdot 15, 2)$$

Since d is undetermined, the conclusion from dimensional analysis is

$$t = \frac{l}{V} f \left(\frac{m}{\rho l^3} \right), \qquad (2 \cdot 15, 3)$$

where the form of the function f must be found by independent considerations. Now, if we keep all the factors except m constant, it is clear that t must be proportional to m since the acceleration varies inversely as m. Accordingly

$$t = \text{const.} \frac{m}{\rho V l^2}. \qquad (2\cdot15, 4)$$

It is usual to replace l^2 by the wing area S, which has the same physical dimensions. Hence we find that t is proportional to

$$\hat{t} = \frac{m}{\rho V S} = \frac{W}{g\rho V S}, \qquad (2\cdot15, 5)$$

and $\hat{t}$ is known as *the unit of aerodynamic time*. We have already seen that this quantity arose in the investigation of gust effects [see equation $(2\cdot14, 6)$]. It is important to note that the times required for similar aircraft to perform similar manœuvres are inversely proportional to the speed of flight and to the air density. This implies that, for equal speeds, the manœuvres of an aircraft occupy more time at high altitudes than at low.* If κ be a conventional mean structural density of the aircraft given by

$$\kappa = \frac{m}{lS}, \qquad (2\cdot15, 6)$$

then the non-dimensional density ratio is

$$\mu = \frac{\kappa}{\rho} = \frac{m}{\rho Sl} = \frac{W}{g\rho Sl}. \qquad (2\cdot15, 7)$$

Equation $(2\cdot15, 5)$ now becomes

$$\hat{t} = \frac{\mu l}{V}. \qquad (2\cdot15, 8)$$

Hence, for similar aircraft and for fixed values of the true speed V and of the density ratio μ, $\hat{t}$ is directly proportional to the linear dimensions. This also follows directly from equation $(2\cdot15, 3)$.

* The same is true if the comparison be made for equal values of the lift coefficient.

An apposite example is provided by the rolling motion of an aircraft induced by application of the ailerons. We see from equation (2·8, 4) that the angular velocity in roll reaches a given fraction of its final value when λt is constant, or when

$$t = \frac{\text{const.}}{\lambda} = \text{const.} \frac{A}{k \rho V}.$$

But $\dfrac{A}{k \rho V}$ is proportional to l, for the moment of inertia A is mass multiplied by square of radius of gyration and k has the dimensions of (length)^4, as noted below equation (2·8, 2).

Chapter 3

THE EQUATIONS OF MOTION OF RIGID AIRCRAFT

3·1   Introductory remarks

The basis for any precise discussion of the motion of an aircraft is the set of dynamical equations which express the Newtonian laws of motion. No aircraft is strictly a rigid body, but it is legitimate for many purposes to treat it as such. Hence the basic dynamical equations are those of a rigid body moving in three dimensions. These equations can be written in various forms according to the set of reference axes adopted, but experience shows that it is usually advantageous to employ 'body axes' which are fixed in the moving body. The dynamical equations appropriate to body axes will be found in any book on higher dynamics, to which the reader might be referred, but we include a proof of them here since we shall obtain incidentally a number of useful results which are not usually given in the text-books. Those readers who do not desire to read the proof may pass immediately to § 3·5.

3·2   Moving axes

Suppose that at time t the components of a vector A referred to the rectangular frame of reference OX, OY, OZ are a, b, c respectively, and suppose further that the frame moves about the fixed point O so that at the instant considered it has the components of angular velocity (p, q, r) about the axes* (see Fig. 3·2, 1, where the arrows indicate the positive senses of the angular velocities).

Then it is required to find the time rates of change of the components of A at time t referred to *fixed axes which coincide with OX, OY, OZ at this particular instant*, given the values of the components a, b, c as functions of time. If this be done for all positions of the frame as it moves, we have what is con-

---

* The reader may refer to Lamb's *Higher Mechanics* or to any book on kinematics or higher dynamics for the proof that angular velocity about an axis is a localized vector quantity.

ventionally called the time rate of change of the vector A referred to the moving axes; it is important for the reader to understand clearly the precise meaning of this conventional terminology.

In order to find the rate of change of the vector let us consider the state of affairs at time $t + dt$. The first step is to find the new position OX', OY', OZ' of the frame of reference. For convenience let X, Y and Z be at unit distance from O. Then during the interval dt the angular velocity q about OY displaces X by the amount $-q \cdot dt$ parallel to OZ, and the angular velocity r about OZ displaces X by the amount $r \cdot dt$ parallel to OY. Hence, if we neglect small quantities of the second order, the coordinates of X' are $(1, r dt, -q dt)$. Similarly the coordinates of Y' and Z' are $(-r dt, 1, p dt)$ and $(q dt, -p dt, 1)$ respectively. Since

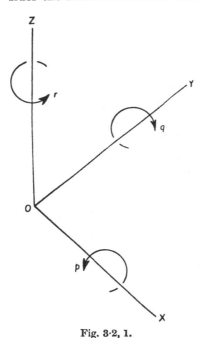

Fig. 3·2, 1.

$$OX' = OY' = OZ' = 1,$$

the foregoing coordinates are also the *direction cosines* of the corresponding axes referred to the original frame OX, OY, OZ.

At time $t + dt$ the components of A along OX', OY', OZ' are $a + \dot{a} dt$, $b + \dot{b} dt$, $c + \dot{c} dt$ respectively. Hence we find that the resolved part of the vector in the direction OX is now

$$(a + \dot{a} dt) \times 1 + (b + \dot{b} dt) \times (-r \cdot dt) + (c + \dot{c} dt) \times (q \cdot dt)$$
$$= a + (\dot{a} - rb + qc) dt,$$

when second order terms are neglected. Thus, during the interval dt the component of A in the fixed direction OX has increased by the amount $(\dot{a} - rb + qc) dt$, and the time rate of change of this component is accordingly

$$\dot{a} - rb + qc. \qquad (3·2, 1)$$

Similarly the time rates of change of the components in the directions OY, OZ are respectively

$$\dot{b} - pc + ra, \tag{3.2, 2}$$

$$\dot{c} - qa + pb. \tag{3.2, 3}$$

With the help of these general results we can answer any question regarding velocities and accelerations referred to moving axes (see § 3.3).

3.3 Components of velocity and acceleration of a point

Suppose that a point P has co-ordinates (x, y, z) referred to the moving frame OX, OY, OZ, where the co-ordinates may vary with time. First let O be fixed as hitherto. Then if we identify the vector A with the displacement of P we find by (3.2, 1) ... (3.2, 3) that the components of the velocity of P referred to the moving axes are

$$u = \dot{x} - ry + qz, \tag{3.3, 1}$$

$$v = \dot{y} - pz + rx \tag{3.3, 2}$$

and
$$w = \dot{z} - qx + py. \tag{3.3, 3}$$

If, however, the origin O has components of velocity (U, V, W) at the instant considered parallel to the instantaneous positions of the moving axes, then clearly the expressions for the components of the velocity of P become

$$u = U + \dot{x} - ry + qz, \tag{3.3, 4}$$

$$v = V + \dot{y} - pz + rx \tag{3.3, 5}$$

and
$$w = W + \dot{z} - qx + py. \tag{3.3, 6}$$

Next, let the components of the acceleration of P be (α, β, γ). We now identify A with the velocity of P and find by (3.2, 1) that

$$\alpha = \dot{u} - rv + qw, \tag{3.3, 7}$$

$$\beta = \dot{v} - pw + ru, \tag{3.3, 8}$$

$$\gamma = \dot{w} - qu + pv. \tag{3.3, 9}$$

In the important case where P is rigidly attached to the frame OX, OY, OZ the co-ordinates (x, y, z) are constant. Hence

$$u = U - ry + qz, \tag{3.3, 10}$$

$$v = V - pz + rx, \tag{3.3, 11}$$

$$w = W - qx + py, \tag{3.3, 12}$$

and the expanded forms of the expressions for the components of acceleration are

$$\alpha = \dot{U} - rV + qW - (q^2 + r^2)x + (qp - \dot{r})y + (rp + \dot{q})z, \quad (3\cdot3, 13)$$

$$\beta = \dot{V} - pW + rU - (r^2 + p^2)y + (rq - \dot{p})z + (pq + \dot{r})x, \quad (3\cdot3, 14)$$

$$\gamma = \dot{W} - qU + pV - (p^2 + q^2)z + (pr - \dot{q})x + (qr + \dot{p})y. \quad (3\cdot3, 15)$$

The foregoing equations would enable us, for instance, to find the velocity and acceleration of any point of an aeroplane in terms of the quantities (U, V, W, p, q, r) and their time rates of change. It will be noted that the components both of velocity and acceleration are linear functions of position.

3·4 Dynamical equations for a rigid body referred to body axes through the centre of mass

When the mass is constant the Newtonian laws of motion are equivalent to the statement that the acceleration of any particle multiplied by its mass is vectorially equal to the resultant of all the forces acting on the particle. Hence, if we define the *reversed effective force* or *inertia force* of the particle to be the reversed acceleration multiplied by the mass, then we can base the analytical formulation of dynamics on the statement that the forces on any particle are always in equilibrium, where it is to be understood that the inertia forces are included among the forces acting. If we consider an extended rigid body we see that the whole system of forces (in the foregoing extended sense) on its particles is in equilibrium. Now the forces due to the actions of the particles of the body on one another (the *internal forces*) occur in equal and opposite pairs because actions and reactions are equal and opposite (Newton's Third Law of Motion). Hence the whole of the internal forces of the body are a system in equilibrium (d'Alembert's principle). Consequently the external forces and inertia forces together are a system in equilibrium. We have merely to write down the conditions of equilibrium of these forces in accordance with the principles of statics in order to obtain the dynamical equations of the rigid body. The conditions for zero force resultant are

$$\Sigma(X_m - \delta m\alpha) = 0, \quad (3\cdot4, 1)$$

$$\Sigma(Y_m - \delta m\beta) = 0, \quad (3\cdot4, 2)$$

and $\quad\quad\quad\quad\quad \Sigma(Z_m - \delta m\gamma) = 0, \quad (3\cdot4, 3)$

where δm is the mass of a particle, (X_m, Y_m, Z_m) is the force acting on it, and the summations extend to all the particles of the body. The conditions for zero couple resultant are

$$\Sigma[y(Z_m - \delta m\gamma) - z(Y_m - \delta m\beta)] = 0, \qquad (3\cdot 4, 4)$$

$$\Sigma[z(X_m - \delta m\alpha) - x(Z_m - \delta m\gamma)] = 0, \qquad (3\cdot 4, 5)$$

$$\Sigma[x(Y_m - \delta m\beta) - y(X_m - \delta m\alpha)] = 0. \qquad (3\cdot 4, 6)$$

Now let the whole system of external forces be equivalent to the force (X, Y, Z) acting at the origin together with the couples (L, M, N) about OX, OY, OZ respectively. Then the dynamical equations become

$$\Sigma\delta m\alpha = \Sigma X_m = X, \qquad (3\cdot 4, 7)$$

$$\Sigma\delta m\beta = \Sigma Y_m = Y, \qquad (3\cdot 4, 8)$$

$$\Sigma\delta m\gamma = \Sigma Z_m = Z, \qquad (3\cdot 4, 9)$$

$$\Sigma\delta m(y\gamma - z\beta) = \Sigma(yZ_m - zY_m) = L, \qquad (3\cdot 4, 10)$$

$$\Sigma\delta m(z\alpha - x\gamma) = \Sigma(zX_m - xZ_m) = M, \qquad (3\cdot 4, 11)$$

$$\Sigma\delta m(x\beta - y\alpha) = \Sigma(xY_m - yX_m) = N. \qquad (3\cdot 4, 12)$$

So far the frame of reference is arbitrary. Now let us take *axes fixed in the rigid moving body, and let the origin be the centre of mass,* so that

$$\Sigma\delta mx = \Sigma\delta my = \Sigma\delta mz = 0. \qquad (3\cdot 4, 13)$$

In order to express concisely the results of substituting the expressions $(3\cdot3, 13) \dots (3\cdot3, 15)$ for (α, β, γ) in the dynamical equations we introduce the following set of quantities which are the dynamical constants of the body for the particular reference axes chosen:

$m = \Sigma\delta m =$ total mass,

$A = \Sigma\delta m(y^2 + z^2) =$ moment of inertia about OX,

$B = \Sigma\delta m(z^2 + x^2) = \qquad ,, \qquad ,, \qquad OY,$

$C = \Sigma\delta m(x^2 + y^2) = \qquad ,, \qquad ,, \qquad OZ,$

$D = \Sigma\delta myz =$ product of inertia about OY and OZ,

$E = \Sigma\delta mzx = \qquad ,, \qquad ,, \qquad OZ \;,,\; OX,$

$F = \Sigma\delta mxy = \qquad ,, \qquad ,, \qquad OX \;,,\; OY.$

By (3·3, 13) ... (3·3, 15) the dynamical equations (3·4, 7) ... (3·4, 9) become in view of (3·4, 13)

$$m(\dot{U} - rV + qW) = X, \qquad (3\cdot4, 14)$$

$$m(\dot{V} - pW + rU) = Y, \qquad (3\cdot4, 15)$$

$$m(\dot{W} - qU + pV) = Z. \qquad (3\cdot4, 16)$$

Next
$$\Sigma\delta m(y\gamma - z\beta)$$

$$= (\dot{W} - qU + pV)\Sigma\delta my - (p^2 + q^2)\Sigma\delta myz + (pr - \dot{q})\Sigma\delta mxy$$

$$+ (qr + \dot{p})\Sigma\delta my^2 - (\dot{V} - pW + rU)\Sigma\delta mz$$

$$+ (r^2 + p^2)\Sigma\delta myz - (rq - \dot{p})\Sigma\delta mz^2 - (pq + \dot{r})\Sigma\delta mzx$$

$$= \dot{p}\,\Sigma\delta m(y^2 + z^2) + qr\Sigma\delta m(y^2 - z^2) + D(r^2 - q^2)$$

$$- E(pq + \dot{r}) + F(pr - \dot{q}).$$

But $\Sigma\delta m(y^2 - z^2) = (C - B)$, and (3·4, 10) accordingly becomes

$$A\dot{p} - (B - C)\,qr + D(r^2 - q^2) - E(pq + \dot{r}) + F(pr - \dot{q}) = L.$$
$$(3\cdot4, 17)$$

Similarly

$$B\dot{q} - (C - A)\,rp + E(p^2 - r^2) - F(qr + \dot{p}) + D(qp - \dot{r}) = M$$
$$(3\cdot4, 18)$$

and

$$C\dot{r} - (A - B)\,pq + F(q^2 - p^2) - D(rp + \dot{q}) + E(rq - \dot{p}) = N.$$
$$(3\cdot4, 19)$$

The three equations of moments can be put in a more concise form by introducing the components of angular momentum (h_1, h_2, h_3) about the co-ordinate axes, which are given by

$$h_1 = Ap - Fq - Er, \qquad (3\cdot4, 20)$$

$$h_2 = -Fp + Bq - Dr, \qquad (3\cdot4, 21)$$

$$h_3 = -Ep - Dq + Cr. \qquad (3\cdot4, 22)$$

Then the equations (3·4, 17) ... (3·4, 19) become*

$$\dot{h}_1 - r\,h_2 + q\,h_3 = L, \qquad (3\cdot4, 23)$$

$$\dot{h}_2 - p\,h_3 + r\,h_1 = M, \qquad (3\cdot4, 24)$$

$$\dot{h}_3 - q\,h_1 + p\,h_2 = N. \qquad (3\cdot4, 25)$$

---

* The equations can be immediately obtained in this form by use of the principle of angular momentum and the method of moving axes as given in § 8·2. On this the reader may consult Lamb's *Higher Mechanics*.

If the co-ordinate axes are the *principal axes of inertia** for the body the three products of inertia are zero. Accordingly the equations of moments become

$$A\dot{p} - (B - C)\,qr = L, \qquad (3\cdot4,\ 26)$$
$$B\dot{q} - (C - A)\,rp = M, \qquad (3\cdot4,\ 27)$$
$$C\dot{r} - (A - B)\,pq = N. \qquad (3\cdot4,\ 28)$$

These are known as Euler's dynamical equations.

The conventional aeroplane possesses a medial fore-and-aft plane about which its masses are symmetrically distributed. If, as is usual, the axes OX and OZ are taken to lie in this plane of symmetry the products of inertia $D = \Sigma\delta myz$ and $F = \Sigma\delta mxy$ are zero, and the equations of motion can be correspondingly simplified. The axes can of course always be chosen so as to make E also vanish, but this is not always convenient. It should be remembered that the masses of a real aircraft are seldom truly symmetrically arranged, so that the products of inertia D and F are not strictly zero, though small.

3·5 General discussion of the external forces on an aircraft

The only external forces of any importance which act upon aircraft in flight are gravity, the aerodynamic forces upon the aircraft due to its own motion through the air and the forces brought into play by the propeller or jet. The latter are discussed separately in Chapter 5. Aerostatic or buoyancy forces are usually negligible for aeroplanes.

The gravitational forces on the particles of the aircraft have a resultant equal to mg which acts vertically downwards through the centre of mass (or c.g.) O. Accordingly, if the downward vertical has direction cosines $(\mathbf{l}, \mathbf{m}, \mathbf{n})$ referred to the body axes of the aircraft, then the components of the gravitational force are

$$X_g = \mathbf{l}mg, \qquad (3\cdot5,\ 1)$$
$$Y_g = \mathbf{m}mg, \qquad (3\cdot5,\ 2)$$
$$Z_g = \mathbf{n}mg, \qquad (3\cdot5,\ 3)$$

and the contributions to the couples are zero. Expressions for $(\mathbf{l}, \mathbf{m}, \mathbf{n})$ in terms of the angular co-ordinates of the aircraft are given in § 3·7.

* Routh's *Elementary Rigid Dynamics*, chap. v.

The full exploration and discussion of the aerodynamic forces is a matter of great complexity which is still incomplete both theoretically and empirically. All that we shall attempt here is to give some guiding ideas and explain certain conventional methods for representing symbolically the aerodynamic actions.

The forces on the aircraft obviously depend on any relative motions within the air through which it moves and these may be important (influence of gusts). However, the influence of such motions is a matter for separate inquiry, and we begin by assuming that the atmosphere is everywhere at rest except in so far as motions are impressed upon it by the aircraft itself.

Both experiment and hydrodynamic theory show that the force system on the aircraft depends in general not only on its instantaneous state of velocity and acceleration (as specified by U, V, W, p, q, r and $\dot{U}, \dot{V}, \dot{W}, \dot{p}, \dot{q}, \dot{r}$) but on the whole history of the motion. That this must be so follows essentially from the fact that the aircraft deposits eddies or vortices in the air as it moves through it, and these vortices, which have a considerable degree of permanence, continue to influence the flow in the neighbourhood of the aircraft. If the air were inviscid and incompressible it would not be possible to create vortices in it,* and the force system would depend simply on the instantaneous state of velocity and acceleration. Clearly the discussion of the motion of a body when the forces on it at any one instant depend on the whole previous history of the motion is in general extremely difficult. However, we shall see that there are important cases where the influence of past motions is negligible, and others in which such influence is calculable with comparative ease because the past motion conforms to some simple type. In a general way it can be said that past history is only of great importance when there are sudden and violent changes of motion or configuration (e.g. sudden opening of a flap), for then powerful vortices are deposited in the air at a small distance from the aircraft, and the induced velocities in its vicinity are therefore large.

We shall begin by considering steady motions, i.e. motions in which U, V, W, p, q, r are all independent of time. Even in this relatively simple case it is rarely possible to predict the

* The same is true of an inviscid compressible fluid provided that during the motion the density is a function of pressure only (barotropic fluid).

aerodynamic forces by pure theory, and recourse must be made to model experiments or the analysis of tests made upon aircraft in flight. Tests in which rotations of the model occur are relatively difficult and are only conducted in the larger aerodynamic laboratories with the help of whirling arms or wind tunnels provided with special spinning balances (see Chapter 8). On account of the difficulty and cost of such tests, attempts are commonly made to estimate the forces due to rotations with the help of experiments in which the rotation is oscillatory in place of steady (see § 8·5) or by calculations based on the results of tests made without rotation (see § 8·3).

When the angular velocities are all zero the technique of measurement of the forces and couples is comparatively simple. In the absence of compressibility and scale effects (see Chapter I) these depend merely on the attitude of the model to the airstream and on the quantities

l = a typical linear dimension (e.g. mean wing chord),

V = resultant relative air speed,

ρ = air density.

Let F be a typical component of force and G a typical component of couple. Then for geometrically similar bodies the non-dimensional quantities $\left(\dfrac{F}{\rho V^2 l^2}\right)$ and $\left(\dfrac{G}{\rho V^2 l^3}\right)$ depend upon attitude alone, so the forces and moments on the full-scale aircraft can be deduced from those measured upon a geometrically similar model.

In investigations concerning both control and stability it is important to know the change in a force or moment brought about by a *small change* in a component of velocity or acceleration. Consider, for example, an aeroplane in steady rectilinear horizontal flight without sideslip. The forces on the aeroplane are then balanced when U and W have certain fixed values and V, p, q and r all vanish. Suppose now that on account of some disturbance the aircraft acquires a small velocity of sideslip, so that V changes from zero to v say. How will the aerodynamic forces and moments depend on v? Since v is restricted to be small it is natural to assume that the change in any force, say the lateral force Y, is *proportional* to v, and equal to v multiplied by the differential coefficient of Y with respect to v.

If the change in Y is δY, then the assumption is that

$$\delta Y = v \frac{\partial Y}{\partial v}, \qquad (3 \cdot 5, 4)$$

where $\dfrac{\partial Y}{\partial v}$ is constant for the given aircraft with the given conditions as regards speed, attitude and atmospheric density. Similarly the increment δL in the rolling moment L, say, would be assumed to be given by

$$\delta L = v \frac{\partial L}{\partial v}, \qquad (3 \cdot 5, 5)$$

with $\dfrac{\partial L}{\partial v}$ constant. This method of representing the changes in the aerodynamic forces and moments is due to Bryan.* The differential coefficients such as $\dfrac{\partial Y}{\partial v}$ are called *aerodynamic derivatives* or simply *derivatives*, and an introductory account of some of them has been given in § 2·5. A shortened notation which has now become standard is explained in § 3·6.

It will be clear from what has been said already that the assumption of the constancy of the aerodynamic derivatives is gravely open to doubt. The question does not admit of direct theoretical attack when the body moving through the air is an aeroplane since no exact and fundamental method for calculating the aerodynamic forces on such a body has yet been evolved. However, guidance can be obtained by considering the simple case of a very thin rigid aerofoil of infinite aspect ratio moving at right angles to its span through an infinite fluid free from obstacles and at rest at infinity, so that the induced motion of the fluid is two-dimensional. It is assumed that the motion of the aerofoil is rectilinear with constant velocity, but that very small two-dimensional movements of translation and rotation are superposed on this; these will be called the deviations. It is further assumed that the mean or initial angle of incidence is very small and skin friction is neglected. The forces and moments are calculated on the hypothesis of Wagner† according to which the circulation round the aerofoil at every instant adjusts itself in such a

---

* G. H. Bryan, *Stability in Aviation*, Macmillan, 1911.

† H. Wagner, 'Ueber die Entstehung des dynamischen Auftriebes von Tragflügeln', *Z.A.M.M.*, vol. 5 (1925), p. 17.

manner that the velocity of the air at the sharp trailing edge is finite.* Details of the calculations will be omitted here.† It will suffice to say that full allowance is made for the influence of the vortices deposited in the wake of the aerofoil. The general results of the investigation‡ can be stated as follows:

(*a*) In general, the aerodynamic forces at any instant depend, strictly, on the whole history of the motion and the concept of a derivative is inapplicable.

(*b*) The derivatives exist and are constant (i.e. independent of time) only in the following circumstances:

(*i*) Where all the velocities of deviation are extremely small in relation to the displacements of deviation, and the accelerations very small in relation to the velocities. The derivatives for this case will be called *slow-motion* or *quasi-static derivatives*.

(*ii*) Where all the disturbances are proportional to the exponential function of time $e^{\lambda t}$, where λ is a real, pure imaginary or complex constant, and the motion of deviation has existed for an infinite time. The derivatives then are functions of λ. For convenience such deviations will be called *exponential*, and the corresponding derivatives *exponential derivatives*.

Slow-motion and exponential derivatives are discussed separately below.

Quasi-static derivatives can be used whenever the changes of circulation occur so slowly that the vortices deposited in the wake are too feeble or remote to influence the aerodynamic forces appreciably. In this case the derivatives are calculated on the assumption that the circulation at every instant has the value corresponding to a state of steady velocity agreeing with that existing at the instant. The criterion for the legitimate use of quasi-static derivatives is that the following non-dimensional quantity shall always be small:

$$\chi = \frac{l\dot{\alpha}}{V\alpha}. \qquad (3\cdot5, 6)$$

---

* This is merely an extension of the Joukowsky hypothesis to conditions of variable motion.

† A valuable review of the earlier work on this subject is given by H. M. Lyon in *R. & M.* 1786 (1937).

‡ W. J. Duncan and A. R. Collar, 'Resistance Derivatives of Flutter Theory', *R. & M.* 1500 (1932).

In this formula l is a convenient linear dimension (say wing chord c), V is the true air speed and α is an angle of incidence, caused either by a change of inclination or by a change of linear velocity (see § 2·5); when the incidence changes are oscillatory, α and $\dot{\alpha}$ must be interpreted as amplitudes. When χ is small the motion may conveniently be called *long period* without thereby implying that it is actually periodic. A more detailed examination of the question* shows that quasi-static derivatives can be used with an error of less than 5 per cent so long as

$$\chi < 0·02. \tag{3·5, 7}$$

When the deviations are exponential the derivatives can be shown† to be functions of the non-dimensional frequency parameter

$$\lambda_1 = \frac{2\pi fc}{V} = \frac{\omega c}{V} \tag{3·5, 8}$$

and of the non-dimensional damping parameter

$$\lambda_2 = \frac{\kappa c}{V}, \tag{3·5, 9}$$

where the coefficient λ occurring in $exp\,(\lambda t)$, to which the deviation is proportional, is given by

$$\lambda = \kappa + i\omega, \tag{3·5, 10}$$

and the frequency
$$f = \frac{\omega}{2\pi}. \tag{3·5, 11}$$

The most important cases are those of simple harmonic motion, where κ is zero, and of non-oscillatory divergence, where ω is zero; in these cases the derivatives in their non-dimensional forms are respectively functions of λ_1 only and of λ_2 only. It was stated above that the exponential derivatives are only strictly applicable when the disturbances have existed for an infinite time. Experiment, however, confirms the obvious suggestion from theory that the derivatives will be sensibly constant provided that the deviation has lasted long enough for the air to have moved a distance relative to the aerofoil amounting to several wing chords. The theoretical investigation

---

* W. J. Duncan, 'Some Notes on Aerodynamic Derivatives', *R. & M.* 2115 (1945).
† Duncan and Collar, *loc. cit.*

only holds for very small angles of incidence, but experiment shows that the derivatives for a simple aerofoil are practically independent of incidence so long as all the incidences occurring in the motion lie within the range for which $\dfrac{dC_L}{d\alpha}$ is sensibly constant. The non-dimensional derivatives for a complete aircraft are not independent of incidence for large changes of incidence, even below the stall. This is largely attributable to the tail unit changing its position relative to the slipstream or wake.

The foregoing discussion leads us to the provisional conclusion that it is legitimate to assume the derivatives for an aeroplane to be constant (i.e. independent of time) when the deviations from a steady state are either of the long period or exponential type. It is fortunate that the kinds of motion which occur in investigations of stability usually conform to one of these types, and therefore the use of constant derivatives in the theory is justified. It must be admitted, however, that the influence of the frequency parameter has hitherto been neglected in most investigations of the stability of aeroplanes, although it has been taken into account in studies of flutter. There can be no doubt that it is incorrect to use the same derivatives for long-period and short-period motions; for instance, the derivatives appropriate to the phugoid oscillation will not strictly be applicable to the rapid longitudinal oscillation (see §§ 2·6 and 2·7).

The aerodynamic derivatives of a self-propelling aircraft (as opposed to a glider) are largely influenced by the propulsive mechanism, both directly and indirectly. The direct effects are those due to the forces acting on the propulsive gear (e.g. airscrew) itself, while the indirect effects are due to the modifications to the airflow in the vicinity of the aircraft brought about by the propulsive apparatus. The effects of the slipstream from a propeller upon immersed parts of the aircraft (e.g. tail surfaces) may be very important (see § 5·6).

3·6   The standard notation for derivatives and their non-dimensional forms

Suppose that R stands for any component (force or couple) of the aerodynamic force on the aircraft, and let s stand for a component of velocity, linear or angular. Then the derivative of R with respect to s is $\dfrac{\partial R}{\partial s}$ and this is conventionally written R_s. Similarly the derivative of R with respect to the acceleration $\dot{s}$ is $\dfrac{\partial R}{\partial \dot{s}}$ and written $R_{\dot{s}}$. Also, if a stands for an angular or other position co-ordinate, either of the aircraft as a whole or of one of its control surfaces, then the derivatives R_a, $R_{\dot{a}}$ and $R_{\ddot{a}}$ may be similarly defined.

Let us consider an aeroplane having the force system $(X_0, Y_0, Z_0, L_0, M_0, N_0)$ when the components of velocity are $(U_0, V_0, W_0, P_0, Q_0, R_0)$ and their rates of change are zero. Then let the velocities be increased by the small quantities (u, v, w, p, q, r) and let the accelerations be $(\dot{u}, \dot{v}, \dot{w}, \dot{p}, \dot{q}, \dot{r})$. The typical force X will then be given by

$$X = X_o + uX_u + vX_v + wX_w + pX_p + qX_q + rX_r$$
$$+ \dot{u}X_{\dot{u}} + \dot{v}X_{\dot{v}} + \dot{w}X_{\dot{w}} + \dot{p}X_{\dot{p}} + \dot{q}X_{\dot{q}} + \dot{r}X_{\dot{r}}, \quad (3\cdot6, 1)$$

and the typical couple L will be given by

$$L = L_0 + uL_u + vL_v + wL_w + pL_p + qL_q + rL_r$$
$$+ \dot{u}L_{\dot{u}} + \dot{v}L_{\dot{v}} + \dot{w}L_{\dot{w}} + \dot{p}L_{\dot{p}} + \dot{q}L_{\dot{q}} + \dot{r}L_{\dot{r}}. \quad (3\cdot6, 2)$$

As another example consider the aileron hinge moment H in a 'longitudinal' motion where the disturbances in v, p and r are absent. If the aileron angle is ξ we shall have

$$H = H_0 + \xi H_\xi + \dot{\xi} H_{\dot{\xi}} + \ddot{\xi} H_{\ddot{\xi}} + uH_u + wH_w + qH_q$$
$$+ \dot{u}H_{\dot{u}} + \dot{w}H_{\dot{w}} + \dot{q}H_{\dot{q}}. \quad (3\cdot6\ 3)$$

The derivatives may be classified in an obvious manner as *force-velocity, force-angular-velocity, moment-velocity* and *moment-angular-velocity derivatives* according to the nature of the dependent and independent variables. Those derivatives which depend on accelerations are naturally called *acceleration derivatives*; when reversed in sign they may also be called *virtual inertias*, and it is easy to see that they always occur in association with the ordinary inertia coefficients (masses, moments and products of inertia) in the equations of motion. The derivatives which depend on angular velocities and accelerations may be called *rotary derivatives*, but this term is usually applied only to those depending on the angular velocities.

Again, the derivatives depending on linear and angular velocities may be classed as *damping derivatives* since they largely govern the rates at which energy is dissipated and motions damped, while those depending on displacements (usually angular) may be called *stiffness derivatives* since they appear in the equations of motion in the same way as elastic stiffnesses.

It is clear that each type of displacement has a *corresponding* force, if we use the last term in a general sense. For instance, displacement or velocity along OX has the corresponding force X, angular displacement or velocity about OX has the corresponding moment L, and so on. Whenever the force and the displacement, velocity or acceleration appearing in the symbol for the derivative correspond in this way the derivative is said to be *direct*. In all other cases it is a *cross derivative* or *coupling derivative*, since it represents symbolically an influence tending to couple two different kinds of motion together. For instance, L_v is a coupling derivative since L is a moment measured about OX and v is a linear velocity; if L_v does not vanish, the velocity of sideslip v brings into play the rolling moment vL_v, so that the distinct kinds of motion sideslip and rolling become coupled together. Coupling derivatives are of the greatest importance in all discussions of stability.

It is often convenient to use derivatives in non-dimensional forms. There is an indefinitely large number of such forms and the choice of that to be adopted should be governed by the following considerations:

(*a*) Simplicity.

(*b*) The non-dimensional form of the derivative should as nearly as possible be constant, i.e. it should depend as little as possible, consistently with (*a*), on the variables which determine the value of the derivative itself. In particular, it is desirable that the non-dimensional form shall be so far as possible independent of the size of the aircraft, of the speed of flight, and of the air density.

With regard to (*b*) we already know (§ 3·5) that in the absence of compressibility and scale effects the non-dimensional quantities $\dfrac{X}{\rho V^2 l^2}$ and $\dfrac{L}{\rho V^2 l^3}$ are independent of l, V and ρ for similar aircraft in similar circumstances of flight. Accordingly these expressions are adopted as the basis of the non-dimensional

TABLE 3·6, 1. NON-DIMENSIONAL FORMS OF
DERIVATIVES

Symbol	Definition	Physical dimension	Non-dimensional form
l	Typical linear dimension	L	—
V	Speed of flight	LT^{-1}	—
ρ	Air density	ML^{-3}	—
θ	An angle	Non-dimensional	—
$\dot{\theta}$	Angular velocity	T^{-1}	—
$\ddot{\theta}$	Angular acceleration	T^{-2}	—
u	A linear velocity	LT^{-1}	$\dfrac{u}{V}$
$\dot{u}$	Linear acceleration	LT^{-2}	$\dfrac{l\dot{u}}{V^2}$
X	A typical force	MLT^{-2}	$\dfrac{X}{\rho V^2 l^2}$
L	A typical couple	$ML^2 T^{-2}$	$\dfrac{L}{\rho V^2 l^3}$
X_θ	Force angular displacement derivative	MLT^{-2}	$\dfrac{X_\theta}{\rho V^2 l^2}$
L_θ	Moment angular displacement derivative	$ML^2 T^{-2}$	$\dfrac{L_\theta}{\rho V^2 l^3}$
$X_{\dot{\theta}}$	Force angular velocity (or force rotary) derivative	MLT^{-1}	$\dfrac{X_{\dot{\theta}}}{\rho V l^3}$
$L_{\dot{\theta}}$	Moment angular velocity (or moment rotary) derivative	$ML^2 T^{-1}$	$\dfrac{L_{\dot{\theta}}}{\rho V l^4}$
$X_{\ddot{\theta}}$	Force angular acceleration derivative	ML	$\dfrac{X_{\ddot{\theta}}}{\rho l^4}$
$L_{\ddot{\theta}}$	Moment angular acceleration derivative	ML^2	$\dfrac{L_{\ddot{\theta}}}{\rho l^5}$
X_u	Force linear velocity derivative	MT^{-1}	$\dfrac{X_u}{\rho V l^2}$
L_u	Moment linear velocity derivative	MLT^{-1}	$\dfrac{L_u}{\rho V l^3}$
$X_{\dot{u}}$	Force linear acceleration derivative	M	$\dfrac{X_{\dot{u}}}{\rho l^3}$
$L_{\dot{u}}$	Moment linear acceleration derivative	ML	$\dfrac{L_{\dot{u}}}{\rho l^4}$

forms of the derivatives given in Table 3·6, 1. These are merely to be regarded as general forms which may, with advantage, be specialized in particular applications such as, e.g., the longitudinal, or the lateral, motion of an aircraft (see §§ 5·2 and 6·2). In particular, the quantity adopted as the typical length l may vary from case to case. It is also to be observed that any quantity of the same dimension may be substituted for l^n in the denominators of the non-dimensional expressions, and it is in fact usually convenient to replace this by Sl^{n-2}, where S is the total wing area. In the case of a hinge moment derivative S would be replaced by the area of the hinged surface or of that part of it lying aft of the hinge.

3·7   The angular coordinates of an aircraft*

An aircraft, considered as a rigid body, possesses six degrees of freedom. Of these, three can be taken as the spatial coordinates of the centre of mass (C.G.) and the remaining three

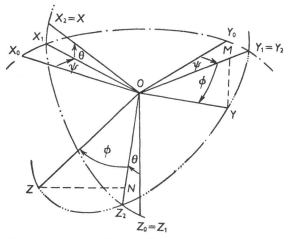

Fig. 3·7, 1. Angular coordinates of a rigid aircraft, showing angles of yaw, pitch and roll.

define the angular orientation. An indefinitely large number of angular coordinate systems might be used, but the one which has become standard in aerodynamics is a modification of a scheme originally due to Euler.

* This section may be omitted at a first reading.

Take the aircraft in a standard position (which is arbitrary) with the body axes in the positions OX_0, OY_0, OZ_0 (see Fig. 3·7, 1). The first step in bringing the aircraft to any other orientation is, in the standard scheme, to rotate (or *yaw*) it about the axis OZ_0 through an angle ψ, which is such that OX now lies at OX_1 in the plane containing OZ_0 and OX_2, the final position of OX. At this stage the body axes occupy the positions OX_1, OY_1, OZ_1, as shown in the figure, where OZ_1 coincides with OZ_0.

The second step consists in rotating the aircraft through an angle θ (the *angle of pitch*) about OY_1, the new position of the transverse axis, until OX reaches its final position. After this displacement the body axes lie at OX_2, OY_2, OZ_2, where Y_2 coincides with Y_1. Lastly the aircraft is rotated in *roll* through the angle ϕ about the axis OX_2 until the axis OY reaches its final position. The final positions of the body axes are OX, OY, OZ, where OX coincides with OX_2. Thus it will be seen that the new angular position is derived from the original by rotation about the axes OZ, OY, OX in that order through angles ψ, θ, ϕ respectively, *where each axis occupies the position into which it is carried by the earlier rotations*. The final position would be different if the rotations occurred in a different order,* but when ψ, θ and ϕ are all infinitesimals of the first order the final position is independent of the order of the rotations, to the first order of infinitesimals.

Now let us find the direction cosines (l_1, m_1, n_1), (l_2, m_2, n_2), (l_3, m_3, n_3) of OX, OY, OZ respectively, referred to the original axes OX_0, OY_0, OZ_0. These can be obtained by spherical trigonometry, but more simply as follows. For convenience take X_0, Y_0, Z_0 at unit distance from 0. On reference to Fig. 3·7, 1 it is clear that the direction cosines of OX_1, OY_1, OZ_1 are $(\cos\psi,\ \sin\psi,\ 0)$, $(-\sin\psi,\ \cos\psi,\ 0)$, $(0, 0, 1)$ respectively. Since OX_1 is the projection of $OX_2' \equiv OX$ upon the plane OX_0Y_0, and OX_2 is inclined at θ to OX_1, it follows that the direction cosines of OX_2 (or OX) are $(\cos\theta\cos\psi,\ \cos\theta\sin\psi,\ -\sin\theta)$. The direction cosines of OY_2 are the same as those of OY_1, i.e. $(-\sin\psi,\ \cos\psi,\ 0)$. Also the projection of OZ_2 on the plane OX_0Y_0 lies along OX_1 and is of length $\sin\theta$. Hence the direction cosines of OZ_2 are $(\sin\theta\cos\psi,\ \sin\theta\sin\psi,\ \cos\theta)$. Let YM be perpendicular

* See Lamb's *Higher Mechanics*, § 3.

to OY_2 and ZN perpendicular to OZ_2. Then the projection of OY upon any line is equal to the sum of the projections of OM and MY upon the line. Now OM and MY are of lengths $\cos\phi$ and $\sin\phi$, and they are parallel to OY_2 and OZ_2 respectively. Hence

$$l_2 = \text{projection of } OY \text{ upon } OX_0$$
$$= \text{projection of } OM \text{ upon } OX_0$$
$$\qquad + \text{projection of } MY \text{ upon } OX_0$$
$$= \cos\phi \times \text{projection of } OY_2 \text{ upon } OX_0$$
$$\qquad + \sin\phi \times \text{projection of } OZ_2 \text{ upon } OX_0$$
$$= -\cos\phi\sin\psi + \sin\phi\sin\theta\cos\psi.$$

Similarly $\qquad m_2 = \cos\phi\cos\psi + \sin\phi\sin\theta\sin\psi$

and $\qquad n_2 = \sin\phi\cos\theta.$

TABLE 3·7, 1. DIRECTION COSINES OF AIRCRAFT AXES

(Rotations in standard order: yaw, pitch, roll)

Axis and suffix	Direction cosine		
	l	m	n
OX 1	$\cos\theta\cos\psi$	$\cos\theta\sin\psi$	$-\sin\theta$
OY 2	$-\cos\phi\sin\psi$ $+\sin\phi\sin\theta\cos\psi$	$\cos\phi\cos\psi$ $+\sin\phi\sin\theta\sin\psi$	$\sin\phi\cos\theta$
OZ 3	$\sin\phi\sin\psi$ $+\cos\phi\sin\theta\cos\psi$	$-\sin\phi\cos\psi$ $+\cos\phi\sin\theta\sin\psi$	$\cos\phi\cos\theta$

Again, the projection of OZ upon any line is the sum of the projections of ON and NZ upon the line. Also ON is of length $\cos\phi$ and is parallel to OZ_2, while NZ is of length $\sin\phi$ and is parallel to OY_2, but opposite in direction. Hence

$$l_3 = \cos\phi \times \text{projection of } OZ_2 \text{ upon } OX_0$$
$$\qquad - \sin\phi \times \text{projection of } OY_2 \text{ upon } OX_0$$
$$= \cos\phi\sin\theta\cos\psi + \sin\phi\sin\psi.$$

Similarly $\qquad m_3 = \cos\phi\sin\theta\sin\psi - \sin\phi\cos\psi,$

and $\qquad n_3 = \cos\phi\cos\theta.$

For ease of reference these results are collected in Table 3·7, 1.

Clearly the state of angular velocity of the aircraft can be expressed alternatively by ϕ, θ, ψ and by p, q, r, the angular velocities about the body axes OX, OY, OZ. There must there-fore be linear relations between these quantities, which can be derived from the geometry of Fig. 3·7, 1. For example, p must be the sum of the components along OX of the angular velocities $\dot{\psi}, \dot{\theta}, \dot{\phi}$ about OZ_0, OY_1 and OX_2 respectively. Now OZ_0 makes the angle $\dfrac{\pi}{2} + \theta$ with OX, OY_1 is perpendicular to OX, while OX_2 coincides with OX. Hence

$$p = \dot{\phi} + \dot{\psi}\cos\left(\frac{\pi}{2} + \theta\right) = \dot{\phi} - \dot{\psi}\sin\theta. \qquad (3\cdot7, 1)$$

In order to find q we require the cosine of the angle between OY and OZ_0. Now OY_1 is perpendicular to the plane OZ_0Z_2, so that the plane OY_1Z_2 is perpendicular to this. Hence YZ_0Z_2 is a spherical triangle right angled at Z_2. Therefore

$$\cos Y\widehat{O}Z_0 = \cos Y\widehat{O}Z_2 \cos Z_2\widehat{O}Z_0$$
$$= \sin\phi\cos\theta.$$

Also the angle between OY and OY_1 is ϕ, while OY is per-pendicular to OX_2. Thus

$$q = \dot{\theta}\cos\phi + \dot{\psi}\sin\phi\cos\theta, \qquad (3\cdot7, 2)$$

and similarly $\qquad r = -\dot{\theta}\sin\phi + \dot{\psi}\cos\phi\cos\theta. \qquad (3\cdot7, 3)$

These relations can also be derived by a method which will be illustrated by proving the foregoing formula for r. The point X has the coordinates $(1, 0, 0)$ referred to the body axes OX, OY, OZ. Hence by $(3\cdot3, 11)$ the component v of the velocity of X in the direction OY (due simply to angular motion of the body) is r. But the component of the velocity of X in the direction OY is $l_2\dot{l}_1 + m_2\dot{m}_1 + n_2\dot{n}_1$, for X, being at unit distance from 0, has the coordinates (l_1, m_1, n_1) referred to the axes OX_0, OY_0, OZ_0. Therefore $\qquad r = l_2\dot{l}_1 + m_2\dot{m}_1 + n_2\dot{n}_1$

$$= (-\cos\phi\sin\psi + \sin\phi\sin\theta\cos\psi)$$
$$\times (-\sin\theta\cos\psi\,.\,\dot{\theta} - \cos\theta\sin\psi\,.\,\dot{\psi})$$
$$+ (\cos\phi\cos\psi + \sin\phi\sin\theta\sin\psi)$$
$$\times (-\sin\theta\sin\psi\,.\,\dot{\theta} + \cos\theta\cos\psi\,.\,\dot{\psi})$$
$$- \sin\phi\cos^2\theta\,.\,\dot{\theta}$$
$$= -\dot{\theta}\sin\phi + \dot{\psi}\cos\phi\cos\theta \qquad (3\cdot7, 4)$$

on reduction.

Equations $(3\cdot7, 1) \ldots (3\cdot7, 3)$ can be solved for the rates of change of the angular co-ordinates and yield

$$\dot{\phi} = p + q\sin\phi\tan\theta + r\cos\phi\tan\theta, \qquad (3\cdot7, 5)$$

$$\dot{\theta} = q\cos\phi - r\sin\phi, \qquad (3\cdot7, 6)$$

$$\dot{\psi} = q\sin\phi\sec\theta + r\cos\phi\sec\theta. \qquad (3\cdot7, 7)$$

It is usually convenient to take the reference axes parallel to the aircraft body axes in undisturbed flight. Then, if the angles θ, ϕ and ψ are small quantities of the first order, the foregoing equations become to the first order

$$\dot{\phi} = p, \qquad (3\cdot7, 8)$$

$$\dot{\theta} = q, \qquad (3\cdot7, 9)$$

$$\dot{\psi} = r. \qquad (3\cdot7, 10)$$

These also follow at once from $(3\cdot7, 1) \ldots (3\cdot7, 3)$.

3·8   Forces and moments due to the inertia of the airscrew*

In any circumstances the airscrew contributes to the inertia forces of the aircraft in the same manner as any other carried mass. However, when the aircraft moves angularly in pitch or yaw new inertia forces and moments come into play. The more important of these are the gyroscopic couples, which are proportional to the angular velocity of the airscrew. For a two-bladed screw the gyroscopic couples vary periodically with a frequency equal to twice the rotational frequency of the screw, but for a screw with more than two blades these couples are constant so long as the angular velocities remain constant. The

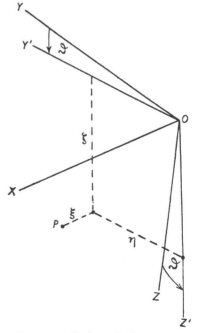

Fig. 3·8, 1. Body and airscrew axes.

* This section may be omitted at a first reading.

forces and couples which we have described are those which
are developed by an airscrew in perfect static and dynamic
balance; if the screw is in any way out of balance additional
forces and moments are introduced. The investigation of these
is beyond the scope of this section.

Take a system of rectangular axes fixed in the aircraft, of
which OX is the axis of rotation of the airscrew* and OY is
lateral, and refer the motions of the aircraft to these in the
usual way (see § 3·2). Also take perpendicular axes OY', OZ'
which are fixed in the airscrew and lie in the plane OYZ. Any
point P of the airscrew has coordinates (x, y, z) referred to the
axes OX, OY, OZ and coordinates (ξ, η, ζ) referred to $OX, OY',$
OZ'. The angle between OY and OY' is ϑ, and it is assumed
that
$$\vartheta = \omega t. \tag{3·8, 1}$$

Thus ω is the (constant) angular velocity of the screw relative
to the aircraft. Clearly the two sets of co-ordinates are connected by the equations

$$x = \xi, \tag{3·8, 2}$$
$$y = \eta \cos \vartheta - \zeta \sin \vartheta, \tag{3·8, 3}$$
$$z = \eta \sin \vartheta + \zeta \cos \vartheta. \tag{3·8, 4}$$

For a given particle of the airscrew ξ, η, ζ are constant, but
y and z vary. By (3·8, 1) we have

$$\dot{x} = 0, \tag{3·8, 5}$$
$$\dot{y} = -\omega \eta \sin \vartheta - \omega \zeta \cos \vartheta$$
$$= -\omega z, \tag{3·8, 6}$$
$$\dot{z} = \omega \eta \cos \vartheta - \omega \zeta \sin \vartheta$$
$$= \omega y. \tag{3·8, 7}$$

Hence by equations (3·3, 4) ... (3·3, 6) the components of velocity of any point of the airscrew referred to the axes $(OX, OY,$
$OZ)$ are

$$u = U - ry + qz, \tag{3·8, 8}$$
$$v = V - (\omega + p)z + rx, \tag{3·8, 9}$$
$$w = W - qx + (\omega + p)y. \tag{3·8, 10}$$

* This is assumed to be parallel to the plane of symmetry of the aircraft,
but need not lie in this plane.

These equations also follow from the consideration that the total components of angular velocity of the screw are $(\omega + p, q, r)$. The components of acceleration of the point in the screw are given by equations (3·3, 7) ... (3·3, 9). Hence

$$
\begin{aligned}
\alpha &= \dot{u} - rv + qw \\
&= \dot{U} - \dot{r}y + \dot{q}z - r\dot{y} + q\dot{z} \\
&\quad - r[V - (\omega + p)z + rx] \\
&\quad + q[W - qx + (\omega + p)y] \\
&= \dot{U} - rV + qW - (q^2 + r^2)x + (qp - \dot{r})y \\
&\quad + (rp + \dot{q})z + 2\omega(rz + qy).
\end{aligned}
$$

Let $(\alpha_0, \beta_0, \gamma_0)$ be the components of acceleration when the airscrew is at rest $(\omega = 0)$. Then [cp. equation (3·3, 13)] the last equation can be written

$$
\alpha = \alpha_0 + 2\omega(rz + qy), \tag{3·8, 11}
$$

and similarly we find that

$$
\beta = \beta_0 - 2\omega py - \omega^2 y \tag{3·8, 12}
$$

and

$$
\gamma = \gamma_0 - 2\omega pz - \omega^2 z. \tag{3·8, 13}
$$

Let (X_s, Y_s, Z_s), (L_s, M_s, N_s) be the *total force and couple applied to the aircraft by the screw due to the inertia forces only*. Then

$$
\left.
\begin{aligned}
X_s &= -\Sigma \delta m \alpha, \\
Y_s &= -\Sigma \delta m \beta, \\
Z_s &= -\Sigma \delta m \gamma, \\
L_s &= -\Sigma \delta m (y\gamma - z\beta), \\
M_s &= -\Sigma \delta m (z\alpha - x\gamma), \\
N_s &= -\Sigma \delta m (x\beta - y\alpha),
\end{aligned}
\right\} \tag{3·8, 14}
$$

where δm is an element of mass of the screw, and the summations extend to all such elements. We have now merely to substitute the expressions for (α, β, γ) in order to obtain the components of force and couple, but certain simplifications in the resulting formulae arise from the fact that the screw is in static and dynamic balance. Since the screw is in static balance its centre of mass must lie upon the axis of rotation OX, and *we shall now take the centre of mass of the screw as the origin O.* Hence

$$\Sigma \delta m x = \Sigma \delta m y = \Sigma \delta m z = 0. \qquad (3 \cdot 8, 15)$$

Also, since the screw is in dynamic balance, OX must be a principal axis of inertia, so that

$$\Sigma \delta m \xi \eta = \Sigma \delta m \xi \zeta = 0. \qquad (3 \cdot 8, 16)$$

Hence by equations $(3 \cdot 8, 2) \dots (3 \cdot 8, 4)$

$$\Sigma \delta m x y = \Sigma \delta m x z = 0. \qquad (3 \cdot 8, 17)$$

Hitherto the position of OY' has not been completely determined, but we shall now take it to be a principal axis of inertia of the screw,* so that

$$\Sigma \delta m \eta \zeta = 0. \qquad (3 \cdot 8, 18)$$

We now introduce the inertia coefficients of the screw as follows:

$$a = \Sigma \delta m (\eta^2 + \zeta^2), \qquad (3 \cdot 8, 19)$$

$$b = \Sigma \delta m (\zeta^2 + \xi^2), \qquad (3 \cdot 8, 20)$$

$$c = \Sigma \delta m (\xi^2 + \eta^2). \qquad (3 \cdot 8, 21)$$

These are constants, but the inertia coefficients referred to the aircraft axes (OX, OY, OZ) are in general not all constant. We have

$$a' = \Sigma \delta m (y^2 + z^2) = \Sigma \delta m (\eta^2 + \zeta^2) = a, \qquad (3 \cdot 8, 22)$$

$$b' = \Sigma \delta m (z^2 + x^2) = \Sigma \delta m (\eta^2 \sin^2 \vartheta + 2\eta \zeta \sin \vartheta \cos \vartheta$$

$$+ \zeta^2 \cos^2 \vartheta + \xi^2)$$

$$= \left(\frac{a+c-b}{2}\right) \sin^2 \vartheta + \left(\frac{a+b-c}{2}\right) \cos^2 \vartheta + \left(\frac{b+c-a}{2}\right),$$

by equations $(3 \cdot 8, 18) \dots (3 \cdot 8, 21)$,

$$= b \cos^2 \vartheta + c \sin^2 \vartheta. \qquad (3 \cdot 8, 23)$$

Similarly $c' = \Sigma \delta m (x^2 + y^2) = b \sin^2 \vartheta + c \cos^2 \vartheta, \qquad (3 \cdot 8, 24)$

$$d' = \Sigma \delta m y z = \tfrac{1}{2}(c - b) \sin 2\vartheta, \qquad (3 \cdot 8, 25)$$

while by $(3 \cdot 8, 17)$ $e' = f' = 0.$

* Since OX is a principal axis there are necessarily two other principal axes in the plane OYZ.

The following results will also be of use:

$$\Sigma\delta mx^2 = \tfrac{1}{2}(b'+c'-a')$$
$$= \tfrac{1}{2}(b+c-a), \quad (3\cdot8,\ 26)$$

$$\Sigma\delta my^2 = \tfrac{1}{2}(a'+c'-b')$$
$$= \tfrac{1}{2}a - \tfrac{1}{2}(b-c)\cos 2\vartheta, \quad (3\cdot8,\ 27)$$

$$\Sigma\delta mz^2 = \tfrac{1}{2}(a'+b'-c')$$
$$= \tfrac{1}{2}a + \tfrac{1}{2}(b-c)\cos 2\vartheta, \quad (3\cdot8,\ 28)$$

$$\Sigma\delta m(z^2-y^2) = (b-c)\cos 2\vartheta. \quad (3\cdot8,\ 29)$$

Now let us return to the expressions for the inertia forces and couples. By (3·8, 11) and (3·8, 14)

$$X_s = -\Sigma\delta m\alpha = -\Sigma\delta m\alpha_0 - 2\omega\Sigma\delta m(rz+qy)$$
$$= -\Sigma\delta m\alpha_0$$
$$= -(\dot{U}-rV+qW)\Sigma\delta m, \quad \text{by } (3\cdot8,\ 15)$$
$$= -\mathscr{M}(\dot{U}-rV+qW), \quad (3\cdot8,\ 30)$$

where $\mathscr{M}$ is the total mass of the airscrew.
Similarly

$$Y_s = -\Sigma\delta m\beta_0$$
$$= -\mathscr{M}(\dot{V}-pW+rU), \quad (3\cdot8,\ 31)$$

and
$$Z_s = -\Sigma\delta m\gamma_0$$
$$= -\mathscr{M}(\dot{W}-qU+pV). \quad (3\cdot8,\ 32)$$

Thus, so far as the forces are concerned, the airscrew behaves as a single particle situated at its centre of mass, a result which could have been deduced from general dynamical principles.

Next, we have

$$L_s = -\Sigma\delta m(y\gamma-z\beta)$$
$$= -\Sigma\delta m(y\gamma_0-z\beta_0) \quad \text{by } (3\cdot8,\ 12) \text{ and } (3\cdot8,\ 13)$$
$$= (q^2-r^2)\Sigma\delta myz - (qr+\dot{p})\Sigma\delta my^2 + (qr-\dot{p})\Sigma\delta mz^2$$
$$= qr(b-c)\cos 2\vartheta - \tfrac{1}{2}(q^2-r^2)(b-c)\sin 2\vartheta - \dot{p}a. \quad (3\cdot8,\ 33)$$

$$M_s = -\Sigma \delta m(z\alpha - x\gamma)$$
$$= -\Sigma \delta m(z\alpha_0 - x\gamma_0) - 2\omega\Sigma \delta mz(rz + qy)$$
$$+ (2\omega p + \omega^2)\,\Sigma \delta mxz$$
$$= (pr - \dot{q})\,\Sigma \delta mx^2 - (rp + \dot{q} + 2\omega r)\,\Sigma \delta mz^2$$
$$- (qp - \dot{r} + 2\omega q)\,\Sigma \delta myz$$
$$= \tfrac{1}{2}(pr - \dot{q})\,(b + c - a)$$
$$- \tfrac{1}{2}(rp + \dot{q} + 2\omega r)\,[a + (b - c)\cos 2\vartheta]$$
$$+ \tfrac{1}{2}(qp - \dot{r} + 2\omega q)\,(b - c)\sin 2\vartheta. \quad (3 \cdot 8,\ 34)$$

$$N_s = -\Sigma \delta m(x\beta - y\alpha)$$
$$= -\Sigma \delta m(x\beta_0 - y\alpha_0) + (2\omega p + \omega^2)\,\Sigma \delta mxy$$
$$+ 2\omega\Sigma \delta my(rz + qy)$$
$$= -(pq + \dot{r})\,\Sigma \delta mx^2 + (qp - \dot{r} + 2\omega q)\,\Sigma \delta my^2$$
$$+ (rp + \dot{q} + 2\omega r)\,\Sigma \delta myz$$
$$= -\tfrac{1}{2}(pq + \dot{r})\,(b + c - a)$$
$$+ \tfrac{1}{2}(qp - \dot{r} + 2\omega q)\,[a - (b - c)\cos 2\vartheta]$$
$$- \tfrac{1}{2}(rp + \dot{q} + 2\omega r)\,(b - c)\sin 2\vartheta. \quad (3 \cdot 8,\ 35)$$

It will be noted that the general expressions for the components of couple contain periodic terms whose frequency is twice that of the rotational frequency of the airscrew. These periodic terms, however, all vanish when $b = c$, i.e. when the airscrew has inertial symmetry about the axis of rotation. An airscrew having three or more equal and evenly spaced blades possesses such symmetry;* accordingly the fluctuating couples only occur with two-bladed screws. Attention may be drawn to the fact that it has not been assumed in the analysis that the axis of the airscrew lies in the plane of symmetry of the aircraft. Thus the results obtained hold good for airscrews mounted on the wings. The rotating parts of engines obviously contribute to the gyroscopic couples, and these contributions are important for gas turbines and for the now obsolete rotary engines; for fixed radial and in-line engines the contributions are small.

* This is most easily seen by considering the ellipse of inertia. With three blades at 120° the ellipse must have three equal radii vectores which are inclined at 120° to one another, and the ellipse must therefore be a circle. A similar argument can be used for four or more blades.

Since the angular velocity ω of the airscrew is much greater than the angular velocities p, q, r of the aircraft, it follows that the important terms in the expressions for the couples are those which are proportional to ω. These are the *gyroscopic* couples of the airscrew, and we see from equations (3·8, 33) ... (3·8, 35) that their values are

$$L_g = 0, \tag{3·8, 36}$$

$$M_g = -\omega r[a + (b - c)\cos 2\vartheta] + \omega q(b - c)\sin 2\vartheta, \tag{3·8, 37}$$

$$N_g = \omega q[a - (b - c)\cos 2\vartheta] - \omega r(b - c)\sin 2\vartheta. \tag{3·8, 38}$$

For a screw with three or more blades these couples are constant, and are given by

$$L_g = 0, \tag{3·8, 39}$$

$$M_g = -\omega r a, \tag{3·8, 40}$$

$$N_g = \omega q a, \tag{3·8, 41}$$

in accordance with elementary dynamical theory. The periodic gyroscopic couples of two-bladed airscrews were first investigated by Lanchester and Melvill Jones.* Lanchester also pointed out the existence of a periodic non-gyroscopic couple due to the periodic change in the moment of inertia of a two-bladed screw about the transverse and normal aircraft axes [see above, equations (3·8, 33) to (3·8, 35)].

3·9   Independence of longitudinal and lateral motions

We shall consider an aircraft which is perfectly symmetrical about the median plane OXZ and shall suppose that it is initially in steady rectilinear flight, with both the vertical and the direction of flight in the plane OXZ. Thus the lateral axis OY is horizontal and there is no sideslip. The components of the velocity of the c.g. in the directions OX and OZ are constant and the angular velocity in pitch is zero.

Suppose that the aircraft is exposed to disturbing agencies of such a nature that after they have ceased to act the motion is still symmetrical. Then the *deviations* from the initial steady state will at this stage consist in the most general case of increments u and w of the components of velocity of the c.g.

---

* *Gyroscopic Action and Propeller Vibration*, Report of the [British] Advisory Committee for Aeronautics, 1914–15.

in the directions OX and OZ respectively and of a displacement θ in pitch with a pitching velocity $\dot{\theta} = q$. On account of the symmetry of the aircraft and of the motion there is no tendency to establish a velocity of sideslip v, an angular velocity in roll p or an angular velocity in yaw r. Hence the motion after a short interval of time will still be symmetrical and this will continue to be true so long as no unsymmetrical disturbance is received. We conclude that motions of the longitudinal-symmetric type do not induce motions of the lateral-anti-symmetric type,* and this is true however large the symmetrical deviations may be, with the proviso in practice that stalling does not occur since this may happen unsymmetrically when the dissymmetry of the disturbance is too small to be detected.

Next, let us suppose that, while the initial state is the same as before, the deviations of the motion are of the lateral-antisymmetric type consisting, in the most general case, of a velocity of sideslip v, an angular velocity in roll p and an angular velocity in yaw r. On account of the symmetry of the aircraft and of the initial motion, it is evident that any influence which the sideslip v has on u, w or q will be the same when v is reversed in sense. Hence there is no first order coupling of these motions with v and by a similar argument they are also uncoupled, to the first order of small quantities, with p and r. Hence, when v, p and r are so small that their squares can be neglected, the lateral motions of deviation can persist without inducing any deviations of the longitudinal type.

We can sum up the results of this argument by saying that, for motions of deviation so small that their squares can be neglected, both the longitudinal and lateral motions persist independently. Moreover, it is evident that, with the foregoing limitation on the magnitude of the deviations, the motion resulting from the most general kind of disturbance will be the sum of longitudinal and lateral motions, each developing independently of the other. When the deviations are not small the longitudinal motion, when pure, can persist alone but a purely lateral motion will, in general, induce longitudinal motion.

* The formal definition of a symmetric deviation is that the displacements and velocities of deviation at points which are mirror images in the plane of symmetry OXZ are themselves mirror images. Antisymmetric deviations at corresponding points to port and starboard are mirror images reversed in direction.

3·10   Coupling of longitudinal and lateral motions

In § 3·9 we have demonstrated the independence of the small deviations from rectilinear uniform flight of the longitudinal-symmetric and lateral-antisymmetric types when the aircraft is perfectly symmetrical about the median plane. Although the conventional aeroplane is superficially symmetrical, the symmetry is usually far from complete. Lack of symmetry may arise from constructional or rigging errors and from minor differences in mass distribution between, e.g. the port and starboard wings; asymmetries of mass distribution are likely to vary as fuel tanks are emptied. A more important lack of symmetry arises in propeller driven aircraft when the gyroscopic moments and slipstream effects are unbalanced. Perfect balance can be obtained with two- and four-propellered aeroplanes when the corresponding propellers on opposite sides are opposite handed and the lack of symmetry with a single contra-rotating pair is negligible. Asymmetries of the kinds mentioned sometimes become apparent in differences of behaviour in rolling and turning to port and starboard.

One not very common but sometimes troublesome coupling between lateral and longitudinal motions is a pitching moment caused by sideslip. This is a 'second order' effect when there is complete symmetry, but not on that account always negligibly small. An explanation can be based on the variation of the downwash along the span of the tailplane. When sideslip occurs the tail unit is displaced laterally in the downwash field of the wings and, in general, the pitching moment due to the tailplane is altered. Shielding of the tailplane by the fin and rudder in sideslip will also give rise to a pitching moment.

3·11   The equations of longitudinal-symmetric motion of a rigid aircraft

We shall now consider the dynamical equations for the *small* deviations of the symmetric type from a state of steady rectilinear flight in equilibrium (see § 3·9). The disposition of the axes OX, OZ, which are fixed in the aircraft, in the undisturbed state are shown in Fig. 3·11, 1.

Since the motion is symmetric v, p and r are zero. We shall suppose that in the steady state the components of velocity of the c.g. of the aircraft referred to the body axes are U and W and that in the disturbed state these become $U+u$ and $W+w$ respectively. The pitching deviation is θ and the angular velocity is $q = \dot{\theta}$. The deviations and their time rates of change are supposed to be so small that their squares and products can be neglected. In the steady state the resultant aerodynamic force, including the propulsive thrust, has the components X, Z which become $X + \Delta X$, $Z + \Delta Z$ in the disturbed motion, while

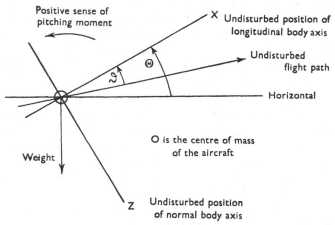

Fig. 3·11, 1. Body axes in undisturbed configuration.

the pitching moment is ΔM. The weight has the components $-mg \sin (\Theta + \theta)$ and $mg \cos (\Theta + \theta)$ in the directions OX, OZ respectively. Accordingly the dynamical equations (3·4, 14), (3·4, 16) and (3·4, 18) become when second order small quantities are neglected

$$m(\dot{u} + qW) = X + \Delta X - mg \sin (\Theta + \theta), \qquad (3·11, 1)$$

$$m(\dot{w} - qU) = Z + \Delta Z + mg \cos (\Theta + \theta), \qquad (3·11, 2)$$

$$B\dot{q} = \Delta M. \qquad (3·11, 3)$$

But in the steady state the first two equations give

$$0 = X - mg \sin \Theta, \qquad (3·11, 4)$$

$$0 = Z + mg \cos \Theta \qquad (3·11, 5)$$

and by subtraction

$$m(\dot{u} + qW) = \Delta X - \theta mg \cos \Theta, \qquad (3\cdot11, 6)$$

$$m(\dot{w} - qU) = \Delta Z - \theta mg \sin \Theta, \qquad (3\cdot11, 7)$$

when θ^2 is neglected. The required dynamical equations are (3·11, 6), (3·11, 7) and (3·11, 3).

As explained in § 3·6 the deviations of the aerodynamic forces and moment will be represented with the help of aerodynamic derivatives, but the acceleration derivatives will be neglected* with the exception of $M_{\dot{w}}$.

Accordingly we shall write

$$\Delta X = uX_u + wX_w + qX_q + X(t), \qquad (3\cdot11, 8)$$

$$\Delta Z = uZ_u + wZ_w + qZ_q + Z(t), \qquad (3\cdot11, 9)$$

$$\Delta M = uM_u + wM_w + \dot{w}M_{\dot{w}} + qM_q + M(t), \qquad (3\cdot11, 10)$$

where $X(t)$, $Z(t)$ and $M(t)$ represent the forces and moment caused by movement of the elevator or external agency. The dynamical equations now become when rearranged

$$\dot{u}m - uX_u - wX_w + \theta mg \cos \Theta + q(mW - X_q) = X(t), \quad (3\cdot11, 11)$$

$$-uZ_u + \dot{w}m - wZ_w + \theta mg \sin \Theta - q(mU + Z_q) = Z(t), \quad (3\cdot11, 12)$$

$$-uM_u - \dot{w}M_{\dot{w}} - wM_w + \dot{q}B - qM_q = M(t). \quad (3\cdot11, 13)$$

We have also the kinematic relations

$$\theta = \int q\, dt, \qquad (3\cdot11, 14)$$

$$U = V \cos \vartheta, \qquad (3\cdot11, 15)$$

$$W = V \sin \vartheta, \qquad (3\cdot11, 16)$$

where V is the true speed of flight.

The choice of the axis OX should be governed by convenience, and it is usually best to take it coincident with the *undisturbed* direction of flight; the axes are then conventionally but misleadingly called *wind axes*. With these axes ϑ and W vanish while Θ becomes the upward inclination of the flight path. For horizontal flight and wind axes Θ, ϑ and W are all zero.

---

* Most of these can be absorbed in the inertial coefficients of the aircraft and affect them only slightly.

3·12   The equations of lateral-antisymmetric motion of a rigid aircraft

We shall obtain the dynamical equations for the small deviations of the lateral type, the initial state of motion being the same as that described in § 3·11 and shown in Fig. 3·11, 1. The equations of motion are (3·4, 15), (3·4, 17) and (3·4, 19). In the undisturbed motion U and W are in general both finite, the initial lateral velocity V is zero as also are all the angular velocities. The non-vanishing deviations are v, p and r.

As a preliminary we must find the inclination of OY to the horizontal in the disturbed motion. Take a point P on OY at positive unit distance from O. The total rotation about OX is

$\phi = \int pdt$ and this is the displacement of P in the direction OZ

due to the rotation; its vertical downward component is $\phi \cos \Theta$. Also the total rotation about OZ is $\psi = \int rdt$ and this is the displacement of P in the direction XO due to the rotation;* its vertical downward component is $\psi \sin \Theta$. The total downward displacement of P, which is the downward inclination of OY to the horizontal in its displaced position, is therefore

$$\phi \cos \Theta + \psi \sin \Theta.$$

Hence, the component of the weight of the aircraft in the direction OY is

$$Y_g = mg(\phi \cos \Theta + \psi \sin \Theta). \qquad (3·12, 1)$$

Since the forces and moments are zero in the undisturbed motion the dynamical equations can be written, when second order quantities are rejected,

$$m(\dot{v} - pW + rU) = \Delta Y + Y_g, \qquad (3·12, 2)$$

$$A\dot{p} - E\dot{r} = \Delta L, \qquad (3·12, 3)$$

$$C\dot{r} - E\dot{p} = \Delta N. \qquad (3·12, 4)$$

---

* These simple expressions for the rotations are valid because OX and OZ only deviate from their undisturbed directions by small angles. It should be noted that OX and OZ are horizontal and vertical respectively only in the special case where Θ is zero.

The increments of the aerodynamic force and moments can be expressed by means of derivatives as follows:

$$\Delta Y = vY_v + pY_p + rY_r + Y(t), \qquad (3\cdot12, 5)$$

$$\Delta L = vL_v + pL_p + rL_r + L(t), \qquad (3\cdot12, 6)$$

$$\Delta N = vN_v + pN_p + rN_r + N(t), \qquad (3\cdot12, 7)$$

where $Y(t)$, $L(t)$ and $N(t)$ represent the influence of the controls. Finally, the dynamical equations become when rearranged

$$\dot{v}m - vY_v - p(mW + Y_p) + r(mU - Y_r)$$
$$- mg(\phi \cos \Theta + \psi \sin \Theta) \qquad = Y(t), \qquad (3\cdot12, 8)$$

$$- vL_v + \dot{p}A - pL_p - \dot{r}E - rL_r = L(t), \qquad (3\cdot12, 9)$$

$$- vN_v - \dot{p}E - pN_p + \dot{r}C - rN_r = N(t). \qquad (3\cdot12, 10)$$

We can always choose OX to be a principal axis of inertia and then E vanishes. For wind axes E is usually small but not strictly zero.

GENERAL REFERENCE FOR CHAPTER 3

A. Robinson and J. A. Laurmann, *Wing Theory*, Cambridge, 1956.

Chapter 4

METHODS FOR SOLVING THE DYNAMICAL EQUATIONS AND FOR INVESTIGATING STABILITY

4·1 Preliminary remarks

The aim of this chapter is to give a clear and brief account of the mathematical methods which are of particular value in solving the kinds of differential equation arising in the dynamics of aircraft and to discuss the theory of stability. Since this book is not a treatise on differential equations, the methods are for the most part described without proof. Singular cases, as when the associated determinantal equations have repeated roots, are seldom or never met in the dynamics of aircraft and are accordingly not discussed here. To sum up, the aim is to be helpful but not exhaustive. Some references to the literature are given at the end of the chapter.

4·2 The equations to be solved

It will be seen that the equations of small deviation of the longitudinal-symmetric type, (3·11, 11) to (3·11, 13), and of the lateral-antisymmetric type, (3·12, 8) to (3·12, 10), can both be written in the form

$$a_{11}\frac{d^2q_1}{dt^2}+b_{11}\frac{dq_1}{dt}+c_{11}q_1+a_{12}\frac{d^2q_2}{dt^2}+b_{12}\frac{dq_2}{dt}+c_{12}q_2$$

$$+a_{13}\frac{d^2q_3}{dt^2}+b_{13}\frac{dq_3}{dt}+c_{13}q_3=Q_1(t), \quad (4\cdot2,\,1)$$

$$a_{21}\frac{d^2q_1}{dt^2}+b_{21}\frac{dq_1}{dt}+c_{21}q_1+a_{22}\frac{d^2q_2}{dt^2}+b_{22}\frac{dq_2}{dt}+c_{22}q_2$$

$$+a_{23}\frac{d^2q_3}{dt^2}+b_{23}\frac{dq_3}{dt}+c_{23}q_3=Q_2(t), \quad (4\cdot2,\,2)$$

$$a_{31}\frac{d^2q_1}{dt^2}+b_{31}\frac{dq_1}{dt}+c_{31}q_1+a_{32}\frac{d^2q_2}{dt^2}+b_{32}\frac{dq_2}{dt}+c_{32}q_2$$

$$+a_{33}\frac{d^2q_3}{dt^2}+b_{33}\frac{dq_3}{dt}+c_{33}q_3=Q_3(t), \quad (4\cdot2,\,3)$$

where q_1, q_2, q_3 are the dynamical variables, the coefficients a, b, c are constants, some of which are zero, and $Q_1(t)$, $Q_2(t)$, $Q_3(t)$ are given functions of the time or zeros.

For longitudinal-symmetric motions the dynamical variables are $q_1 \equiv u$, $q_2 \equiv w$, $q_3 \equiv \theta$ and the coefficients are as given in Table 4·2, 1.

TABLE 4·2, 1. DYNAMICAL CONSTANTS FOR
LONGITUDINAL-SYMMETRIC MOTION

Suffix	a	b	c
11	0	m	$-X_u$
12	0	0	$-X_w$
13	0	$mW - X_q$	$mg \cos \Theta$
21	0	0	$-Z_u$
22	0	m	$-Z_w$
23	0	$-mU - Z_q$	$mg \sin \Theta$
31	0	0	$-M_u$
32	0	$-M_{\dot{w}}$	$-M_w$
33	B	$-M_q$	0

$Q_1(t)$, $Q_2(t)$, $Q_3(t)$ are respectively the longitudinal force, normal downward force and pitching moment produced by the operation of controls or some external agency and are in general functions of the time.

TABLE 4·2, 2. DYNAMICAL CONSTANTS FOR
LATERAL-ANTISYMMETRIC MOTION

Suffix	a	b	c
11	0	m	$-Y_v$
12	0	$-mW - Y_p$	$-mg \cos \Theta$
13	0	$mU - Y_r$	$-mg \sin \Theta$
21	0	0	$-L_v$
22	A	$-L_p$	0
23	$-E$	$-L_r$	0
31	0	0	$-N_v$
32	$-E$	$-N_p$	0
33	C	$-N_r$	0

For lateral-antisymmetric motions the dynamical variables are $q_1 \equiv v$, $q_2 \equiv \int p\,dt = \phi$, $q_3 \equiv \int r\,dt = \psi$, and the coefficients are as given in Table 4·2, 2. The displacement angles ϕ, θ, ψ are all measured with the undisturbed position of the body axes as datum. Here $Q_1(t)$, $Q_2(t)$, $Q_3(t)$ are respectively the lateral force, rolling moment and yawing moment produced by the operation of controls or some external agency and are in general functions of the time.

The dynamical equations, as written at length above, are very cumbrous and a condensed notation must be sought. We take the typical set of terms

$$a_{rs}\frac{d^2q_s}{dt^2} + b_{rs}\frac{dq_s}{dt} + c_{rs}q_s$$

and write them concisely as $A_{rs}(D)q_s$ where

$$A_{rs}(D) \equiv a_{rs}D^2 + b_{rs}D + c_{rs} \qquad (4\cdot2, 4)$$

is a linear differential operator with constant coefficients and

$$D \equiv \frac{d}{dt}. \qquad (4\cdot2, 5)$$

The dynamical equations can now be rewritten

$$\left.\begin{aligned}
A_{11}(D)q_1 + A_{12}(D)q_2 + A_{13}(D)q_3 &= Q_1(t),\\
A_{21}(D)q_1 + A_{22}(D)q_2 + A_{23}(D)q_3 &= Q_2(t),\\
A_{31}(D)q_1 + A_{32}(D)q_2 + A_{33}(D)q_3 &= Q_3(t).
\end{aligned}\right\} \qquad (4\cdot2, 6)$$

Those who are acquainted with matrices will recognize that the last equations can be written still more concisely as

$$A(D)q = Q(t). \qquad (4\cdot2, 7)$$

The condensed notation $(4\cdot2, 4)$ will be used hereafter. In problems where structural distortions have to be taken into account or where the controls are not rigidly locked the number of dependent variables may exceed three, while in simple problems it may be one or two [see, for example, equation $(2\cdot8, 3)$].

The dynamical variables q are not all of the same nature and this implies that the nature of a literal coefficient such as b_{rs} depends on its suffixes. When q_s is a displacement (usually angular) a_{rs} is an *inertia*, b_{rs} a *damping coefficient* and c_{rs} a *stiffness*, where these terms are used in a very general way. However, when q_s is a linear or angular velocity the coefficients a_{rs} will be absent, since time rates of change of accelerations do not appear in dynamical equations, while b_{rs} will be an inertia and c_{rs} a damping coefficient.

4·3 The general nature of the solution

It is shown in treatises on differential equations, and it is indeed obvious, that the general solution of a set of equations of the type (4·2, 6) can be represented as the sum of two parts called the *particular integral* (P.I.) and the *complementary function* (C.F.).* The particular integral is *any* particular solution of the set of equations and should not contain arbitrary constants. The complementary function is the *most general* solution of the set of homogeneous equations obtained on replacing the functions $Q(t)$ by zeros and contains a number n of arbitrary constants of integration. The complementary function is of special importance as it represents the most general free motion.

We begin by considering the complementary function. This can be obtained by assuming

$$\left.\begin{aligned} q_1 &= k_1 e^{\lambda t}, \\ q_2 &= k_2 e^{\lambda t}, \\ q_3 &= k_3 e^{\lambda t}, \end{aligned}\right\} \tag{4·3, 1}$$

where λ and k_1, k_2, k_3 are constants, the latter not all zero. For conciseness we shall write

$$A_{rs}(\lambda) \equiv a_{rs}\lambda^2 + b_{rs}\lambda + c_{rs}, \tag{4·3, 2}$$

so that

$$A_{rs}(D) q_s = k_s A_{rs}(\lambda) e^{\lambda t} \tag{4·3, 3}$$

[see (4·2, 4)].

The homogeneous equations obtained by replacing the functions $Q(t)$ in (4·2, 6) by zeros will accordingly be satisfied if

$$\left.\begin{aligned} A_{11}(\lambda) k_1 + A_{12}(\lambda) k_2 + A_{13}(\lambda) k_3 &= 0, \\ A_{21}(\lambda) k_1 + A_{22}(\lambda) k_2 + A_{23}(\lambda) k_3 &= 0, \\ A_{31}(\lambda) k_1 + A_{32}(\lambda) k_2 + A_{33}(\lambda) k_3 &= 0. \end{aligned}\right\} \tag{4·3, 4}$$

These equations will be incompatible unless

$$\Delta(\lambda) \equiv \begin{vmatrix} A_{11}(\lambda) & A_{12}(\lambda) & A_{13}(\lambda) \\ A_{21}(\lambda) & A_{22}(\lambda) & A_{23}(\lambda) \\ A_{31}(\lambda) & A_{32}(\lambda) & A_{33}(\lambda) \end{vmatrix} = 0. \tag{4·3, 5}$$

* This remains true when the coefficients are functions of *t* instead of constants.

This is called the *determinantal* or *characteristic equation* and when expanded gives an algebraic equation of degree n for λ whose roots, which we here assume to be all distinct, are $\lambda_1, \lambda_2, \dots \lambda_n$. Since $A_{rs}(\lambda)$ is quadratic in λ the maximum possible value of n is, in the present instance, 6 or, in general, twice the number of the dynamical variables; in important cases it is less than this maximum. Let λ_r be one of the roots of the determinantal equation and substitute this for λ in (4·3, 4). These equations are then compatible and yield *unique* values for the *ratios* of the coefficients k. Thus one particular constituent of the c.f. is

$$\left.\begin{aligned}q_1 &= k_{1r}e^{\lambda_r t}, \\ q_2 &= k_{2r}e^{\lambda_r t}, \\ q_3 &= k_{3r}e^{\lambda_r t},\end{aligned}\right\} \tag{4·3, 6}$$

where the ratios $k_{1r} : k_{2r} : k_{3r}$ are characteristic of the constituent and can be found from any *pair* of the equations (4·3, 4). The complete c.f. consists of the sum of n constituents similar to (4·3, 6), each corresponding to a root of the determinantal equation. This is in accordance with the general theorem that the number of arbitrary constants of integration in the solution is equal to the degree in λ of the characteristic equation, for each constituent has coefficients of fixed ratio but arbitrary magnitude. The c.f. is further discussed in § 4·4 where the real form of the constituents corresponding to complex roots of the determinantal equation is given.

The particular integral can always* be expressed as the sum of integrals the integrands of which are products of one of the functions $Q(t)$ and of circular and exponential functions. The p.i. is further considered in § 4·5.

We have seen that the characteristic determinant in (4·3, 5) is, when expanded, at most a sextic in λ which will be written†

$$p_6\lambda^6 + p_5\lambda^5 + p_4\lambda^4 + p_3\lambda^3 + p_2\lambda^2 + p_1\lambda + p_0 = 0. \tag{4·3, 7}$$

The coefficients p can be expressed by determinants whose elements are the dynamical constants a, b, c. To express the

---

* In the absence of repeated roots.

† It is not unusual to write the suffixes in reversed order, so that, e.g., the coefficient of λ^6 is p_0.

coefficients concisely a condensed notation will be used. Thus, for example,

$$| a_j \ \ b_k \ \ c_l | \equiv \begin{vmatrix} a_{1j} & b_{1k} & c_{1l} \\ a_{2j} & b_{2k} & c_{2l} \\ a_{3j} & b_{3k} & c_{3l} \end{vmatrix}, \qquad (4\cdot3,\ 8)$$

so each literal symbol stands for a complete column of a determinant while its suffix is the *second* suffix of the elements in the column. Then we shall have

$$p_6 = |a_1 a_2 a_3|, \qquad (4\cdot3,\ 9)$$

$$p_5 = |a_1 a_2 b_3| + |a_1 b_2 a_3| + |b_1 a_2 a_3|, \qquad (4\cdot3,\ 10)$$

$$p_4 = |a_1 a_2 c_3| + |a_1 c_2 a_3| + |c_1 a_2 a_3| + |a_1 b_2 b_3|$$
$$+ |b_1 a_2 b_3| + |b_1 b_2 a_3|, \quad (4\cdot3,\ 11)$$

$$p_3 = |a_1 b_2 c_3| + |a_1 c_2 b_3| + |b_1 a_2 c_3| + |b_1 c_2 a_3|$$
$$+ |c_1 a_2 b_3| + |c_1 b_2 a_3| + |b_1 b_2 b_3|, \quad (4\cdot3,\ 12)$$

$$p_2 = |a_1 c_2 c_3| + |c_1 a_2 c_3| + |c_1 c_2 a_3| + |b_1 b_2 c_3|$$
$$+ |b_1 c_2 b_3| + |c_1 b_2 b_3|, \quad (4\cdot3,\ 13)$$

$$p_1 = |b_1 c_2 c_3| + |c_1 b_2 c_3| + |c_1 c_2 b_3|, \qquad (4\cdot3,\ 14)$$

$$p_0 = |c_1 c_2 c_3|. \qquad (4\cdot3,\ 15)$$

Since the suffixes always appear in the natural order in these expressions they could be omitted. The rule in forming p_r is that we take every distinct arrangement of the letters a, b, c (with repetitions allowed) whose total *weight* is r, when a, b, c have the weights 2, 1, 0 respectively. With this in mind, the correctness of the expansion will become clear.

For the case of *longitudinal-symmetric motion* we have (see Table 4·2, 1) the columns a_1 and a_2 together with some other dynamical constants zero, and it will be found that the expanded determinant is merely a *quartic* with

$$p_4 = b_{11} b_{22} a_{33}, \qquad (4\cdot3,\ 16)$$

$$p_3 = b_{11}(c_{22} a_{33} + b_{22} b_{33} - b_{23} b_{32}) + c_{11} b_{22} a_{33}, \qquad (4\cdot3,\ 17)$$

$$p_2 = a_{33}(c_{11} c_{22} - c_{21} c_{12}) + b_{11}(c_{22} b_{33} - c_{32} b_{23} - c_{23} b_{32})$$
$$+ c_{11}(b_{22} b_{33} - b_{32} b_{23}) + b_{13}(c_{21} b_{32} - c_{31} b_{22}), \quad (4\cdot3,\ 18)$$

$$p_1 = c_{11}(c_{22} b_{33} - c_{32} b_{23}) - c_{12}(c_{21} b_{33} - c_{31} b_{23}) + c_{13}(c_{21} b_{32} - c_{31} b_{22})$$
$$+ b_{13}(c_{21} c_{32} - c_{31} c_{22}) - c_{23}(b_{11} c_{32} + c_{11} b_{32}), \quad (4\cdot3,\ 19)$$

$$p_0 = c_{13}(c_{21} c_{32} - c_{31} c_{22}) + c_{23}(c_{12} c_{31} - c_{11} c_{32}). \qquad (4\cdot3,\ 20)$$

Next, for *lateral-antisymmetric motion* we have, by Table 4·2, 2, the column a_1 and other dynamical constants zero. Hence we obtain a *quintic equation with the constant term p_0 zero*,

$$p_5 = b_{11}(a_{22}a_{33} - a_{23}a_{32}), \qquad (4·3, 21)$$

$$p_4 = c_{11}(a_{22}a_{33} - a_{23}a_{32}) + b_{11}(a_{22}b_{33} - a_{32}b_{23} + a_{33}b_{22} - a_{23}b_{32}), \qquad (4·3, 22)$$

$$p_3 = c_{11}(a_{22}b_{33} - a_{32}b_{23} + a_{33}b_{22} - a_{23}b_{32}) + b_{11}(b_{22}b_{33} - b_{23}b_{32})$$
$$+ b_{12}(a_{23}c_{31} - a_{33}c_{21}) + b_{13}(a_{32}c_{21} - a_{22}c_{31}), \qquad (4·3, 23)$$

$$p_2 = c_{11}(b_{22}b_{33} - b_{23}b_{32}) + c_{21}(b_{32}b_{13} - b_{33}b_{12}) + c_{31}(b_{12}b_{23} - b_{13}b_{22})$$
$$+ c_{12}(a_{23}c_{31} - a_{33}c_{21}) + c_{13}(a_{32}c_{21} - a_{22}c_{31}), \qquad (4·3, 24)$$

$$p_1 = c_{12}(b_{23}c_{31} - b_{33}c_{21}) + c_{13}(b_{32}c_{21} - b_{22}c_{31}). \qquad (4·3, 25)$$

Accordingly the factor λ may be divided out of the equation and we are again left with a quartic equation (but see § 6·2).

4·4   The complementary function

The first step in calculating the C.F. is to find the n roots of the determinantal equation of the problem, such as (4·3, 5). References to books and papers on this subject are given at the end of the chapter.

Take first the case where λ_r is real. Any pair of the equations (4·3, 4) serves to determine the ratios $k_{1r} : k_{2r} : k_{3r}$ [see equations (4·3, 6)], which are also real. Let $\kappa_{1r}, \kappa_{2r}, \kappa_{3r}$ be a particular and conveniently chosen set of numbers in the correct ratios. Then the constituent corresponding to the root λ_r will be

$$\left. \begin{aligned} q_1 &= k_r \kappa_{1r} e^{\lambda_r t}, \\ q_2 &= k_r \kappa_{2r} e^{\lambda_r t}, \\ q_3 &= k_r \kappa_{3r} e^{\lambda_r t}, \end{aligned} \right\} \qquad (4·4, 1)$$

where k_r determines the amplitude of the constituent as a whole.

When λ_r is complex there will necessarily be a companion root which is its complex conjugate since the coefficients of the determinantal equation are real. Let this pair of roots be $\mu_r \pm i\omega_r$ where μ_r and ω_r are real. When $\mu_r + i\omega_r$ is substituted for λ in a pair of the equations (4·3, 4), which are then solved for the ratios of the coefficients in the ordinary way, the results can be reduced to the form

$$k_{1r} : k_{2r} : k_{3r} = (l_{1r} + im_{1r}) : (l_{2r} + im_{2r}) : (l_{3r} + im_{3r})$$
$$= \alpha_{1r} e^{i\delta_{1r}} : \alpha_{2r} e^{i\delta_{2r}} : \alpha_{3r} e^{i\delta_{3r}}, \qquad (4·4, 2)$$

with the numbers α and δ real, by the rules for the manipulation of complex numbers. If desired, one of the numbers k can be chosen to be unity. The *real* constituents corresponding to the *pair* of roots $\mu_r \pm i\omega_r$ can be taken as the real part of the expressions for the dependent variables q in terms of $\mu_r + i\omega_r$ only. Let R prefixed to an expression indicate that the real part is to be taken and let the arbitrary multiplier for the double constituent as a whole, corresponding to k_r for a real root, be $a_r e^{i\epsilon_r}$, with a_r and ϵ_r real. Then

$$
\begin{aligned}
q_1 &= R a_r \alpha_{1r} e^{\mu_r t} e^{i(\omega_r t + \delta_{1r} + \epsilon_r)} \\
&= a_r \alpha_{1r} e^{\mu_r t} \cos(\omega_r t + \delta_{1r} + \epsilon_r), \\
q_2 &= a_r \alpha_{2r} e^{\mu_r t} \cos(\omega_r t + \delta_{2r} + \epsilon_r), \\
q_3 &= a_r \alpha_{3r} e^{\mu_r t} \cos(\omega_r t + \delta_{3r} + \epsilon_r).
\end{aligned} \qquad (4\cdot4,\,3)
$$

It will be seen that a_r is an arbitrary amplitude factor and ϵ_r an arbitrary phase angle for the double constituent as a whole. An equivalent way of writing the last equations is

$$
\begin{aligned}
q_1 &= b_r \alpha_{1r} e^{\mu_r t} \cos(\omega_r t + \delta_{1r}) + c_r \alpha_{1r} e^{\mu_r t} \sin(\omega_r t + \delta_{1r}), \\
q_2 &= b_r \alpha_{2r} e^{\mu_r t} \cos(\omega_r t + \delta_{2r}) + c_r \alpha_{2r} e^{\mu_r t} \sin(\omega_r t + \delta_{2r}), \\
q_3 &= b_r \alpha_{3r} e^{\mu_r t} \cos(\omega_r t + \delta_{3r}) + c_r \alpha_{3r} e^{\mu_r t} \sin(\omega_r t + \delta_{3r}),
\end{aligned} \qquad (4\cdot4,\,4)
$$

where b_r and c_r are arbitrary amplitude factors related to the former arbitrary constants by the equations

$$
\begin{aligned}
b_r &= a_r \cos \epsilon_r, \\
c_r &= -a_r \sin \epsilon_r.
\end{aligned} \qquad (4\cdot4,\,5)
$$

When we are concerned with a particular free motion of the system it becomes necessary to determine the arbitrary constants of the c.f., which is here the complete solution, from the data defining the state of the system at some particular instant which may be adopted as the origin of time. The number of prescribable elements in the data is just n, the degree of the determinantal equation and the number of arbitrary constants in the general c.f. The initial data are the values of certain displacements and velocities and these serve to determine the unknown coefficients such as k_r, b_r and c_r. Another way of looking at the matter is that the differential equations themselves, considered simply as algebraic equations, together with the prescribable initial conditions, must just suffice to determine all the variables (i.e. the dependent variables and their differential coefficients up to the highest which appear in the differential equations) at the initial instant. Failing this, there will clearly be an indeterminacy.

As an example, let us take the equations of longitudinal-symmetric motion $(3\cdot11, 11) \ldots (3\cdot11, 13)$. We have seen in § $4\cdot3$ that the determinantal equation is a quartic, so the number of prescribable initial conditions is four. In fact, if we prescribe the values at the initial instant of θ, q, u and w, the dynamical equations fix the initial values of $\dot{u}$, $\dot{w}$ and $\dot{q}$ and everything is completely determinate. By inspection of the dynamical equations we could see that the number of assignable variables is 4 and so deduce that the determinantal equation must be a quartic without calculating the determinant.

Clearly, the initial displacements and velocities are related to the unknown amplitude coefficients k, b, c [see equations $(4\cdot4, 1)$ and $(4\cdot4, 4)$] by *linear equations*, for the total expression for any dynamical variable is the sum of n terms corresponding to the several independent constituents. Hence the amplitude coefficients can easily be calculated.

We have used the term constituent to signify the individual free motions represented by equations $(4\cdot4, 1)$, $(4\cdot4, 3)$ or $(4\cdot4, 4)$. These motions are also commonly known as the *free* or *natural modes* of the system.

4·5 The particular integral

Our object here is merely to give a brief account of some of the simpler systematic methods of obtaining a particular integral; in many instances particular integrals can be found by inspection or with particular ease by special devices, but the discussion of these is beyond our scope. We begin with the equation

$$b\frac{dq}{dt} + cq = Q(t), \tag{4·5, 1}$$

where b and c are constants, while $Q(t)$ is a known function of t. A particular integral of the last is

$$q = \frac{e^{-ct/b}}{b} \int_0^t e^{c\tau/b} Q(\tau)\, d\tau. \tag{4·5, 2}$$

This is easily verified, for the equation yields

$$\frac{dq}{dt} = -\frac{cq}{b} + \frac{Q(t)}{b}.$$ (4·5, 3)

Now (4·5, 1) can be written

$$(bD+c)q = Q(t),$$

and the particular integral can be represented formally as

$$q = (bD+c)^{-1}Q(t).$$ (4·5, 4)

By comparison with (4·5, 2) we see that

$$(bD+c)^{-1}Q(t) = \frac{e^{-ct/b}}{b}\int_0^t e^{c\tau/b}Q(\tau)\,d\tau.$$ (4·5, 5)

The importance of this lies in the fact that it enables us to interpret the operator $(bD+c)^{-1}$ whenever it arises, and this is of great utility in solving equations of higher order.

Next let us consider the second order equation with constant coefficients

$$a\frac{d^2q}{dt^2} + b\frac{dq}{dt} + cq = Q(t).$$ (4·5, 6)

Let λ_1, λ_2 be the roots of

$$a\lambda^2 + b\lambda + c = 0,$$ (4·5, 7)

and first suppose that these are real. The P.I. of (4·5, 6) can be written

$$q = (aD^2+bD+c)^{-1}Q(t),$$ (4·5, 8)

and if we assume that D can be treated as a number we get

$$(aD^2+bD+c)^{-1} = \frac{1}{a(\lambda_1-\lambda_2)}\left(\frac{1}{D-\lambda_1} - \frac{1}{D-\lambda_2}\right),$$ (4·5, 9)

provided the roots are not equal. Hence we obtain

$$\begin{aligned} q &= \frac{1}{a(\lambda_1-\lambda_2)}\left[(D-\lambda_1)^{-1}Q(t) - (D-\lambda_2)^{-1}Q(t)\right] \\ &= \frac{1}{a(\lambda_1-\lambda_2)}\left[e^{\lambda_1 t}\int_0^t e^{-\lambda_1\tau}Q(\tau)d\tau - e^{\lambda_2 t}\int_0^t e^{-\lambda_2\tau}Q(\tau)d\tau\right], \end{aligned}$$ (4·5, 10)

by (4·5, 5). The correctness of this solution can be verified at once by differentiation and use of the relations between the roots and coefficients of (4·5, 7). When the roots of (4·5, 7) are not real they must be complex conjugates, since the co-efficients in the equation are real, and may be written

$$\left.\begin{aligned} \lambda_1 &= \mu + i\omega, \\ \lambda_2 &= \mu - i\omega. \end{aligned}\right\} \qquad (4\cdot5,\ 11)$$

Since
$$e^{(\mu+i\omega)\tau} = e^{\mu\tau}(\cos\omega\tau + i\sin\omega\tau), \qquad (4\cdot5,\ 12)$$

equation (4·5, 10) can be reduced to

$$q = \frac{e^{\mu t}}{a\omega}\left[\sin\omega t\int_0^t Q(\tau)e^{-\mu\tau}\cos\omega\tau d\tau\right.$$
$$\left. - \cos\omega t\int_0^t Q(\tau)e^{-\mu\tau}\sin\omega\tau d\tau\right]. \qquad (4\cdot5,\ 13)$$

An equivalent expression is

$$q = \frac{e^{\mu t}}{a\omega}\int_0^t Q(\tau)e^{-\mu\tau}\sin\omega(t-\tau)d\tau, \qquad (4\cdot5,\ 14)$$

where τ is a 'dummy variable' which disappears from the result. It is again easy to verify the correctness of this by use of the rules for the differentiation of definite integrals. Equations (4·5, 8), (4·5, 10) and (4·5, 14) enable us to interpret the inverse of any quadratic function of D.

If we now turn to the set of equations (4·2, 6) and treat D as obeying the laws of ordinary algebra, the solution for q_1 can be written formally

$$q_1\Delta(D) = \begin{vmatrix} Q_1(t) & A_{12}(D) & A_{13}(D) \\ Q_2(t) & A_{22}(D) & A_{23}(D) \\ Q_3(t) & A_{32}(D) & A_{33}(D) \end{vmatrix}, \qquad (4\cdot5,\ 15)$$

where
$$\Delta(D) = \begin{vmatrix} A_{11}(D) & A_{12}(D) & A_{13}(D) \\ A_{21}(D) & A_{22}(D) & A_{23}(D) \\ A_{31}(D) & A_{32}(D) & A_{33}(D) \end{vmatrix}. \qquad (4\cdot5,\ 16)$$

Let the determinant on the right of (4·5, 15) be expanded in the form

$$\Delta_{11}(D)Q_1(t) + \Delta_{21}(D)Q_2(t) + \Delta_{31}(D)Q_3(t),$$

where $\Delta_{11}(D)$ etc., are first minors of $\Delta(D)$. Then the formal solution for q_1 can be written

$$q_1 = \frac{\Delta_{11}(D)}{\Delta(D)} Q_1(t) + \frac{\Delta_{21}(D)}{\Delta(D)} Q_2(t) + \frac{\Delta_{31}(D)}{\Delta(D)} Q_3(t), \quad (4\cdot5,\ 17)$$

and there are similar expressions for q_2 and q_3. In order to interpret the typical operator $\Delta_{11}(D)/\Delta(D)$ we express this, again treating D like a number, as the sum of appropriate partial fractions. For the theory of this the reader should consult a treatise on algebra.* The first step consists in finding the roots of the algebraic equation [see $(4\cdot3, 5)$]

$$\Delta(\lambda) = 0,$$

and then the real factors of $\Delta(\lambda)$. To each real root λ_r corresponds the real linear factor $(\lambda - \lambda_r)$ and to the complex pair $\mu_r \pm i\omega_r$ corresponds the real quadratic factor

$$(\lambda^2 - 2\mu_r \lambda + \mu_r^2 + \omega_r^2). \quad (4\cdot5,\ 18)$$

We can then express $\Delta_{11}(D)/\Delta(D)$ as the sum of terms of the types

$$\frac{l_r}{D - \lambda_r}, \quad (4\cdot5,\ 19)$$

$$\frac{m_r D + n_r}{D^2 - 2\mu_r D + \mu_r^2 + \omega_r^2}, \quad (4\cdot5,\ 20)$$

where the real coefficients l_r, m_r, n_r can easily be found by algebra. The result of operating on $Q_r(t)$ with $(4\cdot5, 19)$ is obtained by use of $(4\cdot5, 5)$. When the operator is $(4\cdot5, 20)$ we first use $(4\cdot5, 14)$ and then apply $(m_r D + n_r)$ to the result. The final result so obtained is

$$\frac{m_r D + n_r}{D^2 - 2\mu_r D + \mu_r^2 + \omega_r^2} Q_r(t)$$

$$= e^{\mu_r t}\left[m_r \int_0^t Q_r(\tau) e^{-\mu_r \tau} \cos \omega_r(t - \tau)\, d\tau \right.$$

$$\left. + \frac{m_r \mu_r + n_r}{\omega_r} \int_0^t Q_r(\tau) e^{-\mu_r \tau} \sin \omega_r(t - \tau)\, d\tau \right]. \quad (4\cdot5,\ 21)$$

* For example, Chrystal's *Algebra*, chap. VIII.

4·6   Impulsive admittances and other special methods

The term 'admittance' was first used with a technical sense in alternating current theory and represented the reciprocal of the impedance, or the generalization of conductance to the case of an alternating current. In mechanics it has come to mean the steady simple harmonic response to an alternating force of unit amplitude while the impulsive admittance is the response of the system at rest to a unit instantaneous impulse.

The method of impulsive admittances permits us to calculate directly the motion of a system whose dynamical equations are linear with constant coefficients under the action of any forces when the system is at rest in its datum configuration at the instant $t = 0$. This is of course a P.I. of the equations of motion and is often the whole solution required. The method has the further advantage that its physical basis is obvious.

The idea behind the method is that any force, constant or variable, acting on the body or system can be considered to be built up from a succession of small impulses, and the effect of the force is just the sum of the effects of the impulses. Suppose that the force is $F(t)$. Then we can suppose that during the interval of length Δt at time t an instantaneous impulse of magnitude $F(t)\Delta t$ is administered to the system. In the limit when Δt tends to zero we obtain the same motion as when the force acts continuously.

Consider one particular displacement or dynamical co-ordinate q of the system and let $\alpha(t)$ be the value of q at time t following the application of unit impulse (at some particular point or spatially distributed in some fixed manner) to the system at rest in its datum configuration at time $t = 0$. Then the action of $F(t)$ between $t = \tau$ and $t = \tau + \Delta\tau$ gives rise to a displacement q at any later time t given by

$$\alpha(t - \tau)\, F(\tau)\Delta\tau.$$

Hence, when we add the effects of all the impulses up to the time t, we obtain*

$$q(t) = \int_0^t \alpha(t - \tau)\, F(\tau)\, d\tau. \tag{4·6, 1}$$

In order to deal systematically with a dynamical system having n degrees of freedom we choose n dynamical co-ordinates. Then

* An integral of this type is often called Duhamel's integral.

we have an array of n^2 impulsive admittances $\alpha(t)$ such that $\alpha_{rs}(t)$ gives the value of q_r at time t following the application of unit impulse at time $t = 0$ in the freedom s. When forces act in several of the freedoms the total displacements are obtained by the summation of expressions similar to (4·6, 1). Thus we obtain

$$q_r(t) = \sum_{s=1}^{n} \int_0^t \alpha_{rs}(t-\tau) Q_s(\tau) \, d\tau, \qquad (4\cdot6, 2)$$

where $Q_s(t)$ is the value of the sth generalized force at time t.

The impulsive admittances are not difficult to calculate since they represent particular free motions, i.e. they are particular cases of the complementary functions of the dynamical equations. If we begin with the equation (4·5, 1), which is of first order with one dependent variable, and regard q as a *velocity* so that b is an inertia, we see that the solution (4·5, 5) can be written

$$q(t) = \int_0^t \frac{e^{-c(t-\tau)/b}}{b} Q(\tau) \, d\tau, \qquad (4\cdot6, 3)$$

and, by comparison with (4·6, 1),

$$\alpha(t) = \frac{e^{-ct/b}}{b}. \qquad (4\cdot6, 4)$$

This makes the velocity q equal to $1/b$ when $t = 0$ which is just that generated by unit impulse and we could have derived $\alpha(t)$ directly from its physical definition.

Again, if we take equation (4·5, 6) and now regard q as a *displacement* and a as an inertia, the solution (4·5, 14) can be written

$$q(t) = \int_0^t \frac{e^{\mu(t-\tau)} \sin \omega(t-\tau)}{a\omega} Q(\tau) \, d\tau. \qquad (4\cdot6, 5)$$

Consequently the impulsive admittance is

$$\alpha(t) = \frac{e^{\mu t} \sin \omega t}{a\omega}. \qquad (4\cdot6, 6)$$

Clearly this gives zero displacement and velocity $1/a$ when $t = 0$; this is just the velocity generated by unit impulse.

The equations (4·2, 1) to (4·2, 3) may be taken to illustrate the general treatment and at first we shall suppose that all the quantities q represent displacements. The first step consists in finding the velocities generated by unit impulse in each freedom in turn. For unit impulse in the first degree of freedom we obtain, on integrating the equations with respect to time over the vanishingly small time during which the impulse acts,

$$\left.\begin{aligned}
a_{11}\frac{dq_1}{dt}+a_{12}\frac{dq_2}{dt}+a_{13}\frac{dq_3}{dt}&=1,\\
a_{21}\frac{dq_1}{dt}+a_{22}\frac{dq_2}{dt}+a_{23}\frac{dq_3}{dt}&=0,\\
a_{31}\frac{dq_1}{dt}+a_{32}\frac{dq_2}{dt}+a_{33}\frac{dq_3}{dt}&=0,
\end{aligned}\right\} \qquad (4\cdot6,\,7)$$

since the displacements and velocities are zero before the impulse begins to act. From these the initial values of the velocities can be found and the initial values of the displacements are all zero. The free motion, or C.F., corresponding to these initial conditions gives

$$\left.\begin{aligned}
q_1&=\alpha_{11}(t),\\
q_2&=\alpha_{21}(t),\\
q_3&=\alpha_{31}(t).
\end{aligned}\right\} \qquad (4\cdot6,\,8)$$

The other admittances are obtained similarly when the unit on the right of equations $(4\cdot6,\,7)$ is moved to the second and third equation in succession.

In dealing with the equations $(3\cdot11,\,11)\,\ldots\,(3\cdot11,\,13)$ of longitudinal-symmetric motion we have to remember that u, w, and q are velocities. Accordingly the equations of impulse are

$$\left.\begin{aligned}
um&=I_1,\\
wm&=I_2,\\
-wM_{\dot w}+qB&=I_3,
\end{aligned}\right\} \qquad (4\cdot6,\,9)$$

where I_1, I_2 are the impulses along OX, OZ and I_3 is the impulsive pitching moment. Hence

$$\left.\begin{aligned}
u&=\frac{I_1}{m},\\
w&=\frac{I_2}{m}\\
\text{and}\qquad q&=\frac{I_3}{B}+\frac{M_{\dot w}}{mB}I_2.
\end{aligned}\right\} \qquad (4\cdot6,\,10)$$

Thus $\alpha_{11}(t)$, $\alpha_{21}(t)$ and $\alpha_{31}(t)$ give the values of u, w, and $\theta*$ in a free motion with the initial values

$$\left.\begin{array}{l} u = \dfrac{1}{m}, \\[2mm] w = 0, \\[2mm] q = 0, \\[2mm] \theta = 0, \end{array}\right\} \quad (4\cdot6, 11)$$

since here $I_1 = 1$, $I_2 = I_3 = 0$. The initial conditions for the other impulsive admittances can be found from (4·6, 10) in a similar manner. For the lateral-antisymmetric motion we derive from equations (3·12, 8) ... (3·12, 10) the equations of impulse

$$\left.\begin{array}{l} vm = I_1, \\[2mm] pA - rE = I_2, \\[2mm] -pE + rC = I_3, \end{array}\right\} \quad (4\cdot6, 12)$$

where I_1 is the impulse along OY while I_2, I_3 are the impulsive moments in roll and yaw respectively.

These equations yield

$$\left.\begin{array}{l} v = \dfrac{I_1}{m}, \\[2mm] p = \dfrac{CI_2 + EI_3}{AC - E^2}, \\[2mm] r = \dfrac{EI_2 + AI_3}{AC - E^2}. \end{array}\right\} \quad (4\cdot6, 13)$$

The initial values of the velocities appropriate to the several admittances are immediately derivable from the last equations.

When the method of impulsive admittances is applied to the dynamical equations in their non-dimensional forms it is obviously convenient to treat the non-dimensional measure of a force as the force itself and the non-dimensional dynamical coefficients as the corresponding true coefficients. This is legitimate on account of the linearity of the equations.

* Alternatively, we may calculate admittances for $q = \dot{\theta}$ in place of θ.

The symbolic method of § 4·5 and the method of impulsive admittances are, of course, equivalent in the sense that the complete solutions (sums of c.f. and p.i.) are always equivalent.

We have remarked that each impulsive admittance represents one component of a particular free motion of the system. Accordingly any impulsive admittance can be expressed as a homogeneous linear combination of the functions representing the corresponding component in the several free modes of the system (see § 4·4). According to K. Mitchell 'the *modal response coefficients* to rolling moment are the amplitudes to which the standardized normal modes are excited when the aeroplane is disturbed from the steady state by the action of an impulsive rolling moment, whose time integral is unit. The modal response coefficient to yawing moment is similarly defined . . . .' Thus Mitchell's method of response coefficients is essentially the same as that of impulsive admittances. However, Mitchell bases his work on the method of the variation of parameters and does not use the terminology of admittances.*

Suppose now that $H(t)$ is some quantity in which we are interested and which is linearly expressible in terms of the deviations q_r by the equation

$$H(t) = \sum_{r=1}^{n} h_r q_r(t) \qquad (4·6, 14)$$

where the coefficients h are constants. Then we may derive from equation (4·6, 2) the relation

$$H(t) = \sum_{r=1}^{n} \int_0^t H_s(t-\tau) Q_s(\tau) d\tau, \qquad (4·6, 15)$$

where
$$H_s(t) = \sum_{r=1}^{n} h_r \alpha_{rs}(t). \qquad (4·6, 16)$$

Thus $H_s(t)$ is effectively an impulsive admittance for the quantity H. We obtain from (4·6, 2) by differentiation with respect to t

$$\dot{q}_r(t) = \sum_{s=1}^{n} \left[\int_0^t \alpha'_{rs}(t-\tau) Q_s(\tau) d\tau + \alpha_{rs}(0) Q_s(t) \right], \qquad (4·6, 17)$$

where
$$\alpha'_{rs}(t) = \frac{d}{dt} \alpha_{rs}(t). \qquad (4·6, 18)$$

* K. Mitchell, 'Lateral Response Theory', *R.A.E. Report No. Aero.* 1925, A.R.C. 7993, *R. & M.* 2297 (1944).

If q_r is a *displacement* $\alpha_{rs}(0)$ is zero and (4·6, 17) becomes

$$\dot{q}_r(t) = \sum_{s=1}^{n} \int_0^t \alpha'_{rs}(t-\tau)\, Q_s(\tau)\, d\tau, \qquad (4\cdot6,\ 19)$$

but if q_r is a velocity $\alpha_{rs}(0)$ is not, in general, zero. Suppose, however, as an example that we consider the case where q_r is the velocity deviation w and where the only applied 'force' is a pitching moment. Then we have

$$w(t) = \int_0^t \alpha_{23}(t-\tau)\, M(\tau)\, d\tau, \qquad (4\cdot6,\ 20)$$

while equations (4·6, 10) show that $\alpha_{23}(0)$ is zero.

Accordingly

$$\dot{w}(t) = \int_0^t \alpha'_{23}(t-\tau)\, M(\tau)\, d\tau \qquad (4\cdot6,\ 21)$$

and we deduce, for example, that the component in the direction OZ of the acceleration at the C.G. is

$$\dot{w}(t) - Uq(t) = \int_0^t [\alpha'_{23}(t-\tau) - U\alpha'_{33}(t-\tau)]\, M(\tau)\, d\tau.$$
$$(4\cdot6,\ 22)$$

When the pitching moment is due to operation of an elevator there will, in general, be a small lift as well as a pitching moment. Then the acceleration contains an extra term equal to the *instantaneous value* of the lift increment due to the elevator divided by the mass of the aircraft.

The operational method of Heaviside was specially devised for dealing with differential equations, or sets of these, having constant coefficients and has the special feature that the solution for given initial conditions is obtained directly, although it is, as usual, necessary to obtain the roots of the determinantal equation as a preliminary. The method based on the Laplace transform is closely related to that of Heaviside and is now generally preferred. These methods are very suitable for application to aircraft dynamics.

Methods based on the use of matrices are also very convenient and well suited to problems on dynamical systems, such as aircraft, whose equations of motion are linear.

Numerical solutions of particular problems of free or forced motion can be obtained by means of analogue or digital computers. These instruments are applicable when the dynamical equations have variable coefficients or are non-linear; step-by-step numerical integration is also applicable in these circumstances.

A list of references will be found at the end of this chapter.

4·7   Response calculations

The object of a response calculation is to find how an aircraft responds to the operation of its controls or to some extraneous force system, such as that caused by a gust. The induced motion or response is thus a special forced motion of the aircraft and analytical expressions for the deviations can be found by the methods already discussed when the deviations are so small that their squares and products can be neglected. For large deviations the equations of motion are non-linear and for any particular case they could be solved approximately step-by-step or with the aid of a differential analyser. In the remainder of this discussion we shall suppose that the deviations are small enough for the linearized dynamical equations to be used.

We shall illustrate the procedure in a response calculation by considering the motion induced by moving the elevators in a prescribed manner. Let us suppose that the aircraft is initially trimmed and flying steadily and let η be the deviation of the elevator angle from the setting for trim. Then η will be some given function of time t and we may find the corresponding pitching moment $M(t)$ about the c.g., calculated on the assumption that the aircraft has continued to fly as in the initial or datum condition. In reality the speed of flight varies and this affects the pitching moment corresponding to any given elevator setting, but we are entitled to neglect this in the linearized treatment as the change in pitching moment is proportional to the product $u\eta$ which is negligible. The response of the aircraft itself gives rise to pitching moments but these are represented by terms on the left-hand side of the dynamical equation (3·11, 13). Hence $M(t)$ as specified above is the applied pitching moment to appear on the right-hand side of equation (3·11, 13). The detailed calculation of the motion of response may with advantage proceed by the method of impulsive admittances, as explained in § 4·6. The first step is to find the roots of the characteristic equation of longitudinal-symmetric motion and then the impulsive admittances corresponding to a unit instantaneous pitching impulse. Lists of integrals which assist in the calculation of responses by the method of impulsive admittances are given in § 15·4.

4·8   The meaning of stability

A given state of equilibrium at rest or in steady motion of a dynamical system is said to be stable when the influence of an imposed disturbing impulse, or force acting throughout a finite interval of time, *ultimately* becomes vanishingly small. Thus stability is an attribute of the free motion which follows a disturbance. No real physical system is stable for disturbances of unlimited magnitude though the stability of a system whose equations of motion are linear is theoretically independent of the magnitude of the disturbances. Usually we are concerned with disturbances which are so small that the equations governing the motions of disturbance are linear. If the solution of these linear equations of disturbance contains no unstable component, the system under investigation is said to be stable for small disturbances. It must be emphasized that stability is a term concerned with the ultimate consequences of a disturbance. Thus it might happen that a disturbance became largely amplified in the early stages of its history in a completely stable system. It must also be emphasized that the concept of stability only applies to systems in equilibrium, at rest or in some regular motion, either free or with prescribed forces. We cannot, for example, discuss the stability of an aircraft which is not trimmed to fly at the speed and in the attitude postulated.

It will be convenient to call the difference in value of a particular dynamical variable in the motion following a disturbance from that in the undisturbed motion a *deviation*. The linearized dynamical equations of deviation contain deviations as dependent variables and are obtained by neglecting the squares and products of the deviations.

4·9   The conditions for stability

We confine attention to deviations satisfying linear differential equations with constant coefficients. Accordingly, each constituent of the motion will be proportional to the real part of $\exp(\lambda t)$, where λ is a root of the determinantal equation. First, when λ is real the constituent will increase steadily with time and tend ultimately to infinity if λ is positive; the condition for stability is that λ shall be negative and the neutral state corresponds to the zero value. Secondly, when λ is complex and equal to $(\mu \pm i\omega)$ the deviations are proportional to $\exp(\mu t) \sin(\omega t + \epsilon)$ where ϵ is a constant. Hence the amplitude of the motion will grow indefinitely when μ is positive and the condition for stability is that μ shall be negative; the neutral state corresponds to μ zero. We may sum up as follows:

The complete conditions for stability are that all the real roots and the real parts of all the complex roots shall be negative.* A neutral or critical condition corresponds to the occurrence of a zero real root or to a zero real part of a complex pair of roots, i.e. to the occurrence of a pair of equal and opposite imaginary roots $\pm i\omega$. Since there are n roots of a determinantal equation of degree n, the number of independent conditions for complete stability is n.

It was shown by Routh that the conditions for complete stability could be expressed as a set of n inequalities to be satisfied by the coefficients in the determinantal equation. Let this equation be written with the coefficient of the highest power of λ positive. Then a necessary, but not in itself sufficient, condition for stability is that all the other coefficients shall also be positive. To prove the necessity of this condition we consider the factorized form of $\Delta(\lambda)$. To any real root λ_r there corresponds the real linear factor $(\lambda - \lambda_r)$, which has positive coefficients when λ_r is negative, as required for stability. The conjugate complex pair of roots $\mu_r \pm i\omega_r$ gives rise to the real quadratic factor

$$(\lambda - \mu_r - i\omega_r)(\lambda - \mu_r + i\omega_r) = (\lambda - \mu_r)^2 + \omega_r^2$$

and this has all its coefficients positive when μ_r is negative. Thus, when all the real roots and all the real parts of the complex roots are negative, all the factors of $\Delta(\lambda)$ have positive

* When repeated roots occur the corresponding constituent will be $\exp(\lambda t)$ multiplied by a polynomial in t. This will tend to zero as t tends to infinity when the real part of λ is negative. Hence repeated roots require no modification of the stated conditions for stability.

coefficients and therefore all the coefficients of $\Delta(\lambda)$ have the same sign as that of the highest power of λ.

Criteria for stability, equivalent to Routh's, have been expressed in a convenient determinantal form by Hurwitz and independently by Frazer. Let the determinantal equation be

$$\Delta(\lambda) \equiv p_n \lambda^n + p_{n-1} \lambda^{n-1} + \ldots + p_1 \lambda + p_0 = 0, \qquad (4 \cdot 9, 1)$$

where p_n is positive. Then the complete set of necessary and sufficient conditions for stability is that all the test functions $T_1 \ldots T_n$ shall be *positive*, where

$$T_1 = p_{n-1}, \qquad\qquad\qquad\qquad (4 \cdot 9, 2)$$

$$T_2 = \begin{vmatrix} p_{n-1} & p_n \\ p_{n-3} & p_{n-2} \end{vmatrix}, \qquad\qquad (4 \cdot 9, 3)$$

$$T_3 = \begin{vmatrix} p_{n-1} & p_n & 0 \\ p_{n-3} & p_{n-2} & p_{n-1} \\ p_{n-5} & p_{n-4} & p_{n-3} \end{vmatrix}, \qquad (4 \cdot 9, 4)$$

and so on. The rule in constructing T_r is:

Begin with p_{n-1} in the top left-hand corner and write p_{n-3}, $p_{n-5}, \cdots p_{n-2r+1}$ below it to form the first column of the determinant, which is of order r. Then beginning with the left-hand element in any row, write down the remaining elements in the row by increasing the suffix by one on passing from any element to its neighbour on the right. A zero is to be substituted for the coefficient when the suffix is less than zero or greater than n. It is apparent on writing out the determinants for T_n and T_{n-1} that

$$T_n = p_0 T_{n-1}, \qquad\qquad\qquad (4 \cdot 9, 5)$$

so p_0 must be positive for stability, as we know already. Finally, the complete set of conditions for stability is as follows:

The test functions $T_1 \ldots T_{n-1}$ and the coefficient p_0 must all be positive, where it is postulated that the determinantal equation is written with p_n positive.

Routh showed that, when the conditions for stability are not all satisfied, the number of roots with real parts positive is equal to the number of changes of sign in the sequence of his test functions beginning with p_n.

Routh's test function R_1 is equal to T_1 and, more generally

$$R_s = \frac{T_s}{T_{s-1}}.$$

When the test functions T have been computed the sequence of functions $R_1 \dots R_n$ can be found from these relations and the number of unstable roots ascertained, but the functions R can be obtained with less labour by Routh's original cross-multiplication scheme as given in his *Advanced Rigid Dynamics*.

Let κ be the reciprocal of λ so that it satisfies the equation

$$p_0 \kappa^n + p_1 \kappa^{n-1} + \dots + p_{n-1} \kappa + p_n = 0. \qquad (4\cdot9, 6)$$

Then a real negative root κ corresponds to a real negative root λ and a complex root κ with its real part negative corresponds to a complex root λ with its real part negative. Consequently the stability criteria appropriate to $(4\cdot9, 6)$ are equivalent to the original criteria. In other words, we can derive an equivalent set of stability criteria by substituting p_{n-r} for p_r throughout the formulae. It is worthy of remark that the test determinant T_r is homogeneous and of degree r in the coefficients and that each of its terms has the same weight* $\frac{1}{2}r(2n-r-1)$. This is related to the fact that the criteria are effectively invariant for substitutions of the form $\lambda' = a\lambda$ with a real and positive.

The special case of the quartic equation,

$$p_4 \lambda^4 + p_3 \lambda^3 + p_2 \lambda^2 + p_1 \lambda + p_0 = 0, \qquad (4\cdot9, 7)$$

is of particular importance in relation to aircraft. In accordance with what has been said, the complete set of conditions for stability is (with p_4 positive)

$$p_3 > 0,$$

$$T_2 = \begin{vmatrix} p_3 & p_4 \\ p_1 & p_2 \end{vmatrix} = p_2 p_3 - p_1 p_4 > 0,$$

$$T_3 = \begin{vmatrix} p_3 & p_4 & 0 \\ p_1 & p_2 & p_3 \\ 0 & p_0 & p_1 \end{vmatrix} = p_1 p_2 p_3 - p_4 p_1^2 - p_0 p_3^2 > 0,$$

$$p_0 > 0.$$

* The weight of a product is the sum of the suffixes of its factors.

However, $$T_3 = p_1 T_2 - p_0 p_3^2,$$

so T_2 must be positive when T_3, p_0 and p_1 are positive. Accordingly, the simplest set of criteria is

$$\left.\begin{array}{l} p_3 > 0, \\ p_1 > 0, \\ p_0 > 0, \\ T_3 = p_1 p_2 p_3 - p_4 p_1^2 - p_0 p_3^2 > 0. \end{array}\right\} \qquad (4\cdot9,\ 8)$$

When the first three of these inequalities are satisfied the fourth implies
$$p_2 > 0, \qquad (4\cdot9,\ 9)$$

but the converse is not true. Very often the quartic equation is written with p_4 unity as

$$\lambda^4 + B\lambda^3 + C\lambda^2 + D\lambda + E = 0. \qquad (4\cdot9,\ 10)$$

Routh's discriminant then becomes

$$T_3 = R \equiv BCD - EB^2 - D^2. \qquad (4\cdot9,\ 11)$$

It is notable that, when the coefficients are all positive, an increase of C necessarily increases T_3 whereas an increase of E reduces it.

It is shown in § 4·10 that the two general criteria

$$p_0 > 0,$$

$$T_{n-1} > 0$$

are critical criteria and have a special importance.

A demonstration by elementary algebra of the necessity and sufficiency of Routh's stability criteria for the quartic is given in the Appendix to this chapter.

4·10   Critical criteria for stability

Let us consider a dynamical system whose stability depends on some parameter or parameters, e.g. an aircraft whose stability depends on the speed of flight, the dihedral angle, fin area, etc. Further, let us suppose that the system is completely stable in some standard state S. Then in the state S all the possible constituent motions of the system are stable, so the real roots of the determinantal equation are all negative and the complex roots all have their real parts negative. Now let a parameter α be varied until a critical condition C is reached such that any further change of α would result in instability. This can occur in two ways:

(a) A real negative root λ reaches the value zero in the state C. This is indicated by the vanishing of p_0. (See further at the end of this section.)

(b) The real part of a conjugate complex pair of roots reaches the value zero in state C. This is indicated by the vanishing of the penultimate test function T_{n-1}, where n is the degree of the determinantal equation.

Case (b) will now be further considered in detail.

For simplicity let us take first a quartic determinantal equation and let the conjugate roots in state C be $\pm i\omega$. Then the equation has a pair of equal and opposite roots, $\pm\beta$ say. Accordingly

$$p_4\beta^4 \pm p_3\beta^3 + p_2\beta^2 \pm p_1\beta + p_0 = 0,$$

which implies that

$$p_4\beta^4 + p_2\beta^2 + p_0 = 0 \qquad (4\cdot10, 1)$$

and

$$p_3\beta^3 + p_1\beta = 0$$

or

$$p_3\beta^2 + p_1 = 0. \qquad (4\cdot10, 2)$$

The last equation gives

$$\beta^2 = -\frac{p_1}{p_3}, \qquad (4\cdot10, 3)$$

and then $(4\cdot10, 1)$ becomes, when cleared of fractions,

$$p_1p_2p_3 - p_4p_1^2 - p_0p_3^2 = 0$$

or

$$T_3 = 0. \qquad (4\cdot10, 4)$$

Since $\beta = i\omega$, equation $(4\cdot10, 3)$ yields the useful result

$$\omega^2 = \frac{p_1}{p_3}. \qquad (4\cdot10, 5)$$

For an equation of degree n we obtain a pair of equations corresponding to $(4\cdot10, 1)$ and $(4\cdot10, 2)$ whose eliminant, as obtained by Sylvester's dialytic method, is

$$T_{n-1} = 0. \qquad (4\cdot10, 6)$$

We thus recognize that the vanishing of T_{n-1} always indicates the occurrence of a pair of equal and opposite roots. When the vanishing of T_{n-1} is the *first* critical occurrence in a stable system it necessarily implies the existence of a conjugate pair of imaginary roots $\pm i\omega$, for the existence of the real pair $\pm\beta$ would imply an *earlier* critical change. We thus recognize the following critical stability criteria:

(a) $$p_0 > 0. \qquad (4\cdot10, 7)$$

When this becomes an equality the determinantal equation has a zero root, separating a stable constituent in the form of a *real subsidence* from an unstable constituent in the form of a *real divergence*. (See Fig. 4·12, 1.)

(b) $$T_{n-1} > 0. \qquad (4\cdot10, 8)$$

When this becomes an equality (in a system hitherto stable) there is a simple harmonic constituent motion corresponding to roots $\pm i\omega$. This separates a stable damped oscillatory constituent from an unstable divergent oscillatory constituent. (See Fig. 4·12, 1.)

The test function T_3 for the quartic is homogeneous and of the third degree in the coefficients of the equation. When divided by p_4^3 it is a homogeneous symmetric function of the roots of the quartic of degree 6, for

$$\frac{p_3}{p_4} = -(\lambda_1 + \lambda_2 + \lambda_3 + \lambda_4),$$

$$\frac{p_2}{p_4} = \lambda_1\lambda_2 + \lambda_1\lambda_3 + \lambda_1\lambda_4 + \lambda_2\lambda_3 + \lambda_2\lambda_4 + \lambda_3\lambda_4,$$

$$\frac{p_1}{p_4} = -(\lambda_1\lambda_2\lambda_3 + \lambda_1\lambda_2\lambda_4 + \lambda_1\lambda_3\lambda_4 + \lambda_2\lambda_3\lambda_4)$$

and $$\frac{p_0}{p_4} = \lambda_1\lambda_2\lambda_3\lambda_4.$$

Moreover, T_3 vanishes whenever the equation has a pair of equal and opposite roots, so it must contain as a factor Π, the continued product of the sums of the roots taken in pairs. It can in fact easily be shown that for the quartic,

$$\frac{T_3}{p_4^3} = (\lambda_1 + \lambda_2)(\lambda_1 + \lambda_3)(\lambda_1 + \lambda_4)(\lambda_2 + \lambda_3)(\lambda_2 + \lambda_4)(\lambda_3 + \lambda_4). \quad (4\cdot10, 9)$$

Similarly it can be shown that for an equation of degree n,

$$T_{n-1} = (-1)^{\frac{n(n-1)}{2}} p_n^{n-1} \Pi, \qquad (4\cdot10, 10)$$

where the index of -1 is the number of factors in Π.

The inequality $p_0 > 0$ is commonly referred to as the condition for 'static stability'. Let us return to the dynamical equations $(4\cdot2, 1)$ to $(4\cdot2, 3)$ and consider the steady deviations corresponding to constant values of Q_1, Q_2 and Q_3. The equations of equilibrium then are

$$\left.\begin{aligned}
c_{11}q_1 + c_{12}q_2 + c_{13}q_3 &= Q_1, \\
c_{21}q_1 + c_{22}q_2 + c_{23}q_3 &= Q_2, \\
c_{31}q_1 + c_{32}q_2 + c_{33}q_3 &= Q_3.
\end{aligned}\right\} \qquad (4\cdot10, 11)$$

The determinant of the coefficients is

$$\begin{vmatrix} c_{11} & c_{12} & c_{13} \\ c_{21} & c_{22} & c_{23} \\ c_{31} & c_{32} & c_{33} \end{vmatrix} = p_0, \qquad (4\cdot10, 12)$$

by $(4\cdot3, 15)$, and, so long as p_0 does not vanish, the deviations q corresponding to finite forces Q are finite. When all the other test functions continue positive, a change of sign of p_0 from positive to negative indicates the onset of instability and the unstable constituent is a non-oscillatory divergence, corresponding to a real positive root of the determinantal equation. In the critical state [Case (a) above] p_0 vanishes and equations $(4\cdot10, 11)$ show that finite values of the deviations can occur in the absence of applied forces. In general, the condition $p_0 > 0$ in isolation gives no information about the stability except that it is not statically neutral. Approach to the neutral condition $p_0 = 0$ is accompanied by increasing sensitiveness to applied forces.

4·11   Methods for investigating stability*

The following is a list of the theoretical methods available for investigating the stability of a dynamical system:

(a) Find all the roots of the determinantal equation and examine the signs of their real parts. This method gives the

* Some references to methods based on 'transfer functions' and the Nyquist criterion will be found at the end of this chapter.

most complete information but is also, in general, the most laborious.

(b) Use Routh's or the equivalent determinantal test-functions (see § 4·9). This requires the computation of the co-efficients of the determinantal equation and of the test functions.

(c) If we know that the system is stable in a certain condition S we need only ascertain whether any critical changes of stability occur in passing from S to the condition under consideration.

(d) In special cases it may be possible to show by means of the equation of energy that a system is stable.

(e) Special theorems are applicable to systems of restricted type. A number of such theorems is given by Routh in his *Advanced Rigid Dynamics*. Methods (d) and (e) are seldom applicable to problems in the stability of aircraft. Method (c) is specially simple, but not always applicable.

4·12 Stability diagrams

Information about stability can often be conveniently presented with the aid of graphs. In the first place let us consider the dependence of the stability of a system upon a single parameter α. We shall give very useful information, covering the stability completely, by plotting against α both the real and imaginary parts of the roots of the determinantal equation corresponding to each constituent motion (see Fig. 4·12, 1, which, for simplicity, refers to a quartic equation. The figure is merely diagrammatic and serves to illustrate possible occurrences). For small values of α the diagram shows two oscillatory constituents corresponding to the roots $\mu_1 \pm i\omega_1$ and $\mu_2 \pm i\omega_2$. The first constituent is unstable for values of α lying between O and A; A corresponds to a critical condition for which the test function T_3 vanishes. At B the frequency for the first constituent becomes zero and the oscillation splits into a pair of subsidences, as shown by the continuous curve $DB'C$ which necessarily has a vertical tangent at the point of bifurcation B'.

The tangent to the curve of μ_1 at B' bisects the vertical chords of $DB'C$ very near to B'. At C the lower branch of $DB'C$ cuts the axis, so C corresponds to a critical condition where p_0 vanishes and to the right of C we have a divergence.

The second constituent oscillation is shown as always stable and not degenerating into subsidences within the range of α covered in the diagram. For values of α lying between A and C the system is completely stable while to the left of A there is an unstable oscillation and to the right of C a divergence.

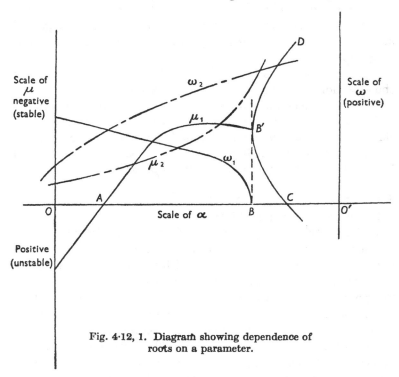

Fig. 4·12, 1. Diagram showing dependence of roots on a parameter.

Next let us consider the dependence of the stability on two parameters α and β. If it is desired to present complete information, as for the case of a single parameter discussed above, it is necessary to have separate diagrams for each constituent. Then, as in Fig. 4·12, 2, we may draw contours of constant μ for the particular constituent motion considered and in particular the critical contour $\mu = 0$. In order to discover the regions of complete stability, if any, we may plot the critical contours for all the constituents in one diagram, as shown in Fig. 4·12, 3 where there are just two constituents. If desired we may also plot contours of constant ω for each constituent.

Another useful diagram shows the plots in the $\alpha\beta$ plane of the two critical conditions $T_{n-1} = 0$ and $p_0 = 0$ which become $R = 0$ and $E = 0$ respectively for the quartic (4·9, 10).

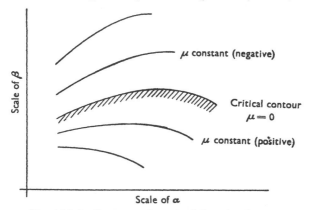

Fig. 4·12, 2. Contours of constant damping factor.

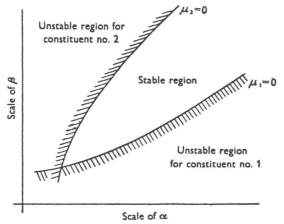

Fig. 4·12, 3. Stable and unstable regions.

Appendix: Routh's criteria of stability for the quartic demonstrated by elementary algebra

A quartic equation with all its coefficients real can always be resolved into *real* quadratic factors as shown in § 4·9. Thus we may always write

$$p_4\lambda^4 + p_3\lambda^3 + p_2\lambda^2 + p_1\lambda + p_0$$
$$\equiv p_4(\lambda^2 + b_1\lambda + c_1)(\lambda^2 + b_2\lambda + c_2). \qquad (1)$$

Moreover, it is necessary and sufficient for stability that b_1, c_1, b_2, c_2 shall all be positive, as follows at once from consideration of the roots of the quadratic factors.

By equating corresponding coefficients on the two sides of (1) we obtain

$$p_3 = p_4(b_1 + b_2), \tag{2}$$
$$p_2 = p_4(b_1 b_2 + c_1 + c_2), \tag{3}$$
$$p_1 = p_4(b_1 c_2 + b_2 c_1), \tag{4}$$
$$p_0 = p_4 c_1 c_2. \tag{5}$$

Hence
$$T_2 = p_3 p_2 - p_4 p_1 \tag{6}$$
$$= p_4^2[b_1 b_2(b_1 + b_2) + b_1 c_1 + b_2 c_2], \tag{7}$$

and
$$T_3 = p_1 T_2 - p_0 p_3^2 \tag{8}$$
$$= p_4^3 b_1 b_2\{(b_1 + b_2)(b_1 c_2 + b_2 c_1) + (c_1 - c_2)^2\} \tag{9}$$
$$= p_4 b_1 b_2\{p_1 p_3 + p_4^2(c_1 - c_2)^2\}. \tag{10}$$

Thus we see that when the system is stable it is necessary that

$$p_3 > 0, \tag{11}$$
$$T_2 > 0, \tag{12}$$
$$T_3 > 0, \tag{13}$$
$$p_0 > 0, \tag{14}$$

where we have supposed the quartic equation to be written with p_4 positive.

Next, let us suppose that the inequalities (11) ... (14) are all satisfied, and let p_4 be positive as before. By (14) the term $p_0 p_3^2$ is positive and by (8), (12) and (13)

$$p_1 > 0. \tag{15}$$

Then, by (6), (11), (12) and (15)

$$p_2 > 0. \tag{16}$$

By (11), (14), (15) and (16) all the coefficients in the quartic must be positive. Hence (10) and (13) yield

$$b_1 b_2 > 0, \tag{17}$$

so b_1 and b_2 have the same sign. But (2) and (11) give

$$b_1 + b_2 > 0, \tag{18}$$

and it follows that
$$b_1 > 0, \tag{19}$$
$$b_2 > 0. \tag{20}$$

By (5) and (14) c_1 and c_2 must have the same sign and by (4) and (15) both must be positive on account of (19) and (20). Thus

$$c_1 > 0, \tag{21}$$
$$c_2 > 0. \tag{22}$$

The inequalities (19) ... (22) show that the system must be stable. Finally, the inequalities (11) to (14) are the necessary and sufficient conditions for stability.

GENERAL REFERENCES FOR CHAPTER 4
SOLUTION OF POLYNOMIAL EQUATIONS

E. T. Whitaker and G. Robinson, *The Calculus of Observations*, Blackie, 1924.

S. Brodetsky and G. Smeal, 'On Graeffe's Method for Complex Roots of Algebraic Equations', *Proc. Camb. Phil. Soc.*, vol. 22 (1924), p. 83.

R. A. Frazer and W. J. Duncan, 'On the Numerical Solution of Equations with Complex Roots', *Proc. Roy. Soc.* A., vol. 125 (1929), p. 68.

H. R. Hopkin, 'Routine Computing Methods for Stability and Response Investigations on Linear Systems', *R.A.E. Tech. Note No. INST.* 954 (1946), *R. & M.* 2392 (1950).

OPERATIONAL AND MATRIX METHODS, MECHANICAL ADMITTANCES

H. S. Carslaw and J. C. Jaeger, *Operational Methods in Applied Mathematics*, Oxford, 1941.

R. V. Churchill, *Modern Operational Mathematics in Engineering*, McGraw Hill, 1944.

R. A. Frazer, W. J. Duncan, and A. R. Collar, *Elementary Matrices*, Cambridge, 1938.

W. J. Duncan, 'Mechanical Admittances and their Applications to Oscillation Problems', A.R.C. Monograph, *R. & M.* 2000 (1947).

STABILITY OF DYNAMICAL SYSTEMS

E. J. Routh, *A Treatise on the Stability of a Given State of Motion*, Macmillan,

E. J. Routh, *Advanced Rigid Dynamics*, Macmillan.   [1877.

A. Hurwitz, 'Über die Bedingungen, unter welchen eine Gleichung nur Wurzeln mit negativen reelen Theilen besitzt', *Mathematische Annalen*, vol. 46, 1895, pp. 273–84.

L. Bairstow and J. L. Nayler, 'Investigations Relating to the Stability of the Aeroplane', and 'Investigation of the Stability of an Aeroplane when in Circling Flight', *R. & M.* 154 (1914).

R. A. Frazer and W. J. Duncan, 'On the Criteria for the Stability of Small Motions', *Proc. Roy. Soc.* A., vol. 124 (1929), p. 642.

W. S. Brown, 'A Simple Method of Constructing Stability Diagrams', *R. & M.* 1905 (1942).

THE NYQUIST CRITERION

Nyquist's methods were originally devised for application to coupled electric circuits but are now finding useful applications to servo-mechanisms. In the hands of experienced users they can be employed to design for required characteristics. For an account the reader may consult

H. W. Bode, *Network Analysis and Feedback Amplifier Design*, Van Nostrand, 1945.

G. S. Brown and D. P. Campbell, *Principles of Servomechanisms*, Wiley, 1948.

A. Porter, *An Introduction to Servomechanisms*, Methuen, 1950.

R. H. Macmillan, *An Introduction to the Theory of Control*, Cambridge, 1951.

Chapter 5

LONGITUDINAL-SYMMETRIC MOTION

5·1 Introduction

We have seen in § 3·9 that one of the possible types of motion of small deviation from a state of rectilinear symmetric flight of a symmetric aircraft is itself symmetric, involving the deviations u, w, of longitudinal and normal velocity respectively and the angle of pitch θ. The equations of motion are (3·11, 11), (3·11, 12) and (3·11, 13), while their solution has been treated in a general way in Chapter 4. In the present chapter the dynamical equations are first transformed into non-dimensional forms and their solutions examined in greater detail. The detailed analysis confirms the existence in typical instances of two typical modes of motion, namely, the rapid incidence adjustment and the phugoid, of which simplified approximate theories are given in § 2·6 and § 2·7 respectively. There is, of course, no necessity to use the non-dimensional forms of the dynamical equations and the analysis proceeds in the same manner whichever form is adopted. The main advantage of the non-dimensional equations is that they are independent of the linear scale of the aircraft. Hence the values of the coefficients are less widely variable and the comparison of different aircraft is facilitated.

For simplicity, we adopt 'wind axes' (see § 3·11) throughout this chapter. Hence we have

$$\left.\begin{aligned} \vartheta &= 0, \\ W &= 0, \\ U &= V, \end{aligned}\right\} \qquad (5\cdot1,\ 1)$$

while Θ is the upward inclination of the flight path to the horizontal (see Fig. 3·11, 1). The condition of balance of the forces normal to the flight path in the undisturbed state is

$$mg \cos \Theta = \tfrac{1}{2}\rho V^2 S C_L. \qquad (5\cdot1,\ 2)$$

It is assumed throughout that the aircraft is rigid and that the controls are fixed or moved in a prescribed manner.

5·2 Non-dimensional forms of the equations of motion

It has been pointed out in § 2·15 that there is a natural unit of time for aircraft motions and this suggests that we should employ a non-dimensional measure of time, namely the ratio of the actual time to the natural unit. Following a method originally proposed by Glauert, we can go further and reduce all the dynamical equations to completely non-dimensional forms.*

We adopt the following basic non-dimensional quantities:

$$\tau = \frac{t}{\hat{t}} = \text{non-dimensional measure of time,} \qquad (5·2, 1)$$

where
$$\hat{t} = \frac{m}{\rho VS} = \frac{W}{g\rho VS} \qquad (5·2, 2)$$

is the natural unit of time, and

$$\mu_1 = \frac{m}{\rho S l_T} = \frac{W}{g\rho S l_T} \qquad (5·2, 3)$$

is the conventional relative density, where l_T is the tail arm, i.e. the distance of the aerodynamic centre of the tailplane aft of the c.g. of the aircraft

$$k_L = \tfrac{1}{2}C_L = \frac{W\cos\Theta}{\rho V^2 S}, \qquad (5·2, 4)$$

$$k' = -k_L \tan\Theta, \qquad (5·2, 5)$$

$$\bar{u} = \frac{u}{V}, \qquad (5·2, 6)$$

$$\bar{w} = \frac{w}{V}. \qquad (5·2, 7)$$

* The notation agrees with that of the Royal Aeronautical Society's Data Sheet 'Aircraft 00·00·02' except that we use $m_{\dot{w}}$ in place of $\bar{m}_{\dot{w}}$ and $\bar{u}, \bar{v}, \bar{w}$ to represent u/V, v/V and w/V respectively.

The aerodynamic derivatives are reduced to non-dimensional forms as follows:

$$\left. \begin{aligned} x_u &= \frac{X_u}{\rho VS}, & x_w &= \frac{X_w}{\rho VS}, \\ z_u &= \frac{Z_u}{\rho VS}, & z_w &= \frac{Z_w}{\rho VS}, \\ x_q &= \frac{X_q}{\rho VSl_T}, & z_q &= \frac{Z_q}{\rho VSl_T}, \\ m_u &= \frac{M_u}{\rho VSl_T}, & m_w &= \frac{M_w}{\rho VSl_T}, \\ m_q &= \frac{M_q}{\rho VSl_T^2}, & m_{\dot{w}} &= \frac{M_{\dot{w}}}{\rho Sl_T^2}. \end{aligned} \right\} \tag{5·2, 8}$$

The non-dimensional moment of inertia coefficient for pitch is

$$i_B = \frac{B}{ml_T^2} = \frac{gB}{Wl_T^2}, \tag{5·2, 9}$$

while the non-dimensional applied force and moment coefficients are

$$\left. \begin{aligned} x(\tau) &= \frac{X(t)}{\rho V^2 S}, \\ z(\tau) &= \frac{Z(t)}{\rho V^2 S}, \\ m(\tau) &= \frac{M(t)}{\rho V^2 Sl_T}, \end{aligned} \right\} \tag{5·2, 10}$$

where t and τ are related by (5·2, 1).

The coefficients in (5·2, 10) are all zero for free motion with fixed controls.

Take equation (3·11, 11), which expresses the balance of longitudinal forces, and divide by $\rho V^2 S$. It will be found, after some reduction, that it can be written

$$\frac{d\bar{u}}{d\tau} - \bar{u}x_u - \bar{w}x_w + \theta k_L - \frac{d\theta}{d\tau}\frac{x_q}{\mu_1} = x(\tau). \tag{5·2, 11}$$

Similarly, the equation (3·11, 12) expressing the balance of normal forces becomes

$$-\bar{u}z_u + \frac{d\bar{w}}{d\tau} - \bar{w}z_w - \theta k' - \frac{d\theta}{d\tau}\left(1 + \frac{z_q}{\mu_1}\right) = z(\tau). \tag{5·2, 12}$$

Finally, the equation (3·11, 13) expressing the balance of pitching moments becomes, when divided by $\rho V^2 S l_T$,

$$-\bar{u}m_u - \frac{d\bar{w}}{d\tau}\frac{m_{\dot{w}}}{\mu_1} - \bar{w}m_w + \frac{1}{\mu_1}\left(\frac{d^2\theta}{d\tau^2}i_B - \frac{d\theta}{d\tau}m_q\right) = m(\tau). \quad (5\cdot2, 13)$$

For the free motion with fixed controls the right-hand sides of these equations are zero. Here, as explained in § 4·3, we assume that all the deviations are proportional to $\exp(\lambda\tau)$ and then the equations can all be satisfied provided that

$$\Delta(\lambda) \equiv \begin{vmatrix} \lambda - x_u & -x_w & -\dfrac{\lambda x_q}{\mu_1} + k_L \\[2ex] -z_u & \lambda - z_w & -\lambda\left(1 + \dfrac{z_q}{\mu_1}\right) - k' \\[2ex] -m_u & -\dfrac{\lambda m_{\dot{w}}}{\mu_1} - m_w & \dfrac{\lambda^2 i_B}{\mu_1} - \dfrac{\lambda m_q}{\mu_1} \end{vmatrix} = 0, \quad (5\cdot2, 14)$$

When the determinant is expanded and divided by the coefficient of λ^4 the equation may be written

$$\lambda^4 + B\lambda^3 + C\lambda^2 + D\lambda + E = 0, \quad (5\cdot2, 15)$$

where

$$B = -(x_u + z_w) - \frac{m_q}{i_B} - \left(1 + \frac{z_q}{\mu_1}\right)\frac{m_{\dot{w}}}{i_B}, \quad (5\cdot2, 16)$$

$$C = (x_u z_w - x_w z_u) + \frac{m_q}{i_B}(x_u + z_w)$$

$$+ \frac{m_{\dot{w}}}{i_B}\left[x_u\left(1 + \frac{z_q}{\mu_1}\right) - \frac{x_q z_u}{\mu_1} - k'\right]$$

$$- \frac{m_w}{i_B}(\mu_1 + z_q) - \frac{x_q m_u}{i_B}, \quad (5\cdot2, 17)$$

$$D = -\frac{m_q}{i_B}(x_u z_w - x_w z_u) + \frac{m_{\dot{w}}}{i_B}(k_L z_u + k' x_u)$$

$$+ \frac{m_w}{i_B}[x_u(\mu_1 + z_q) - x_q z_u - \mu_1 k']$$

$$+ \frac{m_u}{i_B}[\mu_1 k_L - x_w(\mu_1 + z_q) + x_q z_w], \quad (5\cdot2, 18)$$

$$E = \frac{\mu_1 m_w}{i_B}(k_L z_u + k' x_u) - \frac{\mu_1 m_u}{i_B}(k_L z_w + k' x_w). \quad (5\cdot2, 19)$$

9

These general results are subject to some simplification in special cases:

(a) For horizontal flight k' is zero.

(b) In the absence of structural distortion and of compressibility effects m_u is zero [see equation (5·3, 8)].

(c) For tailless aircraft $m_{\dot{w}}$ is negligible.

(d) For modern highly loaded aircraft μ_1 is relatively large while x_q, z_q are commonly small and often neglected.

5·3   The aerodynamic derivatives

The aerodynamic derivatives of longitudinal-symmetric motion, in their non-dimensional forms, are listed in (5·2, 8). We have now to consider their values and we shall at present restrict the discussion to the case of a rigid aircraft and neglect the influence of compressibility. For convenience we shall adopt wind axes.

As pointed out in § 3·5, quasi-static derivatives can be used with an error of less than 5 per cent when the non-dimensional quantity

$$\chi = \frac{c\dot{\alpha}}{V\alpha} \tag{5·3, 1}$$

is less than 0·02 where, for oscillatory motion, α and $\dot{\alpha}$ are to be interpreted as amplitudes. For typical phugoid oscillations this criterion is well satisfied but for the rapid incidence adjustment the limit may be somewhat exceeded. Even so, the errors introduced by neglect of the frequency effect are usually small and we shall here adopt quasi-static derivatives, adding the caution that a special inquiry would be required if χ substantially exceeded the stated limit.

The quasi-static values of the four force-velocity derivatives have already been obtained in equations (2·5, 19) to (2·5, 22). However (see Fig. 2·5, 1), there is the simplification with wind axes that α is zero. Accordingly we obtain the following expressions for the non-dimensional derivative coefficients

$$-x_u = C_D, \tag{5·3, 2}$$

$$-x_w = -\frac{1}{2}\left(C_L - \frac{dC_D}{d\alpha}\right), \tag{5·3, 3}$$

$$-z_u = C_L, \tag{5·3, 4}$$

$$-z_w = \frac{1}{2}\left(C_D + \frac{dC_L}{d\alpha}\right). \tag{5·3, 5}$$

Similarly we obtain from (2·5, 31) and (2·5, 32)

$$m_u = \frac{\bar{c}}{l_T} C_m, \qquad (5·3, 6)$$

$$m_w = \frac{1}{2} \frac{\bar{c}}{l_T} \frac{dC_m}{d\alpha}, \qquad (5·3, 7)$$

where C_m is the standard pitching moment coefficient referred to an axis through the C.G.* As they stand the formulae only apply to gliding flight. The influence of the propulsive system is considered in § 5·6. Since we are concerned with small deviations from a state of equilibrium C_m is zero and (5·3, 6) yields

$$m_u = 0. \qquad (5·3, 8)$$

The physical interpretation of this is that a change of relative air speed without change of incidence merely changes the pitching moment in proportion to V^2 and thus does not change a vanishing pitching moment. This is true only when the aircraft is rigid and the pitching moment coefficient for constant incidence is independent of air speed.

X_q and Z_q. We may calculate the contributions of the tailplane to these derivatives by use of the formulae for x_w and z_w if we regard the angular velocity in pitch as giving rise to a normal downward velocity at the tailplane equal to ql_T. Hence by (5·2, 8) the increment of the longitudinal force is

$$\Delta X' = \rho V S' l_T q x_w'$$

and

$$x_q = \frac{\Delta X'}{\rho V S l_T q} = \left(\frac{S'}{S}\right) x_w'$$

$$= \frac{1}{2}\left(\frac{S'}{S}\right)\left(C_L' - \frac{dC_D'}{d\alpha'}\right) \qquad (5·3, 9)$$

by (5·3, 3). The expression on the right-hand side of the equation is small and the contributions of the other parts of the aircraft are also small. Hence it is usual to neglect x_q in

* If we had used a modified pitching moment coefficient given by

$$C_M = \frac{M}{\frac{1}{2}\rho V^2 S l_T}$$

the factor $\bar{c}/l_T$ would have been absent from (5·3, 6) and (5·3, 7).

calculations of longitudinal stability. We derive similarly

$$z_q = \left(\frac{S'}{S}\right) z'_w$$

$$= -\frac{1}{2}\left(\frac{S'}{S}\right)\left(C'_D + \frac{dC'_L}{d\alpha'}\right) \tag{5·3, 10}$$

by (5·3, 5) for the contribution of tailplane to the derivative.

The contribution of unswept wings to Z_q can be estimated from results given by Glauert in *R. & M.* 1216,* where it is assumed that the aerofoil is thin, that the angular velocity in pitch is steady and that the axis of rotation intersects the wing chord. For a rectangular wing Glauert obtains

$$Z_q = -\rho V Sc\lambda, \tag{5·3, 11}$$

where

$$\lambda = \frac{1}{2}\frac{dC_L}{d\alpha}\left(\frac{3}{4} - h\right) \tag{5·3, 12}$$

and h is the distance of the axis of pitch behind the leading edge, expressed as a fraction of the chord c. We may generalize this result approximately for a tapered wing of span b and obtain

$$Z_q = -\rho V \int_{-b/2}^{b/2} \lambda c^2 \, dy. \tag{5·3, 13}$$

M_q. The contribution to this derivative due to the normal velocity at the tailplane can be found as follows. Clearly

$$\Delta M = l_T \Delta Z'$$

$$= \rho V S' l_T^2 q z'_w$$

when we neglect the small pitching couple on the tailplane, and

$$m_q = \frac{\Delta M}{\rho V S l_T^2 q} = \left(\frac{S'}{S}\right) z'_w = -\frac{1}{2}\left(\frac{S'}{S}\right)\left(C'_D + \frac{dC'_L}{d\alpha'}\right). \tag{5·3, 14}$$

In § 2·6 we obtain the following approximate expression for the damping coefficient,

$$B' = \tfrac{1}{2}\rho V k l'^2 S' \frac{dC'_L}{d\alpha'}.$$

* H. Glauert, 'The Lift and Pitching Moment of an Aerofoil due to a Uniform Angular Velocity of Pitch' (1928).

This is in agreement with the above if we identify l' with l_T, put k equal to unity, and neglect C_D' in comparison with $\dfrac{dC_L'}{d\alpha'}$. It is legitimate to take k as unity if the lift slope $\dfrac{dC_L'}{d\alpha'}$ for the tailplane is found experimentally for the tailplane in situ behind the wing.

The contribution of the wing to M_q can be estimated from Glauert's results as for Z_q and the assumptions are the same. For a rectangular wing

$$M_q = -\rho V S c^2 \mu, \qquad (5{\cdot}3,\,15)$$

where $$\mu = \frac{1}{8}\frac{dC_L}{d\alpha}(1-2h)^2 + \frac{1}{32}\left(2\pi - \frac{dC_L}{d\alpha}\right), \qquad (5{\cdot}3,\,16)$$

and we may extend this approximately to a tapered wing and obtain

$$M_q = -\rho V \int_{-b/2}^{b/2} \mu c^3\, dy. \qquad (5{\cdot}3,\,17)$$

$M_{\dot w}$. A conventional theory of this derivative has been worked out based on the idea that the downwash at the tailplane corresponds to the wing incidence at a time earlier than the instant considered, the interval being l_T/V, i.e. the time needed to traverse the distance between the tailplane and wing. Accordingly we write

$$\epsilon = \frac{d\epsilon}{d\alpha}\left[\alpha - \frac{d\alpha}{dt}\frac{l_T}{V}\right],$$

and the tailplane incidence is

$$\alpha' = \alpha - \epsilon$$
$$= \alpha\left(1 - \frac{d\epsilon}{d\alpha}\right) + \frac{l_T}{V}\frac{d\epsilon}{d\alpha}\frac{d\alpha}{dt}.$$

When the rate of change of wing incidence is entirely attributable to the velocity w we have

$$\frac{d\alpha}{dt} = \frac{\dot w}{V},$$

and by the last equation

$$\frac{d\alpha'}{d\dot{w}} = \frac{l_T}{V^2}\frac{d\epsilon}{d\alpha}.$$

Hence

$$\frac{\partial M}{\partial \dot{w}} = -\tfrac{1}{2}\rho V^2 S' l_T \frac{dC_L'}{d\alpha'}\frac{d\alpha'}{d\dot{w}}$$

$$= -\tfrac{1}{2}\rho S' l_T^2 \frac{dC_L'}{d\alpha'}\frac{d\epsilon}{d\alpha},$$

and

$$m_{\dot{w}} = \frac{\dfrac{\partial M}{\partial \dot{w}}}{\rho S l_T^2}$$

$$= -\frac{1}{2}\left(\frac{S'}{S}\right)\frac{dC_L'}{d\alpha'}\frac{d\epsilon}{d\alpha}. \qquad (5\cdot3, 18)$$

5·4   Shift of the centre of mass

The c.g. of an aircraft alters its position as fuel is consumed and as the other loads are varied. Now the origin of co-ordinates in our system of body axes is always the c.g. and in general the derivatives are dependent on the situation of the origin. We must therefore inquire into this dependence.

This question has two aspects:

(a) On the assumption that the external form of the aircraft is invariable we may obtain the derivatives for the new origin from those for the old origin by means of the principles of statics and kinematics.

(b) Since the aircraft must be trimmed for the new position of the c.g. there will, in general, be a need to retrim and this implies a change of the geometric form of the aircraft in its configuration of equilibrium, e.g. the tailplane or elevator is reset. The geometric changes with their associated alterations of aerodynamic loading bring about changes in the derivatives.

We shall consider these aspects separately and we shall suppose that the *directions* of the axes are fixed. The new c.g. will be taken to lie in the plane of symmetry of the aircraft.

Let accented and unaccented symbols refer to the new and old origin respectively. Then we see from Fig. 5·4, 1 that

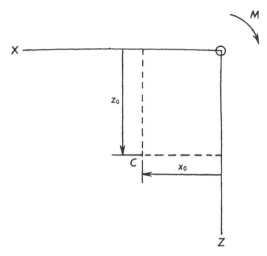

Fig. 5·4, 1. Shift of the centre of mass. O is the original c.g., C is the new c.g.

$$\left.\begin{aligned} \theta' &= \theta, \\ q' &= q, \\ u' &= u + qz_0, \\ w' &= w - qx_0, \end{aligned}\right\} \quad (5·4, 1)$$

$$\left.\begin{aligned} X' &= X, \\ Z' &= Z, \\ M' &= M + Zx_0 - Xz_0. \end{aligned}\right\} \quad (5·4, 2)$$

Since the two forces are unaltered it is only necessary to substitute

$$\left.\begin{aligned} u &= u' - q'z_0, \\ w &= w' + q'x_0, \end{aligned}\right\} \quad (5·4, 3)$$

in the dynamical equations (3·11, 11) and (3·11, 12). Thus

$$uX_u + wX_w + qX_q = (u' - q'z_0) X_u + (w' + q'x_0) X_w + q'X_q$$
$$= u'X_u + w'X_w + q'(X_q + x_0 X_w - z_0 X_u).$$

Hence

$$X'_{u'} = X_u, \tag{5·4, 4}$$

$$X'_{w'} = X_w, \tag{5·4, 5}$$

$$X'_{q'} = X_q + x_0 X_w - z_0 X_u. \tag{5·4, 6}$$

We derive similarly

$$Z'_{u'} = Z_u, \tag{5·4, 7}$$

$$Z'_{w'} = Z_w, \tag{5·4, 8}$$

$$Z'_{q'} = Z_q + x_0 Z_w - z_0 Z_u. \tag{5·4, 9}$$

By the last of equations (5·4, 2)

$$\begin{aligned}
M' &= u M_u + \dot{w} M_{\dot{w}} + w M_w + q M_q + Z x_0 - X z_0 \\
&= (u' - q' z_0) M_u + (\dot{w}' + \dot{q}' x_0) M_{\dot{w}} \\
&\quad + (w' + q' x_0) M_w + q' M_q \\
&\quad + x_0 [u' Z_u + w' Z_w + q'(Z_q + x_0 Z_w - z_0 Z_u)] \\
&\quad - z_0 [u' X_u + w' X_w + q'(X_q + x_0 X_w - z_0 X_u)].
\end{aligned} \tag{5·4, 10}$$

Hence we derive

$$M'_{u'} = M_u + x_0 Z_u - z_0 X_u, \tag{5·4, 11}$$

$$M'_{\dot{w}'} = M_{\dot{w}}, \tag{5·4, 12}$$

$$M'_{w'} = M_w + x_0 Z_w - z_0 X_w, \tag{5·4, 13}$$

$$\begin{aligned}
M'_{q'} = M_q &+ x_0(M_w + Z_q) - z_0(M_u + X_q) \\
&+ x_0^2 Z_w - x_0 z_0(Z_u + X_w) + z_0^2 X_u,
\end{aligned} \tag{5·4, 14}$$

$$M'_{\dot{q}} = x_0 M_{\dot{w}}. \tag{5·4, 15}$$

It must be noted that the expression (5·4, 12) for $M'_{\dot{w}'}$ is incomplete since the derivatives of X and Z with respect to $\dot{w}$ were neglected, but it can be accepted as a sufficient approximation. Similarly the expression (5·4, 15) for $M'_{\dot{q}'}$ is incomplete, but this derivative is not usually retained explicitly.

The only important effect of the retrimming necessitated by the shift of the C.G. is a modification of $M'_{u'}$ which, like M_u, must be zero for a rigid aircraft in the glide, subject to the provisos mentioned in the discussion of M_u in § 5·3.

5·5 Static longitudinal stability

We have seen in § 4·10 that, when the determinantal equation is written with the coefficient of the highest power of λ positive, one necessary condition for stability is

$$p_0 > 0. \qquad (5·5, 1)$$

We also saw that this is a critical condition for stability in the sense that the vanishing of p_0 is the indication of one of the two possible transition states between complete stability and instability. The satisfaction of the inequality (5·5, 1) is called the condition of static stability.

For the longitudinal-symmetric type of motion the determinantal equation in its non-dimensional form is the quartic (5·2, 15). Since the coefficient of the highest power of λ is here positive the condition for static stability is

$$E > 0, \qquad (5·5, 2)$$

and by (5·2, 19)

$$\frac{i_B E}{\mu_1} = m_w(k_L z_u + k'x_u) - m_u(k_L z_w + k'x_w). \qquad (5·5, 3)$$

For a rigid aircraft in a glide and when the influence of the compressibility of the air can be neglected we have m_u zero as shown in § 5·3. In these circumstances (5·5, 3) becomes

$$\frac{i_B E}{\mu_1} = m_w(k_L z_u + k'x_u). \qquad (5·5, 4)$$

By (5·2, 5), (5·3, 2) and (5·3, 4)

$$k_L z_u + k'x_u = k_L(z_u - x_u \tan \Theta)$$

$$= -\frac{C_L}{2 \cos \Theta}(C_L \cos \Theta - C_D \sin \Theta). \qquad (5·5, 5)$$

Let R be the resultant aerodynamic force on the aircraft in the state of equilibrium and γ the (downward) glide angle. Since R balances the weight it acts vertically upward and

$$R = L \cos \gamma + D \sin \gamma$$

$$= L \cos \Theta - D \sin \Theta, \qquad (5·5, 6)$$

for

$$\Theta = -\gamma. \qquad (5·5, 7)$$

Further, since the horizontal component of the resultant aerodynamic force is zero

$$L \sin \Theta + D \cos \Theta = 0. \qquad (5\cdot5, 8)$$

It follows from $(5\cdot5, 6)$ that

$$C_R = C_L \cos \Theta - C_D \sin \Theta, \qquad (5\cdot5, 9)$$

and $(5\cdot5, 5)$ becomes accordingly

$$k_L z_u + k' x_u = \frac{-C_L C_R}{2 \cos \Theta} = -\tfrac{1}{2} C_R^2. \qquad (5\cdot5, 10)$$

Hence $(5\cdot5, 4)$ becomes

$$\frac{i_B \mathbf{E}}{\mu_1} = -\tfrac{1}{2} C_R^2 m_w. \qquad (5\cdot5, 11)$$

In order that $\mathbf{E}$ may be positive it is thus necessary that m_w shall be negative. This is the condition for static stability in the glide, subject to m_u being zero as here postulated. In the same circumstances we have, by $(5\cdot3, 7)$

$$m_w = \frac{\bar{c}}{2l_T} \frac{dC_m}{d\alpha},$$

and finally $\dfrac{dC_m}{d\alpha}$ must be negative for static stability.

Let us now consider how, subject to the assumptions made, the static stability depends on the position of the c.g. We have seen in § 5·4 that when the c.g. is shifted to the point whose co-ordinates are (x_0, z_0) the derivative M_w is changed to

$$M'_w = M_w + x_0 Z_w - z_0 X_w.$$

Hence by $(5\cdot2, 8)$, $(5\cdot3, 3)$ and $(5\cdot3, 5)$

$$m'_w = m_w + \frac{x_0 z_w - z_0 x_w}{l_T}$$
$$= \frac{1}{2l_T} \left[\bar{c} \frac{dC_m}{d\alpha} - x_0 \left(C_D + \frac{dC_L}{d\alpha} \right) - z_0 \left(C_L - \frac{dC_D}{d\alpha} \right) \right]. \qquad (5\cdot5, 12)$$

The coefficient of x_0 in this expression is normally negative and much larger numerically than the coefficient of z_0. Hence a forward shift of the c.g. renders the derivative more negative, i.e. increases the static stability. Conversely a rearward shift of the C.G. reduces the static stability. The coefficient of z_0 is also normally negative, so a downward shift of the C.G. increases the static stability. There is a point on OX which renders the expression on the right-hand side of (5·5, 12) zero. This is called the *neutral point*, or, to be more precise, the elevator-fixed neutral point which is usually called the stick-fixed neutral point. Let this neutral point be at a distance $h_n \bar{c}$ aft of the datum point in the aircraft while the C.G. lies $h\bar{c}$ aft of the datum point. Then the non-dimensional quantity $(h_n - h)$ is called the *stick-fixed C.G. margin* and provides a convenient measure of the static stability with fixed elevator. Further information will be found in Chapter 10 where the stick-free C.G. margin is also discussed.

Gates and Lyon* have given a valuable alternative expression for **E** in the more general case where the non-dimensional force and moment coefficients for constant incidence are not independent of speed.† Before entering on the analysis we shall draw attention to certain simplifications which result from using 'wind axes'. We see from equations (2·5, 7) and (2·5, 8) that with such axes a deviation u gives rise to an increase of speed but to no change of incidence. Hence u is equivalent to

$$\left. \begin{aligned} \Delta V &= u, \\ \Delta \alpha &= 0. \end{aligned} \right\} \qquad (5\cdot5,\ 13)$$

On the other hand a deviation w gives no increment of speed (to the first order of small quantities) but does give an incidence change. Thus w is equivalent to

$$\left. \begin{aligned} \Delta V &= 0, \\ \Delta \alpha &= \frac{w}{V} = \overline{w}\,. \end{aligned} \right\} \qquad (5\cdot5,\ 14)$$

The aim of the analysis will be to express **E** in terms of the resultant force coefficient C_R.

On account of (5·2, 4), (5·2, 5) and the relation

$$C_L = C_R \cos \Theta \qquad (5\cdot5,\ 15)$$

---

* S. B. Gates and H. M. Lyon, 'A Continuation of Longitudinal Stability and Control Analysis. Part I, General Theory', *R. & M.* 2027 (1944).

† This investigation may be omitted at a first reading.

equation (5·5, 3) can be rewritten

$$\frac{2i_B \mathbf{E}}{\mu_1 C_R} = m_u r_w - m_w r_u,$$ (5·5, 16)

where $\qquad\qquad r_u = x_u \sin \Theta - z_u \cos \Theta,$ (5·5, 17)

and $\qquad\qquad r_w = x_w \sin \Theta - z_w \cos \Theta.$ (5·5, 18)

Now $\qquad\qquad R^2 = L^2 + D^2,$ (5·5, 19)

so $\qquad R\dfrac{\partial R}{\partial V} = R\dfrac{\partial R}{\partial u} \text{ by (5·5, 13)} \quad = L\dfrac{\partial L}{\partial u} + D\dfrac{\partial D}{\partial u},$

and $\qquad\qquad \dfrac{\partial R}{\partial V} = \dfrac{\partial L}{\partial u}\cos \Theta - \dfrac{\partial \dot{D}}{\partial u}\sin \Theta,$ (5·5, 20)

since $\qquad\qquad D = -R\sin \Theta.$ (5·5, 21)

But the body axes remain exact wind axes throughout the motion when only u varies. Hence

$$L = -Z,$$

and $\qquad\qquad D = -X,$

so $\qquad\qquad \dfrac{\partial L}{\partial u} = -Z_u,$

and $\qquad\qquad \dfrac{\partial D}{\partial u} = -X_u.$

Accordingly equation (5·5, 20) yields

$$\frac{1}{\rho VS}\frac{\partial R}{\partial V} = x_u \sin \Theta - z_u \cos \Theta = r_u.$$

But $\qquad\qquad \dfrac{\partial R}{\partial V} = \rho VSC_R + \tfrac{1}{2}\rho V^2 S \dfrac{\partial C_R}{\partial V}.$

Hence $\qquad\qquad r_u = C_R + \tfrac{1}{2}V \dfrac{\partial C_R}{\partial V}.$ (5·5, 22)

When only w varies the speed remains fixed and the expressions (5·3, 3), (5·3, 5) for the derivatives remain valid. Hence*

$$r_w = \frac{1}{2}\left(C_L - \frac{\partial C_L}{\partial \alpha}\right)\sin \Theta + \frac{1}{2}\left(C_D + \frac{\partial C_L}{\partial \alpha}\right)\cos \Theta$$

$$= \frac{1}{2}\frac{\partial C_L}{\partial \alpha}\cos \Theta - \frac{1}{2}\frac{\partial C_D}{\partial \alpha}\sin \Theta \quad \text{by (5·5, 8).}$$

* We now must use the symbols for partial differentiation with respect to incidence since the coefficients depend also on speed.

Again, since V is fixed, we have

$$\frac{\partial C_L}{\partial \alpha} = \frac{1}{\frac{1}{2}\rho V^2 S}\frac{\partial L}{\partial \alpha} = \frac{1}{\frac{1}{2}\rho VS}\frac{\partial L}{\partial w} \quad \text{by (5·5, 14)},$$

and similarly

$$\frac{\partial C_D}{\partial \alpha} = \frac{1}{\frac{1}{2}\rho VS}\frac{\partial D}{\partial w}.$$

Therefore

$$r_w = \frac{1}{\rho VS}\left(\frac{\partial L}{\partial w}\cos\Theta - \frac{\partial D}{\partial w}\sin\Theta\right) = \frac{1}{\rho VS}\frac{\partial R}{\partial w}$$

by (5·5, 19), and the relations between R, L and D. Also

$$\frac{\partial R}{\partial w} = \tfrac{1}{2}\rho VS\frac{\partial C_R}{\partial \alpha},$$

and finally

$$r_w = \frac{1}{2}\frac{\partial C_R}{\partial \alpha}, \tag{5·5, 23}$$

where the partial derivative is to be computed for constant V.

The two moment derivatives in (5·5, 16) are obtained as follows. We have

$$M = \tfrac{1}{2}\rho V^2 S\bar{c}C_m,$$

so

$$\frac{\partial M}{\partial u} = \frac{\partial M}{\partial V} = \tfrac{1}{2}\rho V^2 S\bar{c}\frac{\partial C_m}{\partial V},$$

since C_m vanishes in the state of equilibrium. Hence

$$m_u = \frac{1}{2}\frac{\bar{c}}{l_T}V\frac{\partial C_m}{\partial V}. \tag{5·5, 24}$$

Equation (5·3, 7) remains valid when the differential coefficient is written as partial since V does not change with w, so

$$m_w = \frac{1}{2}\frac{\bar{c}}{l_T}\frac{\partial C_m}{\partial \alpha}. \tag{5·5, 25}$$

By use of (5·5, 22) ... (5·5, 25) equation (5·5, 16) can be reduced to

$$\frac{4i_B l_T \mathbf{E}}{\mu_1 \bar{c}C_R^2} = \frac{V}{2C_R}\frac{\partial C_R}{\partial \alpha}\frac{\partial C_m}{\partial V} - \frac{\partial C_m}{\partial \alpha}\left(1 + \frac{V}{2C_R}\frac{\partial C_R}{\partial V}\right). \tag{5·5, 26}$$

This can be more concisely expressed in terms of the total derivative $\dfrac{dC_m}{dC_R}$ calculated for the condition of equilibrium of forces but with elevator fixed. When the aerodynamic forces balance the weight we have

$$C_R V^2 = \frac{W}{\frac{1}{2}\rho S} = \text{constant}$$

and

$$\left(\frac{\partial C_R}{\partial \alpha} d\alpha + \frac{\partial C_R}{\partial V} dV\right) V^2 + 2C_R V dV = 0.$$

The last equation gives

$$\frac{dV}{d\alpha} = -\frac{\dfrac{V}{2C_R}\dfrac{\partial C_R}{\partial \alpha}}{1 + \dfrac{V}{2C_R}\dfrac{\partial C_R}{\partial V}}. \tag{5.5, 27}$$

Then

$$\frac{dC_m}{dC_R} = \frac{\dfrac{\partial C_m}{\partial \alpha} + \dfrac{\partial C_m}{\partial V}\dfrac{dV}{d\alpha}}{\dfrac{\partial C_R}{\partial \alpha} + \dfrac{\partial C_R}{\partial V}\dfrac{dV}{d\alpha}} = \frac{\dfrac{\partial C_m}{\partial \alpha}\left(1 + \dfrac{V}{2C_R}\dfrac{\partial C_R}{\partial V}\right) - \dfrac{V}{2C_R}\dfrac{\partial C_R}{\partial \alpha}\dfrac{\partial C_m}{\partial V}}{\dfrac{\partial C_R}{\partial \alpha}}.$$

$$\tag{5.5, 28}$$

Accordingly equation (5.5, 26) can be written

$$\mathbf{E} = -\frac{\mu_1}{i_B}\frac{\bar{c}}{2l_T}\frac{C_R^2}{2}\frac{\partial C_R}{\partial \alpha}\frac{dC_m}{dC_R}. \tag{5.5, 29}$$

This is the expression given by Gates and Lyon. It shows that when $\dfrac{\partial C_R}{\partial \alpha}$ is positive, as normally, we must have $\dfrac{dC_m}{dC_R}$ negative for static stability. Since no restriction has been imposed on Θ this result holds good for dives, however steep.

It should be emphasized that the foregoing analysis is completely general, in the sense that it establishes the condition for static stability when the force and moment coefficients depend on speed for any cause. In particular, the analysis covers the cases where the coefficients are speed-dependent on account of the compressibility of the air, distortion of the structure or the influence of the propulsive system.

The subject of static stability is considered in further detail in Chapter 10.

5·6   Influence of the propulsive system

When an airscrew is developing positive thrust there is a slipstream extending rearward from the disc of the screw. Within the slipstream the air has an augmented rearward velocity so that its velocity relative to the aircraft is increased; except for a contrarotating pair of propellers there is also some rotary motion about a longitudinal axis in the slipstream. If, as usual, there is a tail wholly or partly immersed in the slipstream the increased velocity over the tail tends to increase its effectiveness and so to increase the static stability. There is, however, a destabilizing effect which is usually more important. This is attributable to an increase in the angle of downwash ϵ at the tail associated with a given wing incidence α, for the factor $\left(1 - \dfrac{d\epsilon}{d\alpha}\right)$ in the contribution of the tail to $\dfrac{dC_m}{d\alpha}$ is reduced (see § 2·6). The destabilizing effect is particularly large for twin tractor airscrews and this is connected with the fact that the slipstreams, in passing over the wings, are flattened into a broad band of small depth within and near which the downwash is particularly intense. It appears that, when the landing flaps are up, the destabilizing effect is least when the screws are opposite handed and rotate so that the blades next the fuselage are rising. However, this may be the worst arrangement with the flaps down and there is therefore no case for departing from the usual arrangement with the screws rotating in the same sense.

There is no satisfactory quantitative theory of the destabilizing effect of the slipstream, but the magnitude of the forward shift of the neutral point for twin and quadruple screws running in the same sense can be estimated from an empirical formula given by Morris and Morrall.* This is based entirely on full-scale measurements of the effect of slipstream and the authors consider that it permits the shift of the neutral point to be estimated to within $\pm 0·02$ of the mean wing chord. The formula can be written

$$\Delta h_n = f(\theta)\,\frac{\overline{V}a_1}{a}\frac{D}{l_T}\sqrt{(T_c)}, \qquad (5·6,\,1)$$

* D. E. Morris and J. C. Morrall, 'Effect of Slipstream on Longitudinal Stability of Multi-engined Aircraft', *R.A.E. Report No. Aero.* 2304, *A.R.C.* 12136, *R. & M.* 2701 (1948).

where Δh_n = forward shift of the neutral point, expressed as a fraction of the mean wing chord.

$f(\theta)$ = a function, defined by a graph, of the angle θ. This is the angle between the line joining the wing root trailing edge to the tail plane root leading edge and the line from the wing root trailing edge to the centre of the wing wake at the tail *in the absence of the slipstream*. $f(\theta)$ has the maximum value 1·6 when $\theta = 15°$ and the value 1·0 when $\theta = 11°$ or $19°$.

$\overline{V}$ = tail volume ratio = $\dfrac{l_T S'}{\overline{c} S}$.

$a_1 = \dfrac{dC'_L}{d\alpha'}$ for the tailplane in situ.

$a = \dfrac{dC_L}{d\alpha}$ for the aircraft as a whole.

D = diameter of propeller.

l_T = tail arm.

T_c = thrust coefficient for one propeller = $\dfrac{T'}{\rho V^2 D^2}$ with T' = thrust of one screw.

Slipstream effect is not the sole influence of the airscrew on the stability of an aircraft and we must next consider what is usually known as *propeller fin effect*. It is shown in the Appendix to this chapter that a rotating propeller acts like a stationary aerofoil when the relative wind is not parallel to the axis of rotation and the lift lies in the plane containing the axis of rotation and the wind velocity vector. Thus, when the angle of incidence of the aircraft is increased so that the upward inclination of the relative wind is increased, the lift on the propeller increases and the propeller is equivalent to a horizontal surface which, in the case of a tractor screw, is well forward of the c.g. and therefore moves the neutral point forward. The equivalent aerofoil produces, when lifting, an upwash in front of itself and a downwash behind; these, by reacting on other parts of the aircraft, may cause a further loss of stability.

Lastly, the airscrew obviously makes a contribution to the derivative X_u. We may with little error assume that the thrust T is in the direction OX and then, for wind axes [see equation

(5·5, 13)], the contribution to X_u is $\dfrac{\partial T}{\partial V}$ so that x_u is increased by $\dfrac{1}{\rho VS}\dfrac{\partial T}{\partial V}$. This can easily be evaluated for an ideal constant-speed airscrew which absorbs constant power and has constant efficiency, for then

$$T V = \text{constant},$$

or $$-C_{AS} = \frac{1}{\rho VS}\frac{\partial T}{\partial V} = -\frac{T}{\rho V^2 S}. \qquad (5\cdot6,\ 2)$$

For horizontal flight T is equal to the drag and then the last equation becomes

$$\frac{1}{\rho VS}\frac{\partial T}{\partial V} = -\tfrac{1}{2}C_D, \qquad (5\cdot6,\ 3)$$

so that, by (5·3, 2),

$$-x_u = (C_D + C_{AS}) = \tfrac{3}{2}\,C_D. \qquad (5\cdot6,\ 4)$$

Results obtained by Bryant and McMillan suggest that for fixed-pitch propellers

$$T V^{\tfrac{1}{2}} = \text{constant},$$

which yields $$\frac{1}{\rho VS}\frac{\partial T}{\partial V} = -\frac{1}{2}\frac{T}{\rho V^2 S}. \qquad (5\cdot6,\ 5)$$

For horizontal flight we should then have

$$-x_u = \tfrac{5}{4}\,C_D. \qquad (5\cdot6,\ 6)$$

When the thrust axis does not pass through the c.g. the dependence of T on u will give rise to a derivative M_u. Suppose that the thrust axis passes *above* the c.g. at the distance r. Then

$$\frac{\partial M}{\partial u} = -r\,\frac{\partial T}{\partial u}. \qquad (5\cdot6,\ 7)$$

We have seen above that $\dfrac{\partial T}{\partial u}$ is negative so M_u will be positive when r is positive. It can be seen from (5·2, 19) and (5·3, 5) that a positive M_u increases the static stability. Hence the static stability can be increased by tilting the axis of a tractor screw downwards to the front.

For jet-propelled aircraft the propeller slipstream is replaced by the jet. In order that the tailplane should not be damaged by heat it is necessary that the jet or jets should pass well below the tailplane and their influence on the velocity over the tail is therefore small. There is no fin effect, so the whole influence of the propulsive system on the longitudinal stability is slight.

5·7   Typical solutions of the dynamical equations of free motion

It is instructive to examine some typical examples of the solutions of the equations of free motion. The systematic solutions here presented in tabular form relate to horizontal flight and are due to Dr S. Neumark. They are based on the following data:

$$\frac{dC_L}{d\alpha} = 4\cdot5, \tag{5·7, 1}$$

$$C_D = 0\cdot02 + 0\cdot06\,C_L^2, \tag{5·7, 2}$$

$$\frac{dC_D}{d\alpha} = \frac{dC_D}{dC_L}\frac{dC_L}{d\alpha} = 0\cdot54\,C_L, \tag{5·7, 3}$$

$$C_{AS} = -\frac{1}{\rho VS}\frac{\partial T}{\partial V} = 0\cdot01, \tag{5·7, 4}$$

so that

$$x_u = -(C_D + C_{AS}) = -(0\cdot03 + 0\cdot06\,C_L^2), \tag{5·7, 5}$$

$$\nu = -\frac{m_q}{i_B} = 3\cdot0, \tag{5·7, 6}$$

$$\chi = -\frac{m_{\dot{w}}}{i_B} = 1\cdot0, \tag{5·7, 7}$$

$$m_u = x_q = z_q = 0. \tag{5·7, 8}$$

The *static stability coefficient*

$$\omega = -\frac{\mu_1 m_w}{i_B} \tag{5·7, 9}$$

is treated as the independent variable and the tables show how the coefficients and roots of the quartic determinantal equation depend on its value. Table 5·7, 1 refers to fast level flight ($C_L = 0\cdot25$) while Table 5·7, 2 refers to slow level flight ($C_L = 1\cdot0$).

In Table 5·7, 1 it will be seen that when ω is negative the roots are all real and one of them is positive. Hence in this condition the aircraft is unstable, in accordance with **E** being negative. When $\omega = 0$ the stability is neutral and there is a point of bifurcation (repeated real root) when $\omega = 0 \cdot 1092$. For all positive values of ω the aircraft has complete longitudinal stability (real roots and real parts of complex roots all negative). For values of ω lying between $0 \cdot 1092$ and $3 \cdot 014$ there are two real roots (corresponding to two subsidences) and one complex pair (corresponding to a damped phugoid oscillation). At $\omega = 3 \cdot 014$ we have another point of bifurcation and for all higher values there are two pairs of complex roots, corresponding respectively to a damped rapid oscillation and a damped phugoid. The behaviour shown in Table 5·7, 2 for the higher lift coefficient is qualitatively similar except that for the largest negative values of ω a damped oscillation replaces a pair of subsidences. Again the aircraft has complete longitudinal stability for all positive values of ω. For an aircraft with a positive c.g. margin of reasonable magnitude ω will be greater than 4 and the solution of the dynamical equations will consist of the typical damped rapid oscillation and damped phugoid.

It is of interest to compare the period of the phugoid oscillation as given by the simple theory of § 2·6 with the results in the tables for infinite static stability. We can easily show that for agreement of the results the coefficient of the imaginary unit in the pair of roots λ_3, λ_4 (the phugoid roots) should be $C_L/\sqrt{2} = 0 \cdot 7071\, C_L$. For Table 5·7, 1 this gives $0 \cdot 1767$ against $0 \cdot 1760$ while for Table 5·7, 2 the formula gives $0 \cdot 7071$ against $0 \cdot 7057$. The slight discrepancies are attributable to the damping of the phugoid oscillation.

5·8 Approximate solution of the dynamical equations of free motion

We have seen that, typically, the quartic equation (5·2, 15) has a pair of complex roots of relatively large modulus, corresponding to the rapid longitudinal oscillation, and a pair of complex roots of relatively small modulus, corresponding to the phugoid oscillation. On account of the large separation of the moduli it follows that an approximation to the larger roots is given by equating to zero the first three terms on the left of

TABLE 5·7, 1.

DEPENDENCE OF ROOTS OF STABILITY QUARTIC ON STATIC STABILITY COEFFICIENT

FAST LEVEL FLIGHT ($C_L = 0.25$)

COEFFICIENTS AND ROOTS IN NON-DIMENSIONAL FORM

Static stability coefficient ω	Coefficients of quartic				Roots of quartic			
	B	C	D	E	λ_1	λ_2	λ_3	λ_4
−2	6·2719	4·9401	0·2359	−0·0625	−5·3586	−0·8420	−0·1586	+0·0873
−1		5·9401	0·2696	−0·03125	−5·1228	−1·0948	−0·1066	+0·0523
−0·5		6·4401	0·2865	−0·01563	−4·9939	−1·2287	−0·0808	+0·0315
0		6·9401	0·3034	0	−4·8554	−1·3709	−0·0456	0
0·1092		7·0493	0·3071	0·00341	−4·8237	−1·4033	−0·0225	
0·5		7·4401	0·3203	0·01563	−4·7048	−1·5243	−0·0214 ± 0·0415i	
1		7·9401	0·3371	0·03125	−4·5383	−1·6930	−0·0203 ± 0·0605i	
2		8·9401	0·3709	0·0625	−4·1260	−2·1084	−0·0187 ± 0·0826i	
3		9·9401	0·4046	0·09375	−3·2360	−3·0004	−0·0177 ± 0·0967i	
3·014		9·9541	0·4051	0·09419	−3·1182		−0·0177 ± 0·0968i	
4		10·9401	0·4384	0·125	−3·1189 ± 0·9940i		−0·0171 ± 0·1066i	
6		12·9401	0·5059	0·1875	−3·1197 ± 1·7291i		−0·0163 ± 0·1203i	
8		14·9401	0·5734	0·25	−3·1201 ± 2·2338i		−0·0159 ± 0·1293i	
10		16·9401	0·6409	0·3125	−3·1203 ± 2·6438i		−0·0157 ± 0·1358i	
12		18·9401	0·7084	0·3750	−3·1204 ± 2·9981i		−0·0156 ± 0·1407i	
15		21·9401	0·8096	0·46875	−3·1204 ± 3·4623i		−0·0155 ± 0·1461i	
20		26·9401	0·9784	0·625	−3·1204 ± 4·1213i		−0·0156 ± 0·1521i	
30		36·9401	1·3159	0·9375	−3·1202 ± 5·1945i		−0·0158 ± 0·1590i	
∞	6·2719	∞	∞	∞	−3·1191 ± $\sqrt{\infty}$. i		−0·0169 ± 0·1760i	

TABLE 5·7, 2.

DEPENDENCE OF ROOTS OF STABILITY QUARTIC ON STATIC STABILITY COEFFICIENT

SLOW LEVEL FLIGHT ($C_L = 1·0$)

COEFFICIENTS AND ROOTS IN NON-DIMENSIONAL FORM

Static stability coefficient ω	Coefficients of quartic				Roots of quartic			
	B	C	D	E	λ_1	λ_2	λ_3	λ_4
-2	6·38	5·6661	1·6283	$-1·00$	$-5·3915$	$-0·6306 \pm 0·5316i$		$+0·2726$
-1		6·6661	1·7183	$-0·50$	$-5·1552$	$-0·6955 \pm 0·3157i$		$+0·1663$
$-0·5$		7·1661	1·7633	$-0·25$	$-5·0260$	$-0·8797$	$-0·5730$	$+0·0987$
0		7·6661	1·8083	$0·00$	$-4·8871$	$-1·1791$	$-0·8138$	0
$0·2$		7·8661	1·8263	$0·10$	$-4·8283$	$-1·2701$	$-0·2001$	$-0·1815$
$0·2369$		7·9030	1·8296	$0·1185$	$-4·8172$	$-1·2864$	$-0·1382$	$-0·1382$
$0·5$		8·1661	1·8533	$0·25$	$-4·7360$	$-1·3993$	$-0·1224 \pm 0·1509i$	
1		8·6661	1·8983	$0·50$	$-4·5690$	$-1·6101$	$-0·1005 \pm 0·2406i$	
2		9·6661	1·9883	$1·00$	$-4·1548$	$-2·0785$	$-0·0733 \pm 0·3323i$	
3		10·6661	2·0783	$1·50$	$-3·2659$	$-2·9999$	$-0·0571 \pm 0·3871i$	
$3·0169$		10·6830	2·0798	$1·5085$	$-3·1331$	$-3·1331$	$-0·0569 \pm 0·3879i$	
4		11·6661	2·1683	$2·00$	$-3·1435 \pm 1·0080i$		$-0·0465 \pm 0·4259i$	
6		13·6661	2·3483	$3·00$	$-3·1557 \pm 1·7447i$		$-0·0343 \pm 0·4791i$	
8		15·6661	2·5283	$4·00$	$-3·1619 \pm 2·2465i$		$-0·0282 \pm 0·5149i$	
10		17·6661	2·7083	$5·00$	$-3·1650 \pm 2·6532i$		$-0·0250 \pm 0·5408i$	
12		19·6661	2·8883	$6·00$	$-3·1665 \pm 3·0045i$		$-0·0235 \pm 0·5607i$	
15		22·6661	3·1583	$7·50$	$-3·1673 \pm 3·4650i$		$-0·0227 \pm 0·5829i$	
20		27·6661	3·6083	$10·00$	$-3·1666 \pm 4·1197i$		$-0·0234 \pm 0·6081i$	
30		37·6661	4·5083	$15·00$	$-3·1638 \pm 5·1883i$		$-0·0262 \pm 0·6368i$	
∞	6·38	∞	∞	∞	$-3·1450 \pm \sqrt{\infty} \cdot i$		$-0·0450 \pm 0·7057i$	

(5·2, 15). Thus, after removal of the factor λ^2, we obtain an approximation to the roots corresponding to the rapid longitudinal oscillation by solving the quadratic equation

$$\lambda^2 + B\lambda + C = 0. \qquad (5·8, 1)$$

Bairstow obtained an approximation $(\lambda^2 + m\lambda + n)$ to the second quadratic factor of the quartic by choosing m and n so that the coefficients of the terms of degree one and zero in the product

$$(\lambda^2 + B\lambda + C)(\lambda^2 + m\lambda + n),$$

shall have their correct values, namely D and E respectively. In this way the approximation

$$\lambda^2 + \left(\frac{D}{C} - \frac{BE}{C^2}\right)\lambda + \frac{E}{C}, \qquad (5·8, 2)$$

to the quadratic factor corresponding to the phugoid oscillation is found. Another method of approximating to the phugoid roots is given in § 6·4.

The accuracy of the roots obtained from the approximate quadratic factors can be improved by various procedures and the simplest of these in principle is Newton's method of successive approximation. Let λ be an approximate root of

$$f(\lambda) = 0, \qquad (5·8, 3)$$

and $\lambda + \Delta\lambda$ the next approximation. Then

$$\Delta\lambda = -\frac{f(\lambda)}{f'(\lambda)} = -\frac{p + iq}{r + is}, \text{ say.}$$

When the denominator is reduced to the real form by multiplication by its conjugate we obtain

$$\Delta\lambda = -\frac{pr + qs}{r^2 + s^2} + i\frac{ps - qr}{r^2 + s^2}. \qquad (5·8, 4)$$

Bairstow has given a method of successive approximation to the quadratic factor itself. Suppose that $(\lambda^2 - \xi\lambda - \eta)$ is an approximate factor of $f(\lambda)$ and let $R_1\lambda + R_0$ be the remainder and $Q_1(\lambda)$ the quotient when $f(\lambda)$ is divided by $(\lambda^2 - \xi\lambda - \eta)$. Further, let $R_3\lambda + R_2$ be the remainder when $Q_1(\lambda)$ is divided by the same quadratic. Then the corrections to be added to ξ and η are given by

$$\delta\xi = \frac{R_1 R_2 - R_0 R_3}{\eta R_3^2 - R_2(R_2 + \xi R_3)} \tag{5·8, 5}$$

and

$$\delta\eta = \frac{R_0(R_2 + \xi R_3) - \eta R_1 R_3}{\eta R_3^2 - R_2(R_2 + \xi R_3)}. \tag{5·8, 6}$$

When a close approximation has already been obtained, r and s need not be recomputed at each stage in Newton's method. In like circumstances R_2 and R_3 need not be recomputed in Bairstow's method. Both the foregoing methods are applicable to equations of any degree.*

Appendix: An elementary investigation of propeller fin effect

In the figure let OZ be fixed in direction and perpendicular to the axis of rotation of the propeller and let P lie upon the

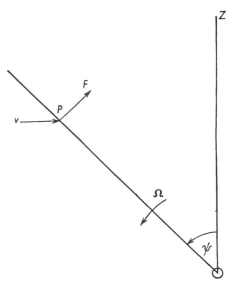

axis of one of the blades at a distance r from the axis of rotation. We consider the influence of a small component v of the relative wind, lying in the disc of the propeller and perpendicular

* A proof of Bairstow's method is given in 'On the Numerical Solution of Equations with Complex Roots', by R. A. Frazer and W. J. Duncan, *Proc. Roy. Soc.* A., vol. 125 (1929), p. 68.

to OZ. It will be seen that v can be resolved into $v \sin \psi$ along the blade and $v \cos \psi$ perpendicular to its axis. The first of these components does not influence the forces on the blade and can be neglected. The second increases the circumferential component of the relative wind at P from $r\Omega(1-a')$, where Ω is the angular velocity and a' is the rotational inflow factor, to $r\Omega(1-a') + v \cos \psi$. Accordingly the triangle of velocities at the element becomes

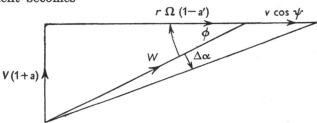

There is an increment $\Delta \alpha$ of the angle of incidence of the blade element given by

$$\Delta \alpha = \frac{v \cos \psi \sin \phi}{W} = \frac{v(1+a)V \cos \psi}{W^2}, \tag{1}$$

where a is the axial inflow factor, while the increment in the relative wind speed is

$$\Delta W = v \cos \psi \cos \phi = \frac{vr\Omega(1-a') \cos \psi}{W}. \tag{2}$$

Next consider the aerodynamic forces on the blade element, which will be supposed to have chord c and radial length dr. We are only concerned at present with the forces in the plane of the disc and we may therefore confine our attention to the circumferential force F per unit radial length of the blade since there is no radial aerodynamic force. Let us put

$$F = \rho c W^2 k_F, \tag{3}$$

where k_F is a non-dimensional force coefficient which, for simplicity, we shall assume to be a function of incidence only. Then the increment in F caused by v is

$$\Delta F = \rho c \left(2 W k_F \, \Delta W + W^2 \frac{dk_F}{d\alpha} \Delta \alpha \right)$$
$$= \rho c v \cos \psi \left[2r\Omega(1-a') \, k_F + V(1+a) \frac{dk_F}{d\alpha} \right]. \tag{4}$$

Therefore the increment of the force Y on the element in the direction of v is

$$\Delta F dr \cos \psi = vc\rho \left[2r\Omega(1-a')\, k_F + V(1+a)\frac{dk_F}{d\alpha} \right] \cos^2 \psi dr.$$

The only variable in this expression as the blade rotates is $\cos^2 \psi$ and the mean value of $\cos^2 \psi$ for a complete revolution is $\frac{1}{2}$. Hence the mean force on the element is

$$dY = \tfrac{1}{2}vc\rho \left[2r\Omega(1-a')\, k_F + V(1+a)\frac{dk_F}{d\alpha} \right] dr.$$

Let B be the number of blades. Then the expression for the standard aerodynamic derivative Y_v is

$$Y_v = -\tfrac{1}{2}B\rho \int c \left[2r\Omega(1-a')\, k_F + V(1+a)\frac{dk_F}{d\alpha} \right] dr, \qquad (5)$$

where the integral covers one blade and the minus sign takes account of the fact that the velocity of the aircraft is that of the relative wind reversed. The derivative is, in fact, negative. Obviously

$$Z_w = Y_v, \qquad (6)$$

since we have not specified the direction of the velocity v in relation to the body axes.

The moment derivatives relative to v and w may be estimated from the corresponding force derivatives by taking into account the appropriate 'arm' of the couple. Likewise, the force and moment derivatives corresponding to the angular velocities in pitch and yaw can be estimated.

[For the blade element theory of airscrews the reader may consult Glauert's *Elements of Aerofoil and Airscrew Theory*, chap. 16. The original paper on propeller fin effect is: R. G. Harris, 'Forces on a Propeller due to Sideslip', *R. & M.* 427 (1918). A review of more recent work is given by E. Priestley, 'Theory of the Propeller "Fin" Effect, including a Review of Existing Theories', *R. & M.* 2030 (1943).]

Chapter 6

LATERAL-ANTISYMMETRIC MOTION

6·1 Introduction

In Chapter 5 we have considered one of the two independent types of motion of small deviation of a symmetric aircraft from a state of steady rectilinear motion in the plane of symmetry, which is vertical. We now pass to the second type of independent motion, called lateral-antisymmetric, in which the deviations consist of a velocity of sideslip v, angular displacement in roll ϕ and angular displacement in yaw ψ. The angular velocities are

$$\left. \begin{array}{l} p = \dot{\phi}, \\ r = \dot{\psi}. \end{array} \right\} \tag{6·1, 1}$$

We have already obtained the dynamical equations as (3·12, 8)...(3·12, 10), and these will be converted into the non-dimensional form (see § 6·2 and the remarks on the non-dimensional equations given in § 5·1).

The detailed analysis shows that in typical instances the motion contains the following constituents:

(a) A damped exponential motion (subsidence) in which the predominant displacement is roll. This corresponds to a numerically large negative real root of the determinantal equation.

(b) An exponential spiral motion in which the predominant components are sideslip and yaw. This motion may be damped or divergent, but the corresponding root of the determinantal equation is numerically small. Hence any divergence is typically gentle.

(c) An oscillation involving all three components of motion. This may be damped or divergent.

The general methods for the solution of the dynamical equations have been discussed in Chapter 4 and the expansion of the determinant of motion for the lateral-antisymmetric case has been given in § 4·3.

Throughout this chapter we assume that the aircraft is rigid and that the controls are fixed. We adopt wind axes, so we have the simplifications

$$\vartheta = 0, \quad W = 0, \quad U = V.$$

6·2   Non-dimensional forms of the equations of motion

We adopt the non-dimensional measure of time as defined by equations (5·2, 1) and (5·2, 2) together with a modified conventional relative density

$$\mu_2 = \frac{m}{\rho S s} = \frac{2W}{g\rho S b}. \tag{6·2, 1}$$

On comparison with (5·2, 3) it will be seen that the tail arm l_T in the expression for μ_1 has been replaced by the semi-span

$$s = \frac{b}{2}. \tag{6·2, 2}$$

We also use k_L and k' as defined by equations (5·2, 4) and (5·2, 5) and the non-dimensional measure of sideslip

$$\bar{v} = \frac{v}{V}. \tag{6·2, 3}$$

The aerodynamic derivatives are reduced to non-dimensional forms as follows:

$$\left.\begin{array}{ll} y_v = \dfrac{Y_v}{\rho V S}, & \\[2mm] y_p = \dfrac{Y_p}{\rho V S s}, & y_r = \dfrac{Y_r}{\rho V S s}, \\[2mm] l_v = \dfrac{L_v}{\rho V S s}, & n_v = \dfrac{N_v}{\rho V S s}, \\[2mm] l_p = \dfrac{L_p}{\rho V S s^2}, & l_r = \dfrac{L_r}{\rho V S s^2}, \\[2mm] n_p = \dfrac{N_p}{\rho V S s^2}, & n_r = \dfrac{N_r}{\rho V S s^2}. \end{array}\right\} \tag{6·2, 4}$$

The non-dimensional moments of inertia in roll and yaw are

$$\left.\begin{array}{l} i_A = \dfrac{A}{ms^2} = \dfrac{gA}{Ws^2}, \\[3mm] i_C = \dfrac{C}{ms^2} = \dfrac{gC}{Ws^2}, \end{array}\right\} \tag{6·2, 5}$$

while the non-dimensional product of inertia is

$$i_E = \frac{E}{ms^2} = \frac{gE}{Ws^2}.$$ (6·2, 6)

Lastly, the non-dimensional forms of the applied forces and moments are

$$\left. \begin{aligned} y(\tau) &= \frac{Y(t)}{\rho V^2 S}, \\[2ex] l(\tau) &= \frac{L(t)}{\rho V^2 Ss}, \\[2ex] n(\tau) &= \frac{N(t)}{\rho V^2 Ss}, \end{aligned} \right\}$$ (6·2, 7)

where t and τ are related by (5·2, 1). These all vanish for free motion with controls fixed.

When equation (3·12, 8), expressing the balance of lateral forces, is divided by $\rho V^2 S$ it can be written, after reduction, as

$$\frac{d\bar{v}}{d\tau} - \bar{v} y_v - \frac{y_p}{\mu_2}\frac{d\phi}{d\tau} - k_L \phi + \left(1 - \frac{y_r}{\mu_2}\right)\frac{d\psi}{d\tau} + k'\psi = y(\tau).$$ (6·2, 8)

The equations (3·12, 9) and (3·12, 10) for the rolling and yawing moments respectively become when divided by $\rho V^2 Ss$ and after reduction

$$-\bar{v} l_v + \frac{i_A}{\mu_2}\frac{d^2\phi}{d\tau^2} - \frac{l_p}{\mu_2}\frac{d\phi}{d\tau} - \frac{i_E}{\mu_2}\frac{d^2\psi}{d\tau^2} - \frac{l_r}{\mu_2}\frac{d\psi}{d\tau} = l(\tau),$$ (6·2, 9)

$$-\bar{v} n_v - \frac{i_E}{\mu_2}\frac{d^2\phi}{d\tau^2} - \frac{n_p}{\mu_2}\frac{d\phi}{d\tau} + \frac{i_C}{\mu_2}\frac{d^2\psi}{d\tau^2} - \frac{n_r}{\mu_2}\frac{d\psi}{d\tau} = n(\tau).$$ (6·2, 10)

The determinantal equation (see § 4·3) corresponding to the foregoing dynamical equations is

$$\Delta(\lambda) \equiv \begin{vmatrix} \lambda - y_v & -\dfrac{\lambda y_p}{\mu_2} - k_L & \lambda\left(1 - \dfrac{y_r}{\mu_2}\right) + k' \\[3ex] -l_v & \dfrac{\lambda^2 i_A}{\mu_2} - \dfrac{\lambda l_p}{\mu_2} & -\dfrac{\lambda^2 i_E}{\mu_2} - \dfrac{\lambda l_r}{\mu_2} \\[3ex] -n_v & -\dfrac{\lambda^2 i_E}{\mu_2} - \dfrac{\lambda n_p}{\mu_2} & \dfrac{\lambda^2 i_C}{\mu_2} - \dfrac{\lambda n_r}{\mu_2} \end{vmatrix} = 0.$$ (6·2, 11)

This when expanded, multiplied by $\left(\dfrac{\mu_2^2}{i_A i_C}\right)$ and the factor λ removed, may be written

$$A\lambda^4 + B\lambda^3 + C\lambda^2 + D\lambda + E = 0, \qquad (6\cdot2, 12)$$

where

$$A = 1 - \frac{i_E^2}{i_A i_C}, \qquad (6\cdot2, 13)$$

$$B = -y_v\left(1 - \frac{i_E^2}{i_A i_C}\right) - \frac{l_p}{i_A} - \frac{n_r}{i_C} - \frac{i_E}{i_C}\frac{l_r}{i_A} - \frac{i_E}{i_A}\frac{n_p}{i_C}, \qquad (6\cdot2, 14)$$

$$C = y_v\left(\frac{l_p}{i_A} + \frac{n_r}{i_C} + \frac{i_E}{i_A}\frac{n_p}{i_C} + \frac{i_E}{i_C}\frac{l_r}{i_A}\right) + \left(\frac{l_p}{i_A}\frac{n_r}{i_C} - \frac{l_r}{i_A}\frac{n_p}{i_C}\right)$$

$$+ \frac{\mu_2 l_v}{i_A}\left\{\frac{i_E}{i_C}\left(1 - \frac{y_r}{\mu_2}\right) - \frac{y_p}{\mu_2}\right\} + \frac{\mu_2 n_v}{i_C}\left\{1 - \frac{y_r}{\mu_2} - \frac{i_E}{i_A}\frac{y_p}{\mu_2}\right\}, \quad (6\cdot2, 15)$$

$$D = -y_v\left(\frac{l_p}{i_A}\frac{n_r}{i_C} - \frac{l_r}{i_A}\frac{n_p}{i_C}\right)$$

$$+ \frac{\mu_2 l_v}{i_A}\left\{\frac{n_p}{i_C}\left(1 - \frac{y_r}{\mu_2}\right) + \frac{n_r}{i_C}\frac{y_p}{\mu_2} - k_L + \frac{i_E}{i_C}k'\right\}$$

$$+ \frac{\mu_2 n_v}{i_C}\left\{-\frac{l_p}{i_A}\left(1 - \frac{y_r}{\mu_2}\right) - \frac{l_r}{i_A}\frac{y_p}{\mu_2} + k' - \frac{i_E}{i_A}k_L\right\}, \qquad (6\cdot2, 16)$$

$$E = \frac{\mu_2 l_v}{i_A}\left(k_L\frac{n_r}{i_C} + k'\frac{n_p}{i_C}\right) - \frac{\mu_2 n_v}{i_C}\left(k_L\frac{l_r}{i_A} + k'\frac{l_p}{i_A}\right). \qquad (6\cdot2, 17)$$

It will be noted that when the product of inertia E vanishes the leading coefficient A becomes unity. In general A differs little from unity.

We have seen that the determinantal equation $(6\cdot2, 11)$ is a quintic whose constant term is zero. Hence any aircraft (without an automatic pilot) has neutral static lateral stability (see § 4·10). In order to understand the significance of this, put λ equal to zero in the dynamical equations of free motion, i.e. make the time rates of change of the deviations zero. We see that equations $(6\cdot2, 8)...(6\cdot2, 10)$ are then satisfied by

$$\bar{v} = 0, \qquad (6\cdot2, 18)$$

$$k'\psi = k_L\phi. \qquad (6\cdot2, 19)$$

Hence there is an undetermined angular displacement. For horizontal flight k' is zero, so ϕ is zero and ψ undetermined.

Thus the azimuth of the flight path is arbitrary and the same conclusion follows for inclined flight since (6·2, 19) is by (5·2, 5) equivalent to

$$\phi \cos \Theta + \psi \sin \Theta = 0. \tag{6·2, 20}$$

The last equation shows that the vertical angular displacement of OY is zero but the horizontal displacement is undetermined.

6·3   The aerodynamic derivatives

Throughout this section, unless the contrary is expressly stated, the aircraft considered have unswept wings which implies that the quarter-chord axis is transverse in plan.

L_v. This is one of the most important derivatives and must be discussed in some detail. We shall begin by considering the derivative for an isolated wing and shall later examine the modifications introduced by the presence of the fuselage and tail.

As already pointed out in § 2·9 a rolling moment is caused by sideslip whenever the wings have a dihedral angle Γ. This angle is measured between the transverse axis OY and a line in the mid-surface of the wing* perpendicular to the line of intersection of the port and starboard wings; when Γ is positive both wing tips are raised relative to their roots. The dihedral angle may vary along the wing span, but it is assumed that the port and starboard wings are mirror images in the plane OXZ.

When the velocity of sideslip is v there is a local increase of incidence on the starboard wing given by

$$\Delta \alpha = \frac{v\Gamma}{V}, \tag{6·3, 1}$$

when Γ is small, as shown in § 2·9. The corresponding increment of lift on a fore-and-aft strip of the wing of width dy and chord c is

$$\tfrac{1}{2}\rho V^2 c \Delta\alpha \frac{dC_L}{d\alpha} dy = \tfrac{1}{2} v\rho V\Gamma \frac{dC_L}{d\alpha} c\, dy,$$

and the increment of rolling moment is

$$-\tfrac{1}{2} v\rho V\Gamma \frac{dC_L}{d\alpha} cy\, dy.$$

---

* This lies half-way between the lower and upper surfaces of the wing.

The total increment of rolling moment on the starboard wing is therefore

$$-\tfrac{1}{2}v\rho V \int_0^s \Gamma \frac{dC_L}{d\alpha} cy\, dy,$$

and the port wing gives an equal contribution (since both $\Delta\alpha$ and y are reversed in sign).

Hence

$$vL_v = -v\rho V \int_0^s \Gamma \frac{dC_L}{d\alpha} cy\, dy$$

and

$$l_v = \frac{L_v}{\rho V S s} = -\frac{1}{Ss}\int_0^s \Gamma \frac{dC_L}{d\alpha} cy\, dy. \qquad (6\cdot3, 2)$$

For a constant and full-span dihedral we derive from the last equation

$$l_v = -\frac{\Gamma}{Ss}\int_0^s \frac{dC_L}{d\alpha} cy\, dy. \qquad (6\cdot3, 3)$$

Assume now for simplicity that $\dfrac{dC_L}{d\alpha}$ is constant along the span

and put

$$\int_0^s cy\, dy = \tfrac{1}{2}Sd, \qquad (6\cdot3, 4)$$

where d is the distance of the centroid of the starboard wing from the plane of symmetry. Then $(6\cdot3, 3)$ becomes

$$l_v = -\tfrac{1}{2}\Gamma\left(\frac{d}{s}\right)\frac{dC_L}{d\alpha}. \qquad (6\cdot3, 5)$$

It is clear that we are making a rough approximation when we treat the effective lift slope as constant over the span and equal to that for the wing as a whole. Experiments indicate that this procedure over-estimates the derivative numerically by about 25 per cent for elliptic and rectangular wings of aspect ratio 6. For an elliptic wing

$$\frac{d}{s} = \frac{4}{3\pi},$$

and if we assign to $\dfrac{dC_L}{d\alpha}$ the value for the complete wing given by lifting line theory, namely

$$\frac{a_0}{1+\dfrac{a_0}{\pi A}},$$

we obtain from (6·3, 5)

$$l_v = -\frac{2}{3\pi}\frac{\Gamma a_0}{1+\dfrac{a_0}{\pi A}},\qquad (6\cdot3,\ 6)$$

where A is the aspect ratio and a_0 the lift slope for infinite aspect ratio. Lifting line theory can easily be applied to the problem of finding the rolling moment on a wing with antisymmetric wing incidence, such as occurs in sideslip when the wing has a constant and full-span dihedral. The result so obtained for an elliptic wing is

$$l_v = -\frac{2}{3\pi}\frac{\Gamma a_0}{1+\dfrac{2a_0}{\pi A}}.\qquad (6\cdot3,\ 7)$$

Thus, when for example, $A = 6$ and $a_0 = 5\cdot5$, the simple theory gives a value for l_v which is about 23 per cent larger numerically than that given by lifting line theory. As the aspect ratio increases the difference in the results decreases and would be zero for infinite aspect ratio.

According to Levačić the value of l_v for straight tapered wings having full-span and constant dihedral, with $A = 6$ and $a_0 = 5\cdot5$, is given closely by the empirical formula

$$-\frac{l_v}{\Gamma} = 0\cdot65 + 0\cdot17\tau,\qquad (6\cdot3,\ 8)$$

where τ is the ratio of tip chord to root chord.* Levačić also considers that the results for other aspect ratios and lift slopes can be obtained by assuming that the derivative for a tapered wing depends on these variables in the same way as for an elliptic wing [see equation (6·3, 7)].

Next, suppose that the central part of the wing, lying between the points $y = \pm ks$, has no dihedral whereas the outer wings

* I. Levačić, 'Rolling Moment due to Sideslip. Part I. The Effect of Dihedral', *R.A.E. Report No. Aero.* 2028, *A.R.C.* 8709 (1945).

have the dihedral Γ (partial-span constant dihedral). For an elliptic wing lifting line theory yields the result that l_v is obtained by multiplying the value of the derivative for full-span constant dihedral by the factor $(1 - k^2)^{\frac{3}{2}}$. The same factor can be applied to straight tapered wings with very fair accuracy when $0.25 \leqslant \tau \leqslant 1.0$.

When an aeroplane is standing on the ground the wings bend downwards under their weight whereas in steady flight they bend upwards by a greater amount under the action of the lift. Hence the effective dihedral in flight is greater than on the ground and may be appreciably greater than the designed figure. Levačić has shown that the increment in l_v associated with an upward bending displacement δ of the wing tips can be estimated from the formula

$$\Delta l_v = -0.88\left(\frac{\delta}{s}\right). \tag{6.3, 9}$$

Sweeping back the wings of an aircraft without dihedral gives rise to a negative L_v and thus has qualitatively the same effect as a positive dihedral. To see how this comes about let us consider first a wing of infinite span set at incidence α_n in a current of speed V and then yawed through the angle Λ (axis of rotation normal to the wind vector and spanwise wing axis). The relative wind then has a component $V \sin \Lambda$ parallel to the axis of the wing and this has no influence on the aerodynamic forces. However, the chordwise component $V \cos \Lambda$ or V_c is fully effective and the lift on an element of wing area ΔS is

$$\Delta L = \tfrac{1}{2}\rho V_c^2 a_0 \alpha_n \Delta S \tag{6.3, 10}$$

where α_n is supposed to be small and measured from the no-lift incidence and a_0 is the lift slope for the unyawed wing. But

$$\alpha_n V_c = V_n \tag{6.3, 11}$$

the component of the relative wind perpendicular to the wing.

Hence $$\Delta L = \tfrac{1}{2}\rho V_c V_n a_0 \Delta S, \tag{6.3, 12}$$

and this equation is valid for all directions of the relative wind, provided only that V_n is a small fraction of V_c.

The foregoing formula for the element of lift can be applied to obtain an approximate expression for the contribution to L_v made by the sweptback wings of an aircraft without dihedral. Let the angle of sweepback of the starboard wing be

11

A and let the angle of sideslip be β. When the angle of incidence is small

$$V_c = V \cos (\Lambda - \beta)$$

while, in the absence of dihedral angle,

$$V_n = V \alpha$$

where α is now measured in the fore-and-aft plane, as usual. Hence equation (6·3, 12) becomes

$$\Delta L = \tfrac{1}{2} \rho V^2 a_0 \alpha \cos (\Lambda - \beta) \Delta S$$
$$= \tfrac{1}{2} \rho V^2 a_0 \alpha (\cos \Lambda + \beta \sin \Lambda) \Delta S \qquad (6\cdot3, 13)$$

when β is small. The part of this depending on β is

$$\tfrac{1}{2} \rho V^2 a_0 \alpha \beta \sin \Lambda \Delta S = \tfrac{1}{2} v \rho V a_0 \sin \Lambda \Delta S = \tfrac{1}{2} v \rho V C_L \tan \Lambda \Delta S,$$

since $C_L = a_1 \alpha$ and $a_1 = a_0 \cos \Lambda$. The last equation follows from the fact that

$$V_c V_n a_0 = V V_n a_1 \text{ (see equation (6·3, 12))}.$$

The contribution to the rolling moment due to sideslip made by an element of the starboard wing of spanwise width dy and of chord c is accordingly

$$- \tfrac{1}{2} \rho v V C_L c y \tan \Lambda \, dy$$

and the whole contribution of the starboard wing is

$$- \tfrac{1}{2} \rho v V \tan \Lambda \int_0^s C_L c y \, dy.$$

The port wing makes an equal contribution since both y and Λ are reversed in sign. Hence we get for the complete wing

$$l_v = - \frac{\tan \Lambda}{Ss} \int_0^s C_L c y \, dy. \qquad (6\cdot3, 14)$$

If as a further approximation we treat C_L as constant and use (6·3, 4) we obtain

$$l_v = - \tfrac{1}{2} C_L \left(\frac{d}{s} \right) \tan \Lambda. \qquad (6\cdot3, 15)$$

There is some experimental evidence tending to show that this formula somewhat underestimates the derivative numerically. It is clear that the argument is approximate since the spanwise component of velocity on a wing of finite span does influence the aerodynamic forces and it is of special importance near the plane of symmetry. It is important to note that the contribution to l_v made by dihedral is independent of lift coefficient [see (6·3, 5)] whereas that made by sweepback is proportional to it. As a working approximation the two effects can be treated as additive.

We have next to consider the influence of the body in modifying the contribution made to l_v by the wings and shall explain the basis of the method of treating this problem which is due to Multhopp.* For simplicity let us regard the fuselage as an infinitely long elliptic cylinder. When there is positive sideslip the relative wind has a component at right angles to the axis of the cylinder and in the direction of the transverse axis of the ellipse in the sense from starboard to port. If the body were isolated this transverse component would give rise to a velocity field symmetrical about the transverse axis of the body and having an upward component in the first and third quadrants but a downward component in the second and fourth quadrants. In accordance with Multhopp's theory we assume that the wings are immersed in the above described velocity field. Thus, quite apart from having any dihedral, the starboard wing will, in a positive sideslip, have increased incidence if it is above the transverse axis of the body but reduced incidence if it is below this axis, while the incidence change on the port wing is reversed. Thus the interference effect of the body is to increase the effective dihedral when the wings are above the transverse body axis and to reduce it when the wings are below this axis; for wings lying entirely in the transverse axial plane of the body the effect is zero. The vertical component of the interference velocity is at a maximum near the body and decreases rapidly as the lateral co-ordinate y increases, and the incidence changes are similarly graded. It results from this that the interference effect increases with the ratio h/b, where h is the vertical depth of the body and b is the span. We cannot here enter into the mathematical details of the theory but shall quote one approximate result for an elliptic wing without dihedral to indicate the magnitude of the effect. Let z be the height of the centre of the body above the mid-plane of the wing and w the maximum width of the elliptic body. When z/h is numerically less than 0·3 the contribution to l_v is nearly

* H. Multhopp, 'Aerodynamics of the Fuselage', *Luftfahrtforschung*, vol. 18, 2/3 (1941), p. 52. R.T.P. translation No. 1220.

proportional to this ratio and we have approximately*

$$l_v = 0 \cdot 12 \left(\frac{z}{h}\right)\left(1 + \frac{w}{h}\right), \qquad (6 \cdot 3, 16)$$

when h/b is $0 \cdot 2$. When h/b has the values $0 \cdot 1$, $0 \cdot 3$, $0 \cdot 4$ the corresponding values of the numerical coefficient are $0 \cdot 035$, $0 \cdot 22$ and $0 \cdot 32$ respectively. When the wing has a dihedral its various elements occupy different vertical positions in the velocity field of the body and the interference effect is altered accordingly. Multhopp makes approximate allowance for the dihedral by supposing the wing to be rotated about a suitably placed axis until the dihedral is reduced to zero. For instance, when h/b is $0 \cdot 2$ and z/h is $0 \cdot 3$, the axis lies at $0 \cdot 23$ of the semi-span from the plane of symmetry.

A horizontal tail surface in general makes a contribution to l_v on account of the presence of the fin and other vertical surfaces. The theory is qualitatively the same as for the body-wing interference but the contribution to l_v is not large on account of the small size of the tailplane.

Lastly we have to mention the contribution to l_v made by the body with its attached vertical surfaces. The most important part of this comes from the fin and rudder and can be estimated from the geometrical particulars and an assumed value of the rate of change of lateral force coefficient with angle of sideslip (see further below under N_v). This contribution may be equivalent to several degrees of wing dihedral.†

We conclude from this discussion that we can obtain a general understanding of the ways in which various parts of the aircraft contribute to l_v and may even make reasonable estimates of the contributions. There is, however, so much uncertainty that an experimental determination of the derivative is greatly to be desired when the stability and response characteristics of an aircraft are to be investigated.

N_v. This highly important derivative is the measure of the yawing moment caused by unit velocity of sideslip. Suppose

---

* Here and elsewhere we use the symbol for the derivative to stand for the particular contribution under consideration.

† I. Levačić, 'Rolling Moment due to Sideslip. Part III. *A*, The Effect of Wing Body Arrangement. *B*, The Effect of the Tail Unit', *R.A.E. Report No. Aero.* 2139, *A.R.C.* 9987 (1946).

that the angle of sideslip is β (positive when the starboard wing tip is forward of the port wing tip). Then

$$v = V\beta,$$

when β is small and, if the corresponding increment of yawing moment (moment about OZ) is ΔN, we have

$$N_v = \frac{\Delta N}{V\beta}. \qquad (6\cdot3,\ 17)$$

Now N is positive when it tends to rotate the aircraft about OZ so that OX moves towards OY, i.e. so that the starboard wing tip is moved aft. Hence the aircraft will have positive 'weathercock stability' when N_v is positive for then the moment ΔN tends to reduce the angle of sideslip β. For satisfactory behaviour it is, in fact, essential that N_v shall be positive and not too small (see further in §§ 6·4 and 6·5).

For simplicity and definiteness let us first consider an isolated vertical fin lying in the plane of symmetry OXZ. Let the profile be symmetrical and let the aerodynamic centre be at a distance l_F aft of OZ (wind axes in use). When the sideslip angle is β the aerodynamic force in the direction OY will be

$$Y = -\tfrac{1}{2}\rho V^2 S_F a\beta, \qquad (6\cdot3,\ 18)$$

where S_F is the fin area and a is the lift slope for the surface. Hence for the isolated fin

$$Y_v = -\tfrac{1}{2}\rho V S_F a \qquad (6\cdot3,\ 19)$$

and $\qquad N_v = \tfrac{1}{2}\rho V S_F l_F a. \qquad (6\cdot3,\ 20)$

Thus N_v will be positive when l_F is positive, i.e. when the fin lies aft of OZ. Consequently a rearwardly placed fin tends to give positive weathercock stability whereas a fin placed forward tends to give negative weathercock stability. If there are two fin surfaces, one aft and one forward of OZ, their contributions to Y_v are of the same sign but those to N_v are of opposite sign. The non-dimensional derivative corresponding to (6·3, 20) is [see (6·2, 4)]

$$n_v = \frac{1}{2}\left(\frac{S_F l_F}{Ss}\right)a = \left(\frac{S_F l_F}{Sb}\right)a, \qquad (6\cdot3,\ 21)$$

where the quantity in the bracket in (6·3, 21) is called the 'fin volume ratio'. If now we consider a complete aircraft, it is convenient to treat the whole vertical surface of the fin, rudder and fuselage aft of the root leading edge of the fin to be fin surface. The formula (6·3, 21) can be applied to estimate the contribution of this surface to n_v but the result should be multiplied by an efficiency factor to allow for interference. The coefficient a must be appropriate to the aspect ratio of the fin surface with due allowance for the influence of the horizontal tail surface as an 'end plate' or fin. A similar method can be used in estimating the contribution of the fin surface to L_v but a suitable vertical arm must replace l_F in the formula (6·3, 20). Moreover, the arm is negative when the fin surface is above OX.

A 'dorsal fin' is now commonly fitted. This consists of an extension of the fin running forward along the top of the fuselage so as to form a long root fillet whose vertical chord gradually tapers to zero at the forward end. This is found to be beneficial, especially by increasing the angle of sideslip at which stalling of the fin occurs.

The body ex tail makes a contribution to n_v which is normally negative. According to an empirical analysis of available data made by Bisgood* the dominant variables are:

(a) The lateral aspect ratio of the body

$$A_B = \frac{h^2}{S_B}, \tag{6·3, 22}$$

where h is the maximum vertical depth of the body and S_B is the area of its projection on the plane of symmetry.

(b) The 'degree of balance' of the body

$$B = 100 \Big/ \sqrt{\left(\frac{S_{FB} l_{FB}}{S_B l_B}\right)}, \tag{6·3, 23}$$

where S_{FB} is the lateral projected area of the body forward of the yawing axis, l_B is the total length of the body and l_{FB} is the length forward of the yawing axis.

Bisgood's results for isolated bodies in two typical cases can be represented closely by formulae:

$$\text{For} \quad A_B = 0·2, \quad n_v = -(0·12 + 0·0008B)\left(\frac{S_B l_B}{Sb}\right). \tag{6·3, 24}$$

$$\text{For} \quad A_B = 0·3, \quad n_v = -(0·15 + 0·0012B)\left(\frac{S_B l_B}{Sb}\right). \tag{6·3, 25}$$

* P. L. Bisgood, unpublished report.

The interference of the wings increases these contributions numerically by roughly 40 per cent, 30 per cent and 20 per cent when the cross sections of the body are rectangular, deep elliptic and circular respectively. The direct contribution of the wings is usually small and stabilizing, but for tailless aircraft and high values of C_L it may be a large part of the total n_v.

Nacelles may make an appreciable contribution to n_v which is usually destabilizing.

As shown in the Appendix to Chapter 5, the airscrew acts like a vertical fin and its contribution to n_v may be important. For tractor screws the effect is destabilizing.

There are so many variables influencing n_v that its value should be obtained by experiment whenever possible.

Y_v. We have briefly referred to this derivative when considering the contribution of the fin surface to N_v, but all vertical surfaces contribute negatively to Y_v. For conventional aircraft of normal proportions y_v appears to have a value near -0.2.

L_p. We shall explain the approximate calculation of this derivative by what is usually called 'strip theory'. In this treatment we suppose the wing divided into parallel strips lying fore-and-aft and calculate the change in aerodynamic load on a typical strip caused by the angular velocity in roll. Finally, we calculate the resulting change in the rolling moment by integration over the wing.

For simplicity we shall suppose that rolling occurs about the wind axis OX. Take a strip of the wing of width dy and chord c distant y from the plane of symmetry OXZ. Then the downward velocity of the strip associated with the rolling velocity p is py and the normal (downward) force dZ on the strip is given by

$$dZ = \rho V c z_w p y\, dy,$$

where z_w is the local non-dimensional derivative [cp. equation (5·2, 8)]. This force makes the contribution

$$dL = \rho V c z_w p y^2 dy,$$

to the rolling moment and the total rolling moment for the wing associated with p is

$$pL_p = p\rho V \int_{-s}^{s} c z_w y^2 dy. \tag{6·3, 26}$$

Hence
$$l_p = \frac{L_p}{\rho V S s^2} = \frac{1}{S s^2}\int_{-s}^{s} c z_w y^2 \, dy. \qquad (6\cdot3,\,27)$$

Let us now assume that z_w has the quasi-static value given by $(5\cdot3,\,5)$, namely
$$z_w = -\frac{1}{2}\left(C_D + \frac{dC_L}{d\alpha}\right),$$

where the drag coefficient and lift curve slope should have the local values appropriate to the strip considered. Accordingly
$$l_p = -\frac{1}{2S s^2}\int_{-s}^{s}\left(C_D + \frac{dC_L}{d\alpha}\right) c y^2 \, dy. \qquad (6\cdot3,\,28)$$

If we assume further that $\left(C_D + \dfrac{dC_L}{d\alpha}\right)$ is constant over the wing

and put
$$\int_{-s}^{s} c y^2 \, dy = S k_x^2, \qquad (6\cdot3,\,29)$$

so that k_x is the radius of gyration of the wing area about an axis parallel to OX at the wing root, we obtain
$$l_p = -\frac{1}{2}\left(\frac{k_x}{s}\right)^2\left(C_D + \frac{dC_L}{d\alpha}\right), \qquad (6\cdot3,\,30)$$

and may assume that the aerodynamic coefficients are those for the complete wing, without fuselage or tail. The values of the derivative given by $(6\cdot3,\,28)$, $(6\cdot3,\,29)$ and $(6\cdot3,\,30)$ are appropriate only to very small values of the frequency parameter (see § 3·5). All the above formulae for the derivative refer to the wing alone, but usually the contributions of other parts of the aircraft are negligible. A very large and high fin may make a not negligible contribution which can be estimated as for the wing.

'Lifting line' theory has been applied to the contribution of the wing to the rotary derivatives L_p, N_p, L_r and N_r and the theory for L_p is very similar to that for L_v, already mentioned.*

* M. Lofts, 'Note on the Calculation of the Wing Contributions to the Lateral Stability Rotary Derivatives', *R.A.E. Report No. Aero.* 2189, *A.R.C.* 10,625 (1947).

For an untwisted wing of elliptic plan form and constant profile shape the expression obtained for l_p can be reduced to

$$l_p = -\frac{1}{8}\frac{a_0}{1+\dfrac{2a_0}{\pi A}}, \tag{6·3, 31}$$

where the influence of drag is neglected. Now for an elliptic wing $k_x = \frac{1}{2}s$ so equation (6·3, 30) yields, when C_D is neglected,

$$l_p = -\frac{1}{8}\frac{dC_L}{d\alpha} \tag{6·3, 32}$$

$$= -\frac{1}{8}\frac{a_0}{1+\dfrac{a_0}{\pi A}}, \tag{6·3, 33}$$

when the value of $\dfrac{dC_L}{d\alpha}$ for the complete wing as given by lifting line theory is substituted. Hence, according to lifting line theory, simple strip theory here over-estimates l_p in the ratio

$$\left(1+\frac{2a_0}{\pi A}\right) \;:\; \left(1+\frac{a_0}{\pi A}\right).$$

However, this cannot be regarded as settling the question since lifting line theory is open to suspicion, especially when the incidence changes near the wing tips are, as here, large and the lift forces near the tips of special importance.

N_p. By an argument similar to that used in the case of L_p we obtain

$$n_p = -\frac{1}{Ss^2}\int_{-s}^{s} cx_w y^2 dy, \tag{6·3, 34}$$

for the contribution of the wing. The quasi-static value of x_w is given by (5·3, 3) and the corresponding value of n_p is

$$n_p = -\frac{1}{2Ss^2}\int_{-s}^{s}\left(C_L - \frac{dC_D}{d\alpha}\right)cy^2 dy. \tag{6·3, 35}$$

If we assume $\left(C_L - \dfrac{dC_D}{d\alpha}\right)$ to be constant we obtain

$$n_p = -\frac{1}{2}\left(\frac{k_x}{s}\right)^2\left(C_L - \frac{dC_D}{d\alpha}\right). \tag{6·3, 36}$$

Again a large and high fin may contribute appreciably to the derivative.

L_r. The following formulae for the contribution of the wing to the non-dimensional derivative are obtained similarly

$$l_r = -\frac{1}{Ss^2}\int_{-s}^{s} cz_u y^2 \, dy \qquad (6\cdot3, 37)$$

$$= \frac{1}{Ss^2}\int_{-s}^{s} C_L cy^2 \, dy, \qquad (6\cdot3, 38)$$

when the quasi-static value of z_u given by $(5\cdot3, 4)$ is substituted. When C_L is assumed constant this becomes

$$l_r = \left(\frac{k_x}{s}\right)^2 C_L. \qquad (6\cdot3, 39)$$

Here also a large vertical fin will contribute to the derivative.

N_r. The contribution of the wings is given by

$$n_r = \frac{1}{Ss^2}\int_{-s}^{s} cx_u y^2 \, dy \qquad (6\cdot3, 40)$$

$$= -\frac{1}{Ss^2}\int_{-s}^{s} C_D cy^2 \, dy, \qquad (6\cdot3, 41)$$

when x_n is given the quasi-static value from equation $(5\cdot3, 2)$. On the assumption that C_D is constant this becomes

$$n_r = -\left(\frac{k_x}{s}\right)^2 C_D. \qquad (6\cdot3, 42)$$

All vertical surfaces make a contribution to N_r and those made by the fin cum rudder are important. These can be calculated in the same manner as the contribution of the tailplane to M_q (see § 5·3).

Y_p. This derivative is small and usually taken to be zero in stability calculations. Evidently a large and high fin and rudder will make a contribution to Y_p, with negative sign, which can be estimated by strip theory.

Y_r. This derivative is also small and is usually neglected. The fin and rudder and other vertical surfaces aft of the c.g. contribute positively to Y_r whereas vertical surfaces forward of the c.g. contribute negatively.

6·4 Typical and approximate solutions of the dynamical equations of free motion

For definiteness we shall take the dynamical equations in their non-dimensional forms (6·2, 8)...(6·2, 10). Typically the solution of the determinantal equation (6·2, 12) consists of:

(a) A numerically large negative real root.

(b) A numerically small real root which is positive or negative according to circumstances.

(c) A pair of complex roots whose real part is small and positive or negative according to circumstances.

We shall now consider these roots and the constituent motions associated with them in turn.

The numerically large real root λ_1 is associated with a motion which is a subsidence in very nearly pure roll. If in equation (6·2, 9) we make $l(\tau)$, $\bar{v}$ and ψ all zero we obtain on multiplication by μ_2/i_A

$$\frac{d^2\phi}{d\tau^2} - \frac{l_p}{i_A}\frac{d\phi}{d\tau} = 0. \qquad (6\cdot4, 1)$$

If we now assume that ϕ is proportional to $\exp(\lambda\tau)$ the equation yields

$$\lambda = \frac{l_p}{i_A}, \qquad (6\cdot4, 2)$$

and this is large and negative, except when the incidence is near the stall. We can take the above as an approximation to the true root λ_1 or alternatively the approximation

$$\lambda = -\frac{\mathbf{B}}{\mathbf{A}}, \qquad (6\cdot4, 3)$$

where $\mathbf{A}$ and $\mathbf{B}$ are given by equations (6·2, 13) and (6·2, 14) respectively; normally the two approximations differ little. The true root can be obtained to any required degree of accuracy by applying Newton's method of successive approximation, beginning with the selected approximate root. When λ_1 has been found the expressions for $\bar{v}$, ϕ and ψ can be obtained, as explained in §§ 4·3 and 4·4.

The numerically small real root λ_2 is given approximately by equating to zero the 'linear' terms in (6·2, 12); thus we obtain the approximation

$$\lambda_2 = -\frac{\mathbf{E}}{\mathbf{D}}. \qquad (6\cdot4, 4)$$

This approximate root can also be refined to any required extent by Newton's process. The predominant components in the motion corresponding to this root are sideslip and yaw and the motion is spiral. In order that the motion shall be stable (a subsidence) it is necessary that E/D shall be positive and, since D is normally positive, this requires that E shall be positive [see equation (6·2, 17)]. For horizontal flight the condition for stability becomes effectively

$$l_v n_r - l_r n_v > 0. \qquad (6·4, 5)$$

Now l_r is normally positive and n_v is positive when there is positive 'weathercock stability'. Hence, for stability, l_v must have the same sign as n_r and be sufficiently large numerically; since n_r is normally negative l_v must also be negative. We see, therefore, that spiral instability can be avoided by providing a sufficiently large positive dihedral in accordance with the inequality

$$\left(\frac{-l_v}{n_v}\right) > \left(\frac{l_r}{-n_r}\right). \qquad (6·4, 6)$$

This inequality will be violated by a sufficient increase of n_v, so too much 'weathercock stability' leads to spiral instability. However, a slight degree of spiral instability is found to be tolerable in practice.

When the real roots λ_1, λ_2 have been found accurately by Newton's process or otherwise, the quadratic factor

$$\lambda^2 - \lambda(\lambda_1 + \lambda_2) + \lambda_1 \lambda_2 \qquad (6·4, 7)$$

of the determinantal equation can be calculated and the remaining quadratic factor obtained by division. The roots of the second quadratic factor can then be calculated in the ordinary way; the roots of the determinantal equation are then all known. Typically, the roots of the second quadratic factor are complex and correspond to an oscillatory motion called the lateral oscillation or 'Dutch roll'. As a rule the real part of the complex pair of roots is not numerically large and when this is so it is possible to estimate the roots directly. Since the theory is of some general interest we shall give an account of it here.

Denote the quartic expression appearing in (6·2, 12) by $\Delta(\lambda)$ and let

$$\frac{1}{n!}\frac{d^n \Delta(\lambda)}{d\lambda^n} = \Delta_n(\lambda). \qquad (6·4, 8)$$

Let $(\mu + i\omega)$ be a complex root of (6·2, 12). Then we have accurately by Taylor's theorem

$$\Delta(\mu) + i\omega\Delta_1(\mu) - \omega^2\Delta_2(\mu) - i\omega^3\Delta_3(\mu) + \omega^4\Delta_4(\mu) = 0.$$

Equate separately to zero the real and imaginary parts of the last equation.*

$$\Delta(\mu) - \omega^2\Delta_2(\mu) + \omega^4\Delta_4(\mu) = 0, \qquad (6\cdot4,\ 9)$$

$$\Delta_1(\mu) - \omega^2\Delta_3(\mu) = 0, \qquad (6\cdot4,\ 10)$$

where in (6·4, 10) we have removed the factor ω, which would not be legitimate if we were dealing with a real root ($\omega = 0$). Equation (6·4, 10) gives

$$\omega^2 = \frac{\Delta_1(\mu)}{\Delta_3(\mu)} = \frac{\mathbf{D} + 2\mu\mathbf{C} + 3\mu^2\mathbf{B} + 4\mu^3\mathbf{A}}{\mathbf{B} + 4\mu\mathbf{A}}, \qquad (6\cdot4,\ 11)$$

so ω can be calculated when μ is known. Substitute in (6·4, 9) the value of ω^2 from (6·4, 11) and clear of fractions

$$\Delta_1(\mu)\Delta_2(\mu)\Delta_3(\mu) - \Delta_4(\mu)\Delta_1(\mu)^2 - \Delta(\mu)\Delta_3(\mu)^2 = 0. \qquad (6\cdot4,\ 12)$$

When expanded this yields a sextic equation for μ which can, however, be written as a cubic in the quantity

$$z = 2\mu(\mathbf{B} + 2\mu\mathbf{A}), \qquad (6\cdot4,\ 13)$$

namely $\qquad\qquad z^3 + 2\mathbf{C}z^2 + Sz + R = 0, \qquad (6\cdot4,\ 14)$

where $\qquad\qquad S \equiv \mathbf{BD} + \mathbf{C}^2 - 4\mathbf{AE} \qquad (6\cdot4,\ 15)$

and $\qquad\qquad R \equiv \mathbf{BCD} - \mathbf{AD}^2 - \mathbf{EB}^2 \qquad (6\cdot4,\ 16)$

is Routh's discriminant (see § 4·9). This can be made the basis for an exact determination of μ and so of the whole root, with the help of (6·4, 11). However, we are concerned with finding an approximation to μ when it is numerically small. It follows from (6·4, 13) and (6·4, 14) that the constant in the equation for μ is R while the coefficient of the first power of μ is $2\mathbf{B}S$. Hence, when μ is small, we have approximately

$$\mu = -\frac{R}{2\mathbf{B}S} = -\frac{\mathbf{BCD} - \mathbf{AD}^2 - \mathbf{EB}^2}{2\mathbf{B}^2\,\mathbf{D} + 2\mathbf{BC}^2 - 8\mathbf{ABE}}. \qquad (6\cdot4,\ 17)$$

* Compare the derivation of the test function T_3 in § 4·10.

If the expression on the right-hand side of the equation is in fact small we may accept it as an approximation to μ and then ω can be found from (6·4, 11). This approximation can usually be applied to the lateral oscillation and to the phugoid oscillation (see § 5·8).

We can obtain an approximation to the period T (in seconds) of the lateral oscillation by putting

$$\omega^2 = \frac{\mathbf{D}}{\mathbf{B}}, \qquad (6\cdot4, 18)$$

which follows from (6·4, 11) when μ is very small. We shall take the case of level flight ($k' = 0$), assume i_E, y_p and y_r zero and μ_2 large. Then we have from (6·2, 14)

$$\mathbf{B} = -y_v - \frac{l_p}{i_A} - \frac{n_r}{i_C}, \qquad (6\cdot4, 19)$$

and approximately from (6·2, 16)

$$\mathbf{D} = \frac{\mu_2}{i_A i_C}(l_v n_p - l_p n_v - i_C k_L l_v) \qquad (6\cdot4, 20)$$

$$= \mu_2 \mathbf{D}' \text{ say}. \qquad (6\cdot4, 21)$$

Further, $$T = \frac{2\pi \hat{t}}{\omega},$$

so $$\frac{T^2}{4\pi^2} = \frac{\hat{t}^2 \mathbf{B}}{\mu_2 \mathbf{D}'} = \frac{bC_L \mathbf{B}}{4g \mathbf{D}'}$$

by (5·2, 2) and (6·2, 1). Thus we obtain the approximation

$$T = \pi \sqrt{\left(\frac{bC_L}{g}\right)} \Big/ \sqrt{\left(\frac{-(y_v i_A i_C + l_p i_C + n_r i_A)}{l_v n_p - l_p n_v - k_L l_v i_C}\right)}. \qquad (6\cdot4, 22)$$

The dominant terms in the numerator and denominator of the fraction under the radical are $-l_p i_C$ and $-l_p n_v$ respectively. When these only are retained we arrive at the approximation

$$T = \pi \sqrt{\left(\frac{bC_L i_C}{gn_v}\right)}. \qquad (6\cdot4, 23)$$

We can easily deduce from the dynamical equation (6·2, 10) for yawing moments that this is the periodic time in seconds for an undamped oscillation in pure yaw.*

---

* In motion about OZ $\psi = -\bar{v}$. On the subject matter of this section, see also the paper by Priestley cited on page 175.

We see from (6·4, 16) that an increase in **C** increases R and therefore has a stabilizing influence on the lateral oscillation. Now when i_E, y_p, y_r are zero and μ_2 is large equation (6·2, 15) shows that

$$\mathbf{C} = \frac{\mu_2 n_v}{i_C}, \qquad (6\cdot4, 24)$$

approximately. Hence **C** can be increased and the lateral oscillation stabilized by an increase of n_v, i.e. by an increase of the area or effectiveness of the fin. This conclusion is fully confirmed by detailed calculations (see § 6·5).

6·5 Importance of the L_v: N_v and N_r: N_v ratios

We have already seen in § 6·4 that the ratio $-L_v/N_v$ must be positive and not too small in order that spiral instability may be avoided and that N_v should be positive and large to ensure that the lateral oscillation shall be stable. Thus an increase in N_v is from one point of view advantageous and from another detrimental. Bryant and Pugsley* pointed out that, with the higher wing loadings now usual, the region of variation of L_v and N_v for which both the spiral motion and lateral oscillation are stable is restricted. In some cases it may not be possible to satisfy both conditions of stability and then experience in flight shows that the lesser evil is to accept the instability of the spiral motion, provided it is not severe.

The facts above mentioned are clearly shown in Figs. 6·5, 1 and 6·5, 2 which are taken with slight modifications from a paper by Priestley.† These diagrams refer to a family of aircraft for which $i_A = 0\cdot09$ and $i_C = 0\cdot16$ and for the first figure C_L is 0·1 while for the second it is 0·5. The spiral stability boundary (**E** = 0) is independent of μ_2 but the oscillatory stability boundary (**R** = 0) depends on μ_2; a family of the latter boundaries appears in each diagram for values of μ_2 ranging from 20 to 100. In Fig. 6·5, 1, which corresponds to the smaller lift coefficient, spiral instability is avoided for any positive value of n_v provided that $-l_v$ is greater than 0·02. Oscillatory instability does not occur unless $-l_v$ is large, the permissible value falling

* L. W. Bryant and A. G. Pugsley, 'The Lateral Stability of Highly Loaded Aeroplanes', *R. & M.* 1840 (1936).

† E. Priestley, 'A Further Investigation of Lateral Stability', *R. & M.* 1989 (1941).

as μ_2 increases, and can always be avoided by a sufficient increase of n_v. For the higher lift coefficient (see Fig. 6·5, 2)

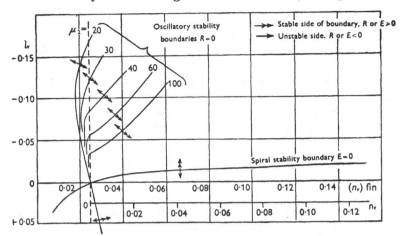

Fig. 6·5, 1. Oscillatory and spiral stability boundaries.
$C_L = 0·1 \quad i_A = 0·09 \quad i_C = 0·16.$

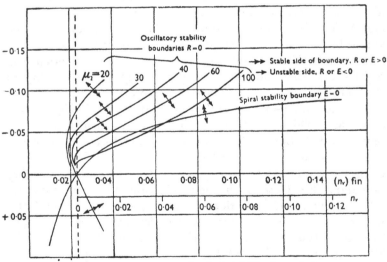

Fig. 6·5, 2. Oscillatory and spiral stability boundaries.
$C_L = 0·5 \quad i_A = 0·09 \quad i_C = 0·16.$

the attainment of complete stability is more difficult and with $\mu_2 = 100$ it is necessary* that $n_v > 0·04$ while the permitted

* We neglect the very small stable region where both l_v and n_v are nearly zero.

range of l_v is small. It will be seen that the boundary $\mathbf{E} = 0$ has been raised while the $R = 0$ boundaries have moved down.

It is a matter of experience that the characteristics of an aircraft in lateral motion, especially with rudder free, are not satisfactory when $-n_r/n_v$ is too small. This ratio controls the rate of damping of a purely yawing oscillation. The ratio can be increased by adding fin area forward of the yawing axis for this increases $-n_r$ and simultaneously decreases n_v.

6·6   Influence of the propulsive system

The slipstream from the airscrew will add to the effectiveness of any fin surface lying within it and may largely affect the lateral trim, especially with free rudder. We have pointed out when discussing N_v that an airscrew acts as a fin which will be destabilizing when forward of the axis of yaw. An elementary account of the theory of propeller fin effect is given in the Appendix to Chapter 5.

Chapter 7

FLAP CONTROLS IN GENERAL

7·1 Introduction

We remarked in § 1·5 that aircraft are usually controlled and manœuvred by producing at will local changes of aerodynamic lift. These changes of lift are caused by some change in the configuration of the surface exposed to the air stream, such as a twist of a wing near the tip (wing warping) as used by the Wright brothers for lateral control. However, by far the most convenient means for producing a local change of lift is to turn a flap, hinged to a wing, tailplane or fin, about its hinge. The aim of this chapter is to examine the characteristics of such hinged flaps in a general way. Throughout this chapter we neglect the compressibility of the air.

In the analysis of flap action there are two main effects to be considered, namely the magnitude of the desired change in the force or moment on the aircraft as a whole corresponding to a given angular deflection of the flap and the magnitude of the hinge moment which must be applied to the flap to keep it deflected. For simplicity let us think first of an untwisted and uniform aerofoil provided with a hinged flap at the rear, with two-dimensional air flow in planes perpendicular to the span. Let α be the angle of incidence referred to a chord line fixed in the main aerofoil and η the angular deflection of the flap; when η is zero the complete aerofoil has its standard undistorted shape.* Then, provided that α and η are not too large numerically, both theory and experiment show that we can express the lift coefficient linearly in terms of these variables thus:

$$C_L = a_1(\alpha - \alpha_0) + a_2\eta, \qquad (7·1, 1)$$

where α_0 is the incidence for no lift with flap undeflected. Likewise the hinge moment coefficient is written

$$C_H = b_0 + b_1\alpha + b_2\eta, \qquad (7·1, 2)$$

where
$$C_H = \frac{H'}{\frac{1}{2}\rho V^2 (Ec)^2}, \qquad (7·1, 3)$$

* Angles are measured in radians unless the contrary is stated.

H' is the hinge moment per unit span, Ec is the flap chord, and b_0 is the value of C_H when α and η are zero. The pitching moment per unit span about the leading edge is M' and the pitching moment coefficient is

$$C_M = \frac{M'}{\tfrac{1}{2}\rho V^2 c^2},\qquad (7\cdot1,\,4)$$

while, according to the theory of thin aerofoils,

$$C_M = C_{MO} - \frac{1}{4}C_L - m\eta,\qquad (7\cdot1,\,5)$$

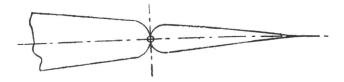

'Plano hinge' at nose

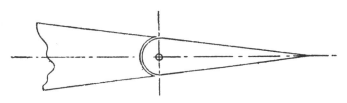

Round nose

Fig. 7·1, 1. Prototype flap controls.

where m is a constant depending on the ratio of flap chord to total chord. Another way of putting (7·1, 5) is that the pitching moment coefficient referred to the quarter-chord point is

$$C_m = C_{MO} - m\eta.\qquad (7\cdot1,\,6)$$

It must, however, be noted that experiment shows C_m not to be strictly independent of C_L, especially for thick aerofoils. However, equation (7·1, 6) remains valid when C_m refers to the actual aerodynamic centre which lies near the quarter-chord point.

The general definition of the hinge moment coefficient for a hinged flap of any plan form on a finite aerofoil is

$$C_H = \frac{H}{\frac{1}{2}\rho V^2 S_f \bar{c}_f},\qquad (7\cdot1,\,7)$$

where H is the total aerodynamic hinge moment, S_f is the plan area of the flap behind the hinge axis and $\bar{c}_f$ is the mean chord of the part of the flap behind the hinge. Alternatively S_f and $\bar{c}_f$ may refer to the whole plan area.

It is convenient to have in mind a prototype flap of the simplest kind. Such a flap is either hinged at the nose and sealed there or has a rounded nose which is part of a circular cylinder having the hinge line as axis (see Fig. 7·1, 1). For the latter the gap at the nose is either sealed or so narrow as to be without influence on the hinge moment. As a rough rule, the gap will be negligible when it is smaller than $\frac{1}{4}$ per cent of the local chord of the main surface. The flap has plane sides and a sharp trailing edge.

7·2   Deductions from the theory of thin aerofoils

A thin aerofoil with deflected flap may, when the hinge is sealed, be regarded as an aerofoil of modified shape and the lift, pitching moment and hinge moment can all be calculated from the theory of thin aerofoils due to Birnbaum and Glauert.* The application of this theory to the aerofoil with deflected flap is due to Glauert† and an extension to multiply hinged flap systems was given by Perring. We shall give here a brief account of Glauert's theory which is valid when the influence of compressibility is negligible. The results of the theory require some correction in practice on account of the influence of the boundary layer, a subject discussed in § 7·6.

In the theory of thin aerofoils the leading edge is taken as origin O and the chord line joining the leading and trailing edges as the axis OX. It is convenient to use an auxiliary position angle θ such that

$$x = \tfrac{1}{2}c(1 - \cos\theta),\qquad (7\cdot2,\,1)$$

---

* See *Elements of Aerofoil and Airscrew Theory*, by H. Glauert. Cambridge, 1926.

† H. Glauert, 'Theoretical Relationships for an Aerofoil with Hinged Flap', *R. & M.* 1095 (1927).

where c is the chord. Then the upward force per unit area, or excess of pressure below the aerofoil over that above it, is $k(\theta)$ given by

$$\frac{k(\theta)}{2\rho V^2} = A_0 \cot\frac{\theta}{2} + \sum_1^\infty A_n \sin n\theta, \qquad (7\cdot2,\,2)$$

where

$$A_0 = \alpha - \frac{1}{\pi}\int_0^\pi \left(\frac{dy}{dx}\right) d\theta, \qquad (7\cdot2,\,3)$$

$$A_1 = \frac{2}{\pi}\int_0^\pi \left(\frac{dy}{dx}\right) \cos\theta\, d\theta, \qquad (7\cdot2,\,4)$$

$$A_n = \frac{2}{\pi}\int_0^\pi \left(\frac{dy}{dx}\right) \cos n\theta\, d\theta, \qquad (7\cdot2,\,5)$$

and y is the ordinate of the centre line of the aerofoil. Moreover the lift coefficient is found by integration to be

$$C_L = \pi(2A_0 + A_1)$$
$$= 2\pi(\alpha - \alpha_0),$$

where, by $(7\cdot2,\,3)$ and $(7\cdot2,\,4)$, the no lift angle is

$$\alpha_0 = \frac{1}{\pi}\int_0^\pi (1 - \cos\theta)\left(\frac{dy}{dx}\right) d\theta. \qquad (7\cdot2,\,6)$$

Also the pitching moment coefficient referred to the leading edge is found to be

$$C_M = -\frac{\pi}{2}(A_0 + A_1 - \tfrac{1}{2}A_2) = -\frac{1}{4}C_L + C_{MO}, \qquad (7\cdot2,\,7)$$

where

$$C_{MO} = \frac{\pi}{4}(A_2 - A_1). \qquad (7\cdot2,\,8)$$

When A_0 does not vanish there is a very large* normal force intensity near the leading edge. This disappears when α is equal to α_i, called the 'ideal angle of attack' by Theodorsen. By equation $(7\cdot2,\,3)$

$$\alpha_i = \frac{1}{\pi}\int_0^\pi \left(\frac{dy}{dx}\right) d\theta, \qquad (7\cdot2,\,9)$$

and the corresponding ideal lift coefficient is

$$C_{Li} = \pi A_1 = 2\int_0^\pi \left(\frac{dy}{dx}\right) \cos\theta\, d\theta. \qquad (7\cdot2,\,10)$$

* Theoretically infinite, in the absence of rounding of the leading edge.

Let us apply these formulae to a thin symmetric aerofoil*
with deflected flap at the rear (see Fig. 7·2, 1). The chord of
the flap is Ec and ϕ is the value of θ corresponding to the hinge
B, so, by (7·2, 1),

$$\cos\phi = -(1-2E), \quad \text{and} \quad \sin\phi = 2\sqrt{[E(1-E)]}.$$
$$(7\cdot2, 11)$$

The angle of incidence α is measured from the chord line AB
of the aerofoil and α' is measured from AC; this is the angle of
incidence to be used in the formulae of thin aerofoil theory and

Fig. 7·2, 1. Skeleton of aerofoil and flap.

the ordinate y is to be measured from AC. Now C is at the
distance $Ec\eta$ below AB produced, when, as we suppose, η is a
very small angle. Hence.

$$\angle BAC = E\eta, \quad \text{and} \quad \alpha' = \alpha + E\eta. \qquad (7\cdot2, 12)$$

Also the values of $\left(\dfrac{dy}{dx}\right)$ for AB and BC are $E\eta$ and $-(1-E)\eta$
respectively. Hence, by (7·2, 6), the no-lift incidence measured
from AC is

$$\alpha_0' = \frac{1}{\pi}\int_0^\phi E\eta(1-\cos\theta)\,d\theta - \frac{1}{\pi}\int_\phi^\pi (1-E)\eta(1-\cos\theta)\,d\theta$$
$$= -\eta\left[(1-E) + \frac{\sin\phi - \phi}{\pi}\right]. \qquad (7\cdot2, 13)$$

Accordingly

$$C_L = 2\pi(\alpha' - \alpha_0')$$
$$= 2\pi\alpha + 2(\pi - \phi + \sin\phi)\eta. \qquad (7\cdot2, 14)$$

On comparison with (7·1, 1) we obtain

$$\left.\begin{aligned} a_1 &= 2\pi, \\ a_2 &= 2(\pi - \phi + \sin\phi). \end{aligned}\right\} \qquad (7\cdot2, 15)$$

* When the aerofoil is not symmetric α_0 and b_0 are not zero, but the
coefficients a_1, a_2, b_1 and b_2 are unaltered.

We also obtain easily

$$\left. \begin{aligned} \alpha_i &= -\frac{\eta(\pi - \phi)}{\pi}, \\ C_{Li} &= 2\eta \sin \phi. \end{aligned} \right\}$$

and

(7·2, 16)

Thus with a rear flap deflected downwards the ideal angle of attack is negative but the corresponding lift coefficient is positive and small.

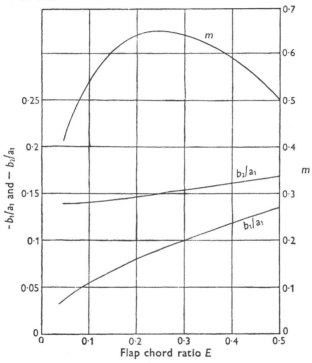

Fig. 7·2, 2. Theoretical characteristics of hinged flaps.

We shall not enter here into the details of the calculations of the pitching and hinge moments but shall content ourselves with quoting Glauert's results, converted to modern notation. The pitching moment coefficient m of equation (7·1, 5) is given by

$$\left. \begin{aligned} m &= \tfrac{1}{2}\sin \phi - \tfrac{1}{4}\sin 2\phi \\ &= \tfrac{1}{2}\sin \phi (1 - \cos \phi), \end{aligned} \right\}$$

(7·2, 17)

while the hinge moment coefficients b_1 and b_2 are given by

$$-\frac{E^2 b_1}{a_1} = \frac{\sin\phi(1-\tfrac{1}{2}\cos\phi)}{2\pi} - \frac{(\pi-\phi)(\tfrac{1}{2}-\cos\phi)}{2\pi}, \quad (7\cdot2, 18)$$

$$-E^2 b_2 = \frac{1}{2\pi}\sin^2\phi + \left(\frac{\pi-\phi}{\pi}\right)\sin\phi - \frac{(\pi-\phi)^2(\tfrac{1}{2}-\cos\phi)}{\pi}. \quad (7\cdot2, 19)$$

Graphs of b_1 and b_2 against E are given in Fig. 7·2, 2.

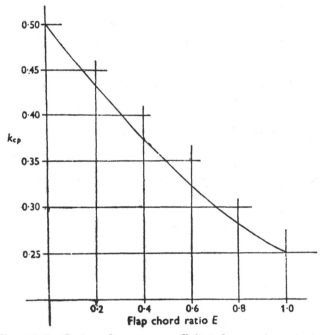

Fig. 7·2, 3. Centre of pressure coefficient for aerodynamic load caused by deflection of flap.

We deduce from the expressions for the pitching moment and lift that the centre of pressure coefficient for the aerodynamic load caused by operating the flap is

$$k_{cp} = \frac{1}{4} + \frac{m}{a_2} = \frac{\pi-\phi+(2-\cos\phi)\sin\phi}{4(\pi-\phi+\sin\phi)}. \quad (7\cdot2, 20)$$

It follows from this that

$$\tfrac{1}{4} < k_{cp} < \tfrac{1}{2}, \quad (7\cdot2, 21)$$

where the lower bound corresponds to a flap hinge at the leading edge ($E = 1$) and the upper bound corresponds to a flap of extremely small chord. Fig. 7·2, 3 gives a graph of k_{cp} against E. It will be seen that for a flap of 15 per cent chord k_{cp} is 0·45 while for a flap of 30 per cent chord k_{cp} is just over 0·4. These results imply that for flaps of ordinary proportions the major part of the lift caused by operating the flap is attributable to pressure changes on the main aerofoil. Another important deduction from this theory is that a flap of given chord deflected through a given angle gives more lift the larger the chord of the main surface to which it is attached.

It is very instructive to compare the performance of a rear flap with that of a nose flap. Let the chord of the nose flap be Fc and let the value of the angle θ corresponding to the hinge position be ψ, so

$$\left.\begin{array}{l} \cos\psi = (1 - 2F), \\ \sin\psi = 2\sqrt{[F(1-F)]}. \end{array}\right\} \tag{7·2, 22}$$

Also let ζ be the angular deflection of the flap, taken positive upwards, and put

$$C_L = a_1\alpha + a_2'\zeta. \tag{7·2, 23}$$

Then a_1 is as before but

$$a_2' = 2(\psi - \sin\psi). \tag{7·2, 24}$$

A comparison with (7·2, 15) shows that the nose flap is very much less effective than the rear flap as a lift producer.* For example, suppose that both are of 15 per cent chord. Then we have $\phi = \pi - \psi$ and according to the theory

$$a_2 = 3·02,$$
$$a_2' = 0·16.$$

On the other hand the nose flap when negatively deflected (i.e. downwards) increases both the ideal angle of attack and the corresponding lift coefficient. For an aerofoil having both nose and rear flaps

$$a_i = -\frac{\zeta\psi + \eta(\pi - \phi)}{\pi}, \tag{7·2, 25}$$

and

$$C_{Li} = 2\eta\sin\phi - 2\zeta\sin\psi. \tag{7·2, 26}$$

* It should be noted that this is not true for supersonic flight speeds.

Hence the nose flap when deflected downward has the same effectiveness as the rear flap in increasing the ideal lift coefficient. It is relevant to remark that the maximum lift coefficient of a given thin aerofoil can be substantially increased by giving a nose flap a downward deflection. Both these effects can be regarded as consequences of the improved fairing of the aerofoil which accompanies downward displacement of the flap for positive wing incidences. The nose flap has the drawback of having an unstable hinge moment, i.e. one which tends to increase numerically any deflection of the flap.

Next let us consider an aerofoil with a flap which is provided with a hinged tab.* By extension of (7·1, 1) and (7·1, 2) we write

$$C_L = a_1(\alpha - \alpha_0) + a_2\eta + a_3\beta, \qquad (7\cdot2, 27)$$

$$C_H = b_0 + b_1\alpha + b_2\eta + b_3\beta, \qquad (7\cdot2, 28)$$

where β is the angle through which the tab is deflected relative to the flap, while the tab hinge moment coefficient is

$$C_{HT} = c_0 + c_1\alpha + c_2\eta + c_3\beta, \qquad (7\cdot2, 29)$$

where
$$C_{HT} = \frac{H'_T}{\frac{1}{2}\rho V^2 (E_T c)^2}, \qquad (7\cdot2, 30)$$

H'_T is the tab hinge moment per unit span and $E_T c$ is the tab chord. The theoretical values of a_3, c_1 and c_3 can be obtained from those for a_2, b_1 and b_2 respectively on substitution of E_T for E. The theoretical values of the remaining coefficients, as first obtained by Perring, are given by

$$E^2 b_3 = -\frac{\pi - \phi}{2\pi} \sin\phi_T (2\cos\phi - \cos\phi_T)$$

$$-\frac{\sin\phi\sin\phi_T}{2\pi} - \frac{\pi - \phi_T}{\pi}\sin\phi(1 - \tfrac{1}{2}\cos\phi)$$

$$+\frac{(\pi - \phi)(\pi - \phi_T)}{\pi}(\tfrac{1}{2} - \cos\phi)$$

$$-\frac{(\cos\phi - \cos\phi_T)^2}{4\pi}\log_e\left\{\frac{1 - \cos(\phi_T + \phi)}{1 - \cos(\phi_T - \phi)}\right\}, \qquad (7\cdot2, 31)$$

---

* W. G. A. Perring, 'Theoretical Relationships for an Aerofoil with a Multiply Hinged Flap System', *R. & M.* 1171 (1928).

and

$$E_T^2 c_2 = \frac{\pi - \phi_T}{2\pi} \sin \phi (\cos \phi - 2 \cos \phi_T)$$

$$- \frac{\sin \phi \sin \phi_T}{2\pi} - \frac{\pi - \phi}{\pi} \sin \phi_T (1 - \tfrac{1}{2} \cos \phi_T)$$

$$+ \frac{(\pi - \phi)(\pi - \phi_T)}{\pi} (\tfrac{1}{2} - \cos \phi_T)$$

$$- \frac{(\cos \phi - \cos \phi_T)^2}{4\pi} \log_e \left\{ \frac{1 - \cos(\phi_T + \phi)}{1 - \cos(\phi_T - \phi)} \right\}, \qquad (7 \cdot 2, 32)$$

where ϕ_T is related to E_T by the formulae of (7·2, 11).

We next turn our attention to aerofoils of finite aspect ratio. Now, provided as usual that the incidences and flap angles are small, the lift, pitching moment and hinge moment coefficients will be linear functions of these angles, but it remains to inquire how the coefficients a_1, b_1, a_2, b_2, etc. depend on aspect ratio and other factors. Following Glauert we shall treat this question on the basis of the Lanchester-Prandtl theory of the finite aerofoil ('lifting-line theory'), and at first we shall confine attention to uniform unyawed and untwisted aerofoils of rectangular plan form provided with full-span flaps of constant chord.

The relevant basic ideas of the Lanchester-Prandtl theory are:

(a) The pressure distribution for any narrow fore-and-aft strip of the wing depends only upon the profile, local angle of incidence, resultant air speed and air density.

(b) The effective local velocity vector is the vector resultant of the general stream velocity and the velocity ('downwash') induced by the vortices which trail from the aerofoil. The downwash is assumed to be uniform over the chord.

(c) When no part of the wing is lifting the trailing vortices are absent and the pressure distribution at any strip depends only on the general stream velocity and air density. An important particular consequence of this is that the no-lift angle for an untwisted uniform aerofoil is independent of aspect ratio.

Now let us take a finite wing with the flap set at some fixed angle. The lift coefficient is given by

$$C_L = a_1(\alpha - \alpha_0) + a_2 \eta,$$

where we must expect a_1 and a_2 to depend on aspect ratio. This equation shows that the no-lift incidence with flap deflected is

$$\alpha_n = \alpha_0 - \frac{a_2}{a_1}\eta, \qquad (7\cdot2, 33)$$

and, in accordance with principle (c) above, this is independent of aspect ratio for all values of η. Consequently a_2/a_1 is independent of aspect ratio. Next we take the pitching moment coefficient, which, for aerofoils of the type considered, can be expressed by

$$C_M = C_{MO} - \frac{1}{4}C_L - m\eta.$$

Let the incidence be α_n so C_L is zero. According to principle (c) the pressure distribution at all strips is independent of aspect ratio and the same is therefore true of the pitching moment. Consequently C_{MO} and m are independent of aspect ratio. The hinge moment coefficient is

$$C_H = b_0 + b_1\alpha + b_2\eta,$$

and when the incidence is α_n

$$C_H = b_0 + b_1\alpha_0 + b\eta, \qquad (7\cdot2, 34)$$

where $$b = \frac{a_1 b_2 - a_2 b_1}{a_1}. \qquad (7\cdot2, 35)$$

We deduce by the same argument as before that b must be independent of aspect ratio. Likewise $(b_0 + b_1\alpha_0)$ is invariant or b_0 is invariant if α is measured from the no-lift incidence.

It appears from equations $(7\cdot2, 3)\ldots(7\cdot2, 5)$ that only the coefficient A_0 depends on incidence. Hence, by $(7\cdot2, 2)$, the change in pressure distribution with incidence over the entire chord is in proportion to $\cot\dfrac{\theta}{2}$. Thus the changes with incidence of lift, pitching moment and hinge moment are all proportional, and, according to the Lanchester-Prandtl theory, this is true for every strip of the wing irrespective of aspect ratio. Accordingly $\dfrac{dC_M}{dC_L}$ and $\dfrac{dC_H}{dC_L}$ are independent of aspect ratio. The first of these results has already been noted and the second shows that

$$\left(\frac{\partial C_H}{\partial \alpha}\right)\bigg/\left(\frac{\partial C_L}{\partial \alpha}\right), \quad \text{or} \quad b_1/a_1,$$

is independent of aspect ratio. To sum up, the seven quantities α_0, C_{MO}, a_2/a_1, b_1/a_1, b, m and $(b_0 + b_1 \alpha_0)$ are all independent of aspect ratio for uniform untwisted rectangular wings with full-span flaps of constant chord. Table 7·2, 1, which is derived from Glauert's results, shows the dependence of a_1 on aspect ratio A for uniform rectangular wings as given by lifting line theory. In the table a_0 is the value of $\dfrac{dC_L}{d\alpha}$ for infinite aspect ratio.

TABLE 7·2, 1.

$\dfrac{A}{a_0}$	$\dfrac{a_1}{a_0}$
0·5	0·587
0·75	0·675
1·0	0·729
1·25	0·767
1·5	0·794
1·75	0·815

An aerofoil carrying a part span flap which is deflected is, in effect, a twisted aerofoil. Now it can be shown that the no-lift angle of a twisted aerofoil is not, in general, independent of aspect ratio. The same is, in general, true even for a full span flap when the aerofoil or flap is tapered. Hence the deductions made above about the invariance of certain quantities cease to be valid for part span flaps or tapered surfaces. However, the 'invariants' are usually but little dependent on aspect ratio.

The values of the aerofoil and flap derivatives as given by the theory of thin aerofoils, or 'potential theory' in general, are found to be numerically larger than those measured. This is attributable to the influence of the boundary layer, a subject briefly discussed in § 7·6. The experimental values of the flap derivatives are of the order of 20 per cent below the theoretical when the trailing edge angle is about 10° and the deficiency is considerably greater for the tab derivatives. Unsealing the gap at the hinge is found to decrease both the effectiveness and efficiency of a flap control, i.e. the hinge moment corresponding to a given lift increment is increased although the hinge moment for a given flap deflection is reduced.

There are various means for adjusting the hinge moments of flap controls without making drastic alterations of shape or hinge position. A slight adjustment of the vertical height of the whole flap relative to the main surface and the adjustment of shrouds are two such measures. One of the simplest and most valuable means of adjustment consists in fixing a length of cord, say about $\frac{3}{16}$ in. in diameter, along the control near the trailing edge; the length of the cord is varied until the desired result is obtained. It is found that with the cord attached on the upper or lower surface or on both there is an algebraic decrease of both b_2 and b_1; thus, for the normal case where these coefficients are both negative, attachment of cord increases them numerically. Hence addition of cord is a convenient means for correcting aerodynamic balance which has been carried to excess. Cord on the upper surface alone gives also a positive hinge moment coefficient b_0 tending to cause the control to float down; this effect is reversed when the cord is applied on the lower surface alone. A ridge of metal or wood may be substituted for the cord. The effects of cords or ridges are related to changes in the circulation about the main surface (see § 7·6). It is found that a cord or ridge fitted only on the lower surface becomes ineffective for upward displacement of the control larger than about 5°. Likewise a cord or ridge fitted only on the upper surface becomes ineffective for a downward deflection of about the same magnitude.

7·3 Non-linear hinge moment curves

The theory of thin aerofoils leads to the conclusion that the lift and hinge moment are linear functions of incidence and flap angle provided that these are small and that the flap conforms to the prototype as described in § 7·1. This theory, however, does not warrant the conclusion that hinge moments must vary linearly with these angles for all nose shapes, especially when the hinge is set back from the nose and the gap is unsealed, unless indeed the angles are infinitesimal. In fact experiment shows that the hinge moment curves for controls with heavily set back hinges and noses which, in certain settings of the control, protrude into the air stream are very far from linear.

An example is provided in Fig. 7·3, 1 which refers to a single Frise aileron. It will be seen that for a small range of movement round the neutral setting the slope of the curve is reversed. Hence, in this region, the hinge moment characteristic is unstable. In practice, when a pair of ailerons are coupled, there should be no unstable region (see § 9·5). In general, however, hinge moment characteristics which are non-linear for small

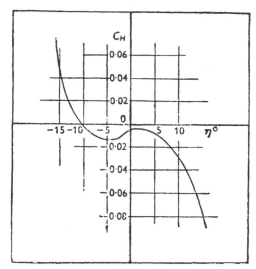

Fig. 7.3, 1. Hinge moment coefficient for single Frise aileron.

flap deflections are most objectionable. For large flap deflections there will always be departures from linearity and likewise for large incidences of the main surface.

7·4   The influence of response on hinge moments

When a flap control is moved the aircraft responds and the motion of response reacts on the hinge moment. This effect can be investigated quantitatively by means of equation (7·1, 2).* Suppose for simplicity that during the application of the control $\frac{d\alpha}{d\eta}$ is constant, where α is the incidence of the main surface to which the flap is attached. Then we derive that the

* We neglect here the lag in the build-up of the hinge moment (Wagner effect).

increment in the hinge moment coefficient is

$$\Delta C_H = \left(1 + \frac{b_1}{b_2}\frac{d\alpha}{d\eta}\right) b_2 \Delta\eta$$

$$= K b_2 \Delta\eta, \qquad (7\cdot4, 1)$$

where K is called the 'response factor' and is given by

$$K = 1 + \frac{b_1}{b_2}\frac{d\alpha}{d\eta}. \qquad (7\cdot4, 2)$$

If there were no response, i.e. if $\dfrac{d\alpha}{d\eta}$ were zero, the response factor would be unity. Thus K gives the ratio of the actual increment of hinge moment to that which would occur in the absence of response with the same flap deflection. Hence it is $K b_2$ rather than b_2 which determines the heaviness of the control.

For an ordinary flap control which is appreciably under-balanced aerodynamically b_1/b_2 is positive but $\dfrac{d\alpha}{d\eta}$ is usually negative. For example, when the aileron on one side goes down (η positive) the wing to which it is attached acquires an upward velocity which reduces the angle of incidence (see § 2·5). Hence the response factor is less than unity and the response of the aircraft to the application of the control gives rise to a hinge moment which assists the pilot to apply the control. If the response effect were too large we should have instability (K negative) but this would be most abnormal.

7·5 Aerodynamic balancing of flap controls

The moment which must be applied to a flap in order to deflect it through an angle adequate for manœuvring tends to be awkwardly large for an unassisted pilot when the aircraft is large or flies fast. Hence one of the basic problems in the design of flap controls consists in so arranging matters that the hinge moments are reduced. This is called aerodynamic balancing. An essential requirement is that the control shall in no circumstance become over-balanced aerodynamically for, if this occurred, the control would be unstable and swing round through a large angle if released by the pilot. In practice this imposes a limit to the closeness of aerodynamic balance although the difficulty may be largely avoided by the use of special devices, such as the spring tab (see § 9·4).

The principal methods of aerodynamic balancing are as follows:

(a) Set back hinge.

(b) Horn balance.

(c) Sealed nose balance.

(d) Geared balance tab.

(e) Special profile shape.

Two or more of these methods may be used on a single control. The spring tab, which is a valuable means for reducing hinge moments precisely when they tend to be large, is discussed separately in § 9·4.

The above-mentioned methods of balancing will now be briefly considered in turn.

(a) *Set back hinge*

The simplest method of aerodynamic balancing consists in placing the control hinge axis some distance aft of the nose. Any degree of balance can, in theory, be attained in this manner, but the variability of b_2 with small changes of shape and of surface condition makes it inadvisable to attempt very close balancing by this means, on account of the danger of overbalance. However, if the control with forward hinge has a b_2 of say $-0·75$ this can be safely reduced (numerically) to say $-0·2$ by setting back the hinge axis through something like 30 per cent of the flap chord.

We can obtain the basic relations on the influence of hinge position on the derivatives by an argument based on kinematics and statics alone. Take the main surface and flap in a certain datum configuration, placed in a uniform airstream of fixed characteristics and let the hinge be at a datum position in the chord line. For a given incidence of the main surface the aerodynamic forces on the flap will be statically equivalent to a force N, normal to the chord and through the hinge axis, a tangential or chordwise force T and a hinge moment H. The force T is small and will at present be neglected but its influence would not be negligible if the hinge axis had a large vertical offset from the chord line. N is positive downwards and H positive in the T.E. down sense. In general, N and H will depend on the flap deflection η and on z, the normal downward displacement of the flap* and we shall express them by means of derivatives as follows:

$$N = N_0 + z\frac{\partial N}{\partial z} + \eta\frac{\partial N}{\partial \eta}, \tag{7.5, 1}$$

$$H = H_0 + z\frac{\partial H}{\partial z} + \eta\frac{\partial H}{\partial \eta} \tag{7.5, 2}$$

Now, with the flap in its datum configuration, let us shift the hinge axis rearward along the flap chord through the distance r, and then let the flap be deflected through the angle η. This raises the original hinge by the amount $r\eta$, i.e. $z = -r\eta$. Hence N and H, still referred to the original hinge axis, are given by

$$N = N_0 + \eta\left(\frac{\partial N}{\partial \eta} - r\frac{\partial N}{\partial z}\right), \tag{7.5, 3}$$

$$H = H_0 + \eta\left(\frac{\partial H}{\partial \eta} - r\frac{\partial H}{\partial z}\right). \tag{7.5, 4}$$

By statics the normal force N' and hinge moment H' referred to the new hinge axis are given by

$$N' = N, \tag{7.5, 5}$$

$$H' = H - Nr. \tag{7.5, 6}$$

Hence by (3) and (4)

$$\frac{\partial N'}{\partial \eta} = \frac{\partial N}{\partial \eta} - r\frac{\partial N}{\partial z}, \tag{7.5, 7}$$

$$\frac{\partial H'}{\partial \eta} = \frac{\partial H}{\partial \eta} - r\left(\frac{\partial H}{\partial z} + \frac{\partial N}{\partial \eta}\right) + r^2\frac{\partial N}{\partial z}. \tag{7.5, 8}$$

Thus the hinge moment derivative varies parabolically with hinge position. This simple fact will be obscured when the conventional non-dimensional coefficient b_2 is calculated on account of the division by the area of the flap aft of the hinge, since this varies as the hinge is shifted.

Suppose next that λ is some parameter affecting N and H but not involving change of position of flap relative to main surface. Then we derive immediately from (7.5, 5) and (7.5, 6) that

$$\frac{\partial N'}{\partial \lambda} = \frac{\partial N}{\partial \lambda} \tag{7.5, 9}$$

and

$$\frac{\partial H'}{\partial \lambda} = \frac{\partial H}{\partial \lambda} - r\frac{\partial N}{\partial \lambda}. \tag{7.5, 10}$$

* The reason for introducing this will become apparent later.

In particular, if we identify λ with the angle of incidence α we derive

$$\frac{\partial H'}{\partial \alpha} = \frac{\partial H}{\partial \alpha} - r\frac{\partial N}{\partial \alpha}. \qquad (7\cdot5, 11)$$

Hence this derivative varies linearly with hinge position. Again this simple result is masked in the conventional non-dimensional coefficient b_1.

Similar arguments apply to the damping derivatives. Thus corresponding to $(7\cdot5, 8)$ we have

$$\frac{\partial H'}{\partial \dot{\eta}} = \frac{\partial H}{\partial \dot{\eta}} - r\left(\frac{\partial H}{\partial \dot{z}} + \frac{\partial N}{\partial \dot{\eta}}\right) + r^2\frac{\partial N}{\partial \dot{z}}, \qquad (7\cdot5, 12)$$

while $(7\cdot5, 11)$ corresponds to

$$\frac{\partial H'}{\partial \dot{\alpha}} = \frac{\partial H}{\partial \dot{\alpha}} - r\frac{\partial N}{\partial \dot{\alpha}}. \qquad (7\cdot5, 13)$$

All the above results apply to flaps in two and three dimensions.

In equation $(7\cdot5, 8)$ we shall normally have $\dfrac{\partial H}{\partial \eta}$ and $\dfrac{\partial N}{\partial \eta}$ negative while $\dfrac{\partial H}{\partial z}$ will be numerically smaller than $\dfrac{\partial N}{\partial \eta}$. Thus the total coefficient of r will be positive and, when r is small, the term in r^2 will be negligible. Hence, when r is small and positive, $\dfrac{\partial H'}{\partial \eta}$ will be numerically less than $\dfrac{\partial H}{\partial \eta}$ so some measure of aerodynamic balance is attained. At the same time $\dfrac{\partial H'}{\partial \alpha}$ will become less negative, i.e. b_1, which is negative for a forward hinge, is reduced numerically by shifting the hinge aft.

(b) Horn balance

A horn balance is a local protuberance of the control surface lying forward of the hinge axis. Thus, a control with a horn balance has, in effect, a heavily set back hinge over a small part of its span and a measure of aerodynamic balance is therefore secured. The horn may lie behind the main surface (shielded horn) or be exposed to the airstream (unshielded horn). It is found that it is not difficult to make b_1 zero or positive by use of an unshielded horn without incurring aerodynamic overbalance. This kind of balance is often used on elevators and rudders where it is advantageous to have b_1 zero or positive.

(c) Sealed nose balance

In this scheme the control has a forward projecting plate or tongue ahead of the hinge and the gap between the forward edge of this and the main surface is sealed, say by a loose fold of impervious fabric. The space above the tongue is open to the air above the main surface through a slit parallel to the hinge axis and the space below the tongue is similarly open beneath. When the flap is deflected downward the circulation round the main surface is increased and the excess of the pressure below over that above is also increased. Consequently the pressures on the tongue assist the angular movement and a measure of aerodynamic balance is attained. An increase of incidence of the main surface likewise contributes a positive hinge moment to the tongue so there will be a positive contribution to b_1.

There is evidence to show that b_1 and b_2 are more sensitive to changes of Mach number with sealed nose balance than with plain forward balance.

(d) Geared balance tab

Suppose that the control flap is provided with a tab at the rear and that the tab is so geared that it is deflected upwards relative to the flap when the latter moves downward. Then the displacement of the tab brings into play a moment assisting the deflection of the control and a measure of aerodynamic balance is attained.

The arrangement can be conveniently investigated quantitatively by consideration of the work done by the aerodynamic loads when the deflection of the flap is increased by a small angle $\delta\eta$. Let the gearing of the tab be such that

$$\delta\beta = k\delta\eta, \qquad (7\cdot5, 14)$$

where k will in fact be negative for a balance tab. The hinge moments for the flap and tab are given respectively by

$$H = qS_f\bar{c}_f(b_0 + b_1\alpha + b_2\eta + b_3\beta)$$

$$= q(\mathscr{B}_0 + \mathscr{B}_1\alpha + \mathscr{B}_2\eta + \mathscr{B}_3\beta) \qquad (7\cdot5, 15)$$

and $\qquad H_t = qS_t\bar{c}_t(c_0 + c_1\alpha + c_2\eta + c_3\beta)$

$$= q(\mathscr{C}_0 + \mathscr{C}_1\alpha + \mathscr{C}_2\eta + \mathscr{C}_3\beta), \qquad (7\cdot5, 16)$$

where q is the dynamic pressure $\frac{1}{2}\rho V^2$. The work done by the aerodynamic loads in the small displacement is

$$\delta W = H\delta\eta + H_t\delta\beta$$
$$= (H + kH_t)\,\delta\eta$$

by (7·5, 14). Hence the effective hinge moment is

$$H_e = \frac{dW}{d\eta} = H + kH_t$$

or

$$\frac{H_e}{q} = (\mathscr{B}_0 + k\mathscr{C}_0) + (\mathscr{B}_1 + k\mathscr{C}_1)\alpha + [\mathscr{B}_2 + k(\mathscr{B}_3 + \mathscr{C}_2) + k^2\mathscr{C}_3]\,\eta.$$

$$(7\cdot5, 17)$$

We see that $\mathscr{B}_2$ is replaced by

$$\mathscr{B}_2 + k(\mathscr{B}_3 + \mathscr{C}_2) + k^2\mathscr{C}_3,$$

and the term in k^2 is very small. Now $\mathscr{B}_3$ is in fact negative and so is $\mathscr{C}_2$ when the tab is not aerodynamically balanced. Hence the effective $\mathscr{B}_2$ is increased algebraically or reduced numerically when k is negative. A balance tab reduces somewhat the effectiveness of the control as follows from (7·2, 27) since a_2 and a_3 have the same sign while β and η have opposite signs.

When k is positive we have an *anti-balance tab*. Such a tab has been used in conjunction with a large horn balance to secure an effective positive b_1 without aerodynamic over-balance.

Outline diagrams of the gearing of balance and anti-balance tabs are shown in Figs. 7·5, 1 and 7·5, 2 respectively. In these diagrams the flap is hinged at A and the tab at B while C is the trailing edge. D is fixed on the wing or other main surface and is connected by a rigid link to E on the tab. The control rod is attached to the flap at F.

(e) *Special profile shape*

The best known control flap of special profile is the Frise aileron, which is shown in Fig. 7·5, 3. Here the balancing effect is largely attributable to the fact that when the aileron moves up the nose protrudes below the wing and there is here a region of high velocity and therefore of suction. This suction assists the deflection. The attainment of satisfactory balance depends on the gearing together of the two ailerons in opposition and is very sensitive to the nose shape. Each individual aileron has a highly non-linear hinge moment curve (see § 7·3).

The hinge moment characteristics of a flap can be largely modified by changing the shape, particularly near the trailing edge (see §§ 7·2 and 7·6). It has been found, for example, that a thick flap with a bevelled trailing edge has a numerically small b_2. This is not regarded as a satisfactory method of balancing as the hinge moment characteristics are very sensitive to the thickness of the boundary layer. Bulging the surfaces of

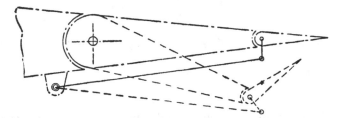

Fig. 7·5, 1. Diagram of gearing for balance tab.

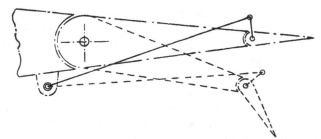

Fig. 7·5, 2. Diagram of gearing for anti-balance tab.

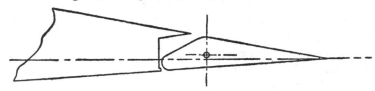

Fig. 7·5, 3. Frise aileron.

the flap (making them convex outwards) tends to reduce the hinge moment but likewise at the expense of making the hinge moment very sensitive to the state of the boundary layer; hence this form of flap is very objectionable. Anomalous variation of b_2 with speed has in some instances been traced to variation of convexity with speed, attributable to varying pressure difference between the outside and inside of the flap. The pressure difference depends acutely on the venting of the flap. This trouble can be avoided by making the surfaces sufficiently stiff.

7·6   Influence of the boundary layer

The influence of the boundary layer is altogether neglected in the theory of thin aerofoils, to which reference has been made in § 7·2. In this theory it is assumed that the flow is everywhere irrotational except in an infinitely thin wake extending backwards from the trailing edge and the circulation is determined by the condition that there shall be stagnation along this edge. In fact, an aerofoil in motion is surrounded by a region called the boundary layer where there is vorticity and retardation of the flow relative to the aerofoil. At the forward stagnation point the boundary layer is of zero thickness but its thickness increases towards the trailing edge where the upper and lower boundary layers meet and merge into the wake. This thickening of the boundary layer towards the rear implies that the velocity of the air relative to the surface at a given small normal distance from the surface decreases near the trailing edge. Hence a flap at the rear is bathed in retarded air and the forces upon it are considerably affected. Moreover, this effect is greater the smaller the ratio of flap chord to total chord and reaches a maximum for tabs of very small chord.

The pressure difference across the boundary layer is quite negligible over the forward part of the aerofoil where the boundary layer is thin, for the pressure gradient normal to the surface is small. Even near the trailing edge where the boundary layer is relatively thick the difference between the pressure at the surface at any point P and at a point on the normal at P just outside the boundary layer is small. Hence there is little error in taking the pressure at the surface of the aerofoil or flap, which determines the lift and hinge moment, to be equal to that in the irrotational flow just outside the boundary layer. This pressure is, however, not the same as it would be in the absence of the boundary layer, for the effective surface of the aerofoil is moved outwards by an amount equal to the 'displacement thickness' of the boundary layer.* For a given air stream

---

* The displacement thickness is δ given by

$$\delta = \int\left(1 - \frac{u}{u_0}\right) dy,$$

taken across the boundary layer, where u is the velocity parallel to the surface, u_0 the value of u just outside the boundary layer and y is normal to the surface.

and angle of incidence the excess of the pressure below the aerofoil over that immediately above the aerofoil is mainly determined by the circulation and this in turn depends on the state of the boundary layer. If this is everywhere thin there will be a comparatively small departure of the circulation from that appropriate to irrotational or 'potential' flow with stagnation at the trailing edge but the departure may be large when the boundary layer is thick.

When the lift and circulation are steady the total vorticity passing into the wake per second is zero. For definiteness let us think of an aerofoil at rest in a stream moving from left to right. Then the vorticity* in the upper boundary layer is clockwise whereas that in the lower boundary layer is anti-clockwise. Hence in the steady state the total clockwise vorticity discharged into the wake per second by the upper boundary layer is numerically equal to the total anti-clockwise vorticity discharged per second by the lower boundary layer. With only slight error we can substitute for this the condition that the velocity just outside the boundary layer at the trailing edge is the same in magnitude above and below the aerofoil. This then is the condition which replaces the Joukowsky condition of stagnation at the trailing edge. It is clear that any agency which affects the boundary layers, and especially if it affects them differentially, will influence the circulation, lift and hinge moment. One important factor is the position in the chord of the transition from laminar to turbulent flow in the upper and lower boundary layers. In general, the further forward is the transition the thicker is the boundary layer at the rear and the greater is the modification of the circulation. Cords or strips near the trailing edge have a powerful influence in modifying the circulation.

The conception that, even in the presence of a relatively thick boundary layer, the pressure distribution all over the aerofoil and flap is mainly determined by the circulation has led Preston† to conclude that the ratio of the actual hinge moment derivative b_2 to the value $(b_2)_T$ given by potential theory should

---

* The vorticity is twice the rate of spin of the fluid.

† See, e.g., 'The Effect of the Boundary-layer Thickness on the Normal Force Distribution of Aerofoils, with Particular Reference to Control Problems', by A. S. Batson and J. H. Preston, *R. & M.* 2008 (1942).

be a function of the ratio $a_1/(a_1)_T$; likewise $b_1/(b_1)_T$ should be a function of the same ratio. However, some more recent work by Bryant and Batson at the National Physical Laboratory indicates that for a particular trailing edge angle (about 9°) the value of $b_2/(b_2)_T$ is independent of $a_1/(a_1)_T$ and that the ratios $b_1/(b_1)_T$ and $b_2/(b_2)_T$ are approximately equal.

It is found that the sensitiveness of the hinge moment derivatives to changes in the condition of the boundary layer depends largely on the profile shape of the flap. The sensitiveness is least when the trailing edge angle is small and the surfaces are concave ('hollow ground'). Convex surfaces render the derivatives particularly sensitive to the effects of small manufacturing errors in the shape as well as to the state of the boundary layer.

7·7 Mass balancing of controls

Mass-balancing of control flaps is often necessary for the prevention of flutter or of 'snaking' and its presence largely influences the general dynamical behaviour of the controls. For example, mass-balance of the elevator, or rather the lack of it, influences the stick-free manœuvre margin (see § 10·7). Broadly, mass-balancing is a means for eliminating the inertial coupling between the motion of the flap and of some motion of the aircraft structure. This is explained in some detail below.

For simplicity let us first consider a control flap hinged to a surface which is being accelerated uniformly in a direction perpendicular to its plane and let the system be in a vacuum, so that aerodynamic forces are absent. In order to hold the flap undeflected it would, in general, be necessary to apply a hinge moment proportional to the acceleration; this moment would equilibrate the hinge moments due to the inertia forces. However, if it were possible to redistribute the mass of the flap or to add balance masses we could reduce this hinge moment to zero. Then the flap would be *mass balanced for normal translatory motion of the main surface* and it would not tend to deflect when the main surface had any kind of motion purely normal to the surface and uniform over it. Evidently the condition for mass balance in this simple case is

$$\Sigma x \delta m = 0, \qquad (7·7, 1)$$

where x is measured from the hinge line and parallel to the surface and the summation covers the entire flap. Clearly this equation implies that, with the surface horizontal, the total hinge moment due to the weight of the flap is zero. The flap in this condition would usually be said to be statically balanced.

With the flap statically balanced, as above, the hinge moment would not necessarily be zero for an acceleration parallel to the main surface, or, what amounts to the same thing, for normal acceleration with the flap already deflected. If z be measured normal to the plane of the flap the condition for static balance in fore-and-aft acceleration is

$$\Sigma z \delta m = 0. \qquad (7 \cdot 7, 2)$$

For complete static balance equations $(7 \cdot 7, 1)$ and $(7 \cdot 7, 2)$ must both be satisfied (c.g. of flap on hinge axis).

In general the surface to which the flap is attached does not have a purely translatory motion and the condition for absence of hinge moment depends on the nature of the motion. As an important simple case, let the main surface move as a rigid body about a fixed hinge line in its own plane and perpendicular to the flap hinge, and let y be measured from the hinge line of the main surface. Let the flap, which we suppose to be flat and thin, be in its undeflected position and consider an element of mass δm with co-ordinates x, y. If ϕ is the angular movement of the main surface the normal acceleration of the element is $y\ddot{\phi}$ and the hinge moment due to it is therefore $-xy\delta m\ddot{\phi}$. Hence the total inertial hinge moment is

$$H_i = -\ddot{\phi}\Sigma xy\delta m, \qquad (7 \cdot 7, 3)$$

and the condition for mass balance is accordingly

$$\Sigma xy\delta m = 0. \qquad (7 \cdot 7, 4)$$

This expresses the condition that the *product of inertia* of the flap for the two hinge axes is zero.

The next important case is where the main surface moves as a rigid body about an axis parallel to the hinge axis of the flap. Let θ be the angular movement of the main surface and d the distance of its axis of rotation forward of the flap hinge. Then the inertial hinge moment is

$$H_i = -\ddot{\theta}\Sigma x(d+x)\,\delta m, \qquad (7 \cdot 7, 5)$$

where we have taken x to be positive when measured aft of the flap hinge. The product of inertia or *inertial coupling coefficient* is here
$$P = \Sigma x(d+x)\,\delta m, \qquad (7\cdot7, 6)$$
and this is normally positive for a flap without balance masses. Let us consider the effectiveness of a unit mass in helping to reduce this to zero. The addition to P is $x(d+x)$ and this is negative only when x is negative and numerically less than d. *Thus a balance mass reduces P only when it lies between the two axes of rotation.* If the balance mass is on an arm lying forward of the flap hinge there is a limit, namely d, to the length of arm; if this limiting length is exceeded the mass will act as an anti-balance mass. The negative maximum value of $x(d+x)$ occurs when x is $-d/2$. Hence a mass is most effective in giving mass balance when it lies halfway between the parallel axes, i.e. at half the limiting length ahead of the flap hinge.

Mass balance for the most general system and kinds of motion can be investigated most conveniently by means of Lagrange's dynamical equations. Suppose that the two kinds of displacement concerned (exemplified above by the rotations about the two hinge axes) are measured by the generalized co-ordinates q_1 and q_2. With the system in a given configuration, the kinetic energy T is a homogeneous quadratic function of the generalized velocities and will contain a term $P_{12}\dot{q}_1\dot{q}_2$. Then P_{12} is the 'product of inertia' or inertial coupling coefficient for the pair of co-ordinates considered and for the given configuration of the system. When we confine attention to small movements about a given datum configuration of the system P_{12} will be a constant and the general expression for it is
$$P_{12} = \Sigma\left(\frac{\partial x}{\partial q_1}\frac{\partial x}{\partial q_2} + \frac{\partial y}{\partial q_1}\frac{\partial y}{\partial q_2} + \frac{\partial z}{\partial q_1}\frac{\partial z}{\partial q_2}\right)\delta m, \qquad (7\cdot7, 7)$$
where x, y, z are the co-ordinates of the element of mass δm, and the summation covers all the particles which partake of *both* motions. The expression within the brackets under the summation sign has a simple physical interpretation which it is important to understand. Let the displacement of δm per unit increment of q_1 be represented by the vector ρ_1 and similarly let ρ_2 refer to q_2. Then
$$\frac{\partial x}{\partial q_1}\frac{\partial x}{\partial q_2} + \frac{\partial y}{\partial q_1}\frac{\partial y}{\partial q_2} + \frac{\partial z}{\partial q_1}\frac{\partial z}{\partial q_2} = \rho_1\cdot\rho_2, \qquad (7\cdot7, 8)$$

that is, the *scalar product* of the vectors ρ_1 and ρ_2 or the product of one of the vectors into the projection of the other vector upon it. We notice, in particular, that the contribution of δm to P_{12} vanishes when either ρ_1 or ρ_2 vanishes and whenever they are perpendicular. In the design of a mass balance system we usually have the problem of arranging that an added mass shall make a negative contribution to P_{12}. This can occur only when ρ_1 is oppositely directed to the projection upon it of ρ_2, a criterion which should always be kept in mind. Suppose, for example, that it was desired to provide elevators with a remote balance mass, suitably geared to the elevators. We shall take the co-ordinates q_1 and q_2 to be rotation of the elevator η and downward displacement of the tail z respectively. Then, in order that the contribution to P_{12} shall be negative, the mass must move up when the T.E. of the elevator moves down. But suppose now that q_2 is a displacement in a particular vertical oscillatory mode of the aircraft with a transverse node near the tail. Then the direction of the displacement of the mass when the elevator moves will, for a negative contribution to P_{12}, have to be reversed if it is shifted across the nodal line.

It is preferable that balance masses be distributed along the leading edge of a control flap having a set-back hinge so that each strip perpendicular to the hinge is independently mass-balanced rather than be concentrated on balance arms. The advantages are that the mass-balancing will then be effective irrespective of torsion of the flap and of variations in the distribution of the normal motion of the main surface.

Tabs for trimming or aerodynamic balance need not be mass-balanced, always provided that there is no appreciable back-lash. The mass-balancing of spring tabs is specially intricate and will not be discussed here. However, it appears probable that mass-balancing of such tabs can be dispensed with so far as flutter prevention is concerned provided that the ratio of tab chord to flap chord is small and that the tab is extremely light.

In the whole of the above discussion we have neglected all aerodynamic forces on the flap. However, aerodynamic forces may contribute effectively to products of inertia, and any such contribution is called a *virtual product of inertia*. In general, if the generalized force in the equation corresponding to generalized co-ordinate q_s contains a term $-\mathscr{P}_{rs}\ddot{q}_r$ of aerodynamic origin, then $-\mathscr{P}_{rs}$ is an acceleration derivative and $\mathscr{P}_{rs}$ is a virtual product of inertia when r and s differ or a virtual moment of inertia when r and s are the same. Virtual inertias are approximately independent of air speed although dependent to some extent on the Reynolds and Mach numbers and on the frequency parameter. For an ordinary flap control the presence of the virtual product of inertia requires that the total structural product of inertia shall be negative in order that the total effective product of inertia shall be zero. Thus, complete effective mass balance is attained when there is a slight degree of structural over mass balance.

Chapter 8

THE MEASUREMENT OF AERODYNAMIC DERIVATIVES

8·1 Introduction

We have introduced the idea of aerodynamic derivatives in § 2·5 and have considered them further in a general way in §§ 3·5 and 3·6. The particular derivatives of longitudinal symmetric motion have been discussed in § 5·3 and those of lateral-antisymmetric motion in § 6·3. In the present chapter we shall explain in outline how the important derivatives can be determined experimentally.

8·2 Basis of the measurements

By definition an aerodynamic derivative is the rate at which some component of the aerodynamic force or moment on a particular body varies with a quantity representing position, velocity or acceleration, linear or angular. Thus the determination of a derivative involves directly or indirectly the measurement of a given force or moment in two or more distinct states of the body concerned, as regards configuration, velocity or acceleration. There are two broad methods of making the measurements, the 'static' and the 'dynamic'. In the static method each measurement is made with the body in a configuration and with velocities relative to the air which are fixed during the measurement. The derivative is given by the limit of the ratio of change in force or moment to corresponding change of the kinematic variable and is of the type which we have called 'quasi-static'. The static method cannot be applied to all derivatives and, as pointed out in § 3·5, quasi-static derivatives are not, in general, applicable in unsteady motions. In the dynamic method we use the equation of motion in which the derivative occurs to determine its value and select the experimental conditions so as to obtain results of the highest accuracy in the most convenient way. For example, we determine L_p by experiments in which the sole degree of freedom of the body is in roll and use the dynamical equation of rolling moments (see § 8·4). The number of degrees of freedom in dynamic tests should be as small as possible and all 'direct' derivatives can be obtained from tests with only a single freedom. Cross derivatives can be measured in single freedom tests by various special devices, e.g. the use of strain gauges or the comparison of the results of tests made in succession with differing fixed axes of rotation.

In planning experiments to measure derivatives it must be remembered that they are dependent on the following factors:

(*a*) Reynolds number.

(*b*) Mach number.

(*c*) The state of the boundary layer and, in particular, the situation of the transition from laminar to turbulent flow.

(*d*) The non-dimensional frequency parameter.

(*e*) The attitude and configuration of the body, especially when the characteristic on which the derivative depends is non-linear. As examples we may take the angle of incidence and the angular setting of a control flap.

Further, it is to be remembered that the value of a derivative may be sensitive to *small* changes of shape of the body under test. Here we may mention the influence of surface roughness and waviness and the sensitiveness of hinge moments to small changes of gap, nose shape, etc. In model tests it is necessary to make sure that the model is so rigid that its distortions have a negligible effect while in full-scale tests important distortions must be measured.

8·3   Quasi-static derivatives

The values of many aerodynamic derivatives for the zero value of the frequency parameter can be obtained from ordinary wind tunnel tests in which the model is fixed during each measurement. We have seen in § 5·3 that the quasi-static values of the non-dimensional derivatives x_u, x_w, z_u, z_w, m_u and m_w can be obtained from the measured values of C_L, C_D and C_m and of their rates of change with incidence. These rates of change may be found by graphical differentiation or sometimes by the use of semi-empirical formulae. For example, if we assume that

$$C_D = C_{DZ} + kC_L^2, \qquad (8·3, 1)$$

where C_{DZ} and k are independent of α, it follows that

$$\frac{dC_D}{d\alpha} = 2kC_L\frac{dC_L}{d\alpha}, \qquad (8\cdot3, 2)$$

and $\dfrac{dC_D}{d\alpha}$ can be calculated when k and $\dfrac{dC_L}{d\alpha}$ have been determined.

The derivatives with respect to velocity of sideslip v can easily be obtained from tests in which the model is stationary. Let the model be yawed through the angle β (positive when the starboard wing tip moves forward). Then the wind velocity has a component $V\sin\beta$ perpendicular to the plane of symmetry of the model and to port. But, if we regard the air as stationary, this is equivalent to a velocity of sideslip in the positive sense OY given by

$$v = V\sin\beta. \qquad (8\cdot3, 3)$$

Thus by plotting forces and moments against $\sin\beta$ (or β if the angle is always small) we can determine the derivatives with respect to sideslip.

The aerodynamic derivatives connected with control surfaces are of great importance. Let ξ be the angular setting of the control, C_L the lift coefficient on the surface to which it is attached and C_H the hinge moment coefficient. In the absence of yaw these coefficients are functions of α and of ξ and in the usual notation

$$\left.\begin{aligned} a_1 &= \frac{\partial C_L}{\partial\alpha}, \\[2mm] a_2 &= \frac{\partial C_L}{\partial\xi}, \\[2mm] b_1 &= \frac{\partial C_H}{\partial\alpha}, \\[2mm] b_2 &= \frac{\partial C_H}{\partial\xi}. \end{aligned}\right\} \qquad (8\cdot3, 4)$$

Thus, a_1 and b_1 are found from a series of tests in which ξ is fixed and α varied, while a_2 and b_2 are obtained from tests in which α is fixed and ξ varied. It is not unusual for b_1 and b_2 to depend largely on ξ.

8·4   Measurement of rotary derivatives in wind tunnels

We shall begin by discussing the determination of L_p by the forced oscillation method. A diagram of the experimental arrangement is given in Fig. 8·4, 1.

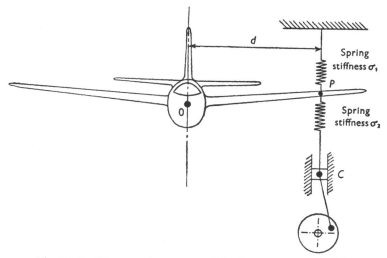

Fig. 8·4, 1. Diagram of arrangement for the measurement of L_p.

The model is mounted so that it can rotate freely about a fixed axis in the plane of symmetry, shown at O in the figure. A point P on one wing is connected to a fixed point through a helical spring of stiffness σ_1 and to a crosshead C through a helical spring of stiffness σ_2. The crosshead is made to perform a simple harmonic oscillation given by

$$z = h \sin \omega t, \qquad\qquad (8·4, 1)$$

where z is the downward displacement of C. In the motion of the model v and r are constantly zero and the dynamical equation (3·12, 9) becomes

$$A\ddot{\phi} - L_p\dot{\phi} = L(t), \qquad\qquad (8·4, 2)$$

where ϕ is the angle of roll measured from the position of equilibrium when the crosshead is at mid-stroke and $L(t)$ is the applied rolling moment. Suppose that P moves down the distance ϕd when the displacement of C is z. Then the stretch of the lower spring is $(z - \phi d)$ and the increase in tension is

14

$\sigma_2(z - \phi d)$ while the increase in tension in the upper spring is $\sigma_1 \phi d$. Hence the net downward force applied at P is

$$\sigma_2 z - (\sigma_1 + \sigma_2) \phi d,$$

and the applied moment is by (8·4, 1)

$$L(t) = dh\sigma_2 \sin \omega t - (\sigma_1 + \sigma_2) \phi d^2. \qquad (8\cdot4, 3)$$

It is now convenient to rewrite the dynamical equation (8·4, 2) as

$$a\ddot{\phi} + b\dot{\phi} + c\phi = k \sin \omega t \qquad (8\cdot4, 4)$$

where*
$$\left. \begin{aligned} a &= A, \\ b &= -L_p, \\ c &= d^2(\sigma_1 + \sigma_2), \\ k &= dh\sigma_2. \end{aligned} \right\} \qquad (8\cdot4, 5)$$

Suppose that the motor driving the crosshead has been run steadily until the model has settled down to an oscillation given by

$$\phi = \phi_1 \sin \omega t + \phi_2 \cos \omega t. \qquad (8\cdot4, 6)$$

Substitute in (8·4, 4) and equate separately the coefficients of $\sin \omega t$ and of $\cos \omega t$ on the two sides of the equation

$$(c - a\omega^2)\phi_1 - b\omega\phi_2 = k,$$

$$b\omega\phi_1 + (c - a\omega^2)\phi_2 = 0.$$

These equations yield

$$\phi_1 = \frac{k(c - a\omega^2)}{(c - a\omega^2)^2 + b^2\omega^2}, \qquad (8\cdot4, 7)$$

$$\phi_2 = \frac{-kb\omega}{(c - a\omega^2)^2 + b^2\omega^2}. \qquad (8\cdot4, 8)$$

We may write (8·4, 6) alternatively as

$$\phi = \phi_0 \sin(\omega t - \epsilon) \qquad (8\cdot4, 9)$$

where ϕ_0 is the resultant amplitude of the forced oscillation and ϵ is the angle by which the displacement in roll lags behind

---

* If the c.g. of the model does not lie on the axis of rotation, c will also contain a gravitational term.

the applied couple. Evidently

$$\phi_1 = \phi_0 \cos \epsilon$$

and

$$\phi_2 = -\phi_0 \sin \epsilon.$$

Square these, add and use (8·4, 7) and (8·4, 8). We derive

$$\phi_0^2 = \frac{k^2}{(c - a\omega^2)^2 + b^2\omega^2} \qquad (8\cdot4, 10)$$

and

$$\tan \epsilon = \frac{b\omega}{c - a\omega^2}. \qquad (8\cdot4, 11)$$

When we know the value of k and measure ϕ_0, ϵ and ω we can determine two of the quantities a, b, c from the last equations if we know the value of one of them, say c, which can be found from the spring stiffnesses or by a static calibration. There are some particularly convenient ways of conducting the experiment which we shall next explain.

Suppose that we keep k constant and vary ω until ϕ_0 is a maximum; this is the condition of *resonance*. Then we have by (8·4, 10)

$$\frac{d}{d\omega}\left(\frac{k}{\phi_0}\right)^2 = 4a^2\omega\left(\omega^2 - \frac{c}{a} + \frac{b^2}{2a^2}\right) = 0$$

or

$$\omega_r^2 = \frac{c}{a} - \frac{b^2}{2a^2}, \qquad (8\cdot4, 12)$$

and the *resonance frequency* is

$$f_r = \frac{\omega_r}{2\pi}. \qquad (8\cdot4, 13)$$

Let ϕ_r be the value of ϕ_0 at resonance. Then, by (8·4, 10) and (8·4, 12)

$$\left(\frac{k}{\phi_r}\right)^2 = b^2\left(\frac{c}{a} - \frac{b^2}{4a^2}\right). \qquad (8\cdot4, 14)$$

We shall see shortly that the value of ω for the *free* oscillations of the system with the cross-head fixed is given by

$$\omega_n^2 = \frac{c}{a} - \frac{b^2}{4a^2}, \qquad (8\cdot4, 15)$$

and if b were zero this would become

$$\omega_{n0}^2 = \frac{c}{a}. \qquad (8\cdot4, 16)$$

It appears from (8·4, 14) and (8·4, 15) that

$$-L_p = b = \frac{k}{\omega_n \phi_r}, \qquad (8·4, 17)$$

and the derivative can be found from this equation. In practice there is a correction to allow for the mechanical damping in the system. When the system is lightly damped, i.e. when b^2/ac is small, the values of ω given by (8·4, 12), (8·4, 15) and (8·4, 16) differ only slightly and need not be distinguished except when the aim is to attain high accuracy. It will be noted that in this method we do not need to measure the phase difference ϵ but only the maximum amplitude ϕ_r and the corresponding value of ω.

A more accurate variation of the last method can be used when there is some device for ascertaining when the applied couple and the response are in quadrature ($\epsilon = 90°$). It will be seen from (8·4, 11) and (8·4, 16) that this occurs when

$$\omega = \omega_{n0}, \qquad (8·4, 18)$$

and then from (8·4, 10)

$$-L_p = b = \frac{k}{\omega_{n0} \phi_0}. \qquad (8·4, 19)$$

The derivative can also be found from the logarithmic decrement of the free oscillations of the model* which occur after it has been pulled from the position of equilibrium and with the crosshead C fixed. The dynamical equation is (8·4, 4) with k made zero and to solve it we put ϕ proportional to $\exp(\lambda t)$. The equation is satisfied provided that λ is a root of

$$a\lambda^2 + b\lambda + c = 0.$$

For the experimental system we shall have in practice

$$b^2 < 4ac,$$

so

$$\lambda = -\mu \pm i\omega_n, \qquad (8·4, 20)$$

where

$$\mu = \frac{b}{2a}, \qquad (8·4, 21)$$

* Strictly the value of the derivative for a damped oscillation differs from that for an S.H.M. of the same frequency but in practice the difference is inappreciable.

and ω_n is given by (8·4, 15). The most general real form of the solution corresponding to the values of λ given by (8·4, 20) is

$$\phi = \phi_0 e^{-\mu t} \sin(\omega_n t + \eta), \qquad (8·4, 22)$$

where η is a phase angle. Hence the free motion is a damped oscillation of periodic time

$$T = \frac{2\pi}{\omega_n}. \qquad (8·4, 23)$$

Now successive maxima on the same side of the position of equilibrium will occur with a constant time interval T and the ratio of the adjacent maxima will be, by (8·4, 22),

$$R = e^{-\mu T}$$

$$\therefore \quad \mu = \frac{1}{T} \log_e\left(\frac{1}{R}\right)$$

or $\qquad -L_p = b = \frac{2a}{T} \log_e\left(\frac{1}{R}\right) = \frac{4 \cdot 605 a}{T} \log_{10}\left(\frac{1}{R}\right). \qquad (8·4, 24)$

It follows from the foregoing that if we plot the logarithms of the maxima against time or number of swings we should obtain a straight line from whose slope the derivative can be found. If the negative slope increases with time the presence of a 'solid' frictional resistance is indicated. To eliminate this we must find by trial a constant f such that, when this is added to all the observed maxima, the sums yield a straight logarithmic plot from whose slope the derivative can be found as before. It can be shown that the numerically constant frictional resisting moment F is given by

$$F = fc \tanh \frac{\mu\pi}{2\omega_n}. \qquad (8·4, 25)$$

The value of the moment of inertia a which appears in (8·4, 24) can be derived from that of the stiffness c by the relation

$$\frac{c}{a} = \mu^2 + \omega_n^2. \qquad (8·4, 26)$$

The derivative L_p can be measured for a motion of continuous rotation when a rolling balance is available. This piece of apparatus enables us to keep the model rotating steadily in roll about a fixed axis and at the same time to measure the torque required to maintain the rotation, namely $-pL_p$. The influence of angle of incidence on the derivative is readily found by this means and the phenomena of *auto-rotation* can be exhibited. It is found that for incidences near and above the stall L_p vanishes for a certain value of p and may even become positive. Usually it is necessary to give the model an initial spin to set the auto-rotation going (see also § 11·5).

An arrangement similar to that described above can be used to determine M_q but the measured quantity is a linear combination of M_q and $M_{\dot{w}}$. Let the model be arranged to oscillate in pitch about a spanwise axis at right angles to the airstream whose velocity is V. Take wind axes in the model and consider the state of affairs when the model has pitched through the angle θ. The velocity deviations are evidently

$$u = V \cos \theta - V = 0$$

to the first order of small quantities, while to the same order

$$w = V \sin \theta = V\theta.$$

Hence the equation of motion (3·11, 13) becomes

$$- Vq M_{\dot{w}} - V\theta M_w + \dot{q}B - q M_q = M(t),$$

which can be written

$$B\ddot{\theta} + \dot{\theta}(-M_q - VM_{\dot{w}}) - \theta V M_w = M(t). \qquad (8\cdot4, 27)$$

Thus an oscillation test can be made to yield a value for $(M_q + VM_{\dot{w}})$ and to obtain M_q it is necessary to estimate $M_{\dot{w}}$. The derivatives cannot be separated by changing V since M_q is (approximately) proportional to V while $M_{\dot{w}}$ is independent of it.

The combination $(N_r - VN_{\dot{v}})$ can be obtained from experiments in which the model oscillates about the normal axis OZ. Since $N_{\dot{v}}$ is negligible we have a convenient method for the determination of N_r.

Considerably more complicated experimental arrangements are needed for the measurement of cross derivatives such as N_p. We shall begin by describing one possible scheme for determining this derivative:

Restrain the model in yaw and force an oscillation in roll. Measure the yawing moment $N(t)$ with a strain gauge. By equation (3·12, 10) we have

$$N(t) = -\dot{p}E - pN_p, \qquad (8\cdot4, 28)$$

since r and v are always zero. The product of inertia E must be measured or reduced to zero by mass balancing the model and N_p can then be obtained from the last equation. In a similar manner we can measure L_r by forcing an oscillation in yaw and measuring the rolling moment $L(t)$ with a strain gauge.

The sum of L_r and N_p can be obtained from experiments in which the model oscillates about an inclined axis through O lying in the plane of symmetry. Let the axis make the angle γ with the forward axis OX and $\frac{\pi}{2} - \gamma$ with the downward axis OZ and let the instantaneous angular velocity be ω. Then we have

$$p = \omega \cos \gamma, \\ r = \omega \sin \gamma, \Big\} \tag{8.4, 29}$$

and the aerodynamic damping moment about OX is

$$- pL_p - rL_r = - \omega(L_p \cos \gamma + L_r \sin \gamma).$$

The component of this moment about the axis of rotation is

$$- \omega(L_p \cos^2 \gamma + L_r \cos \gamma \sin \gamma).$$

Similarly the aerodynamic damping moment about OZ contributes a moment about the axis of rotation equal to

$$- \omega(N_p \cos \gamma \sin \gamma + N_r \sin^2 \gamma),$$

and the total damping coefficient is

$$- L_p \cos^2 \gamma - (N_p + L_r) \cos \gamma \sin \gamma - N_r \sin^2 \gamma. \tag{8.4, 30}$$

If this be measured by one of the methods described in relation to L_p and if L_p and N_r are known, the quantity $(N_p + L_r)$ can be found. The optimum value of γ is $45°$. Alternatively, $(N_p + L_r)$ and $(L_p + N_r)$ can be obtained from a pair of tests with $\gamma = \pm 45°$.

8·5 Measurement of derivatives on whirling arms

A whirling arm consists of a single or double cantilever which can rotate at a controlled speed about a vertical axis. The arm is preferably mounted concentrically in a closed room having circular cylindrical walls which may be provided with evenly spaced baffle plates, to reduce the swirl. The model under test is attached to the arm at or near its outer end. Hence the c.g. of the model has a horizontal linear velocity and the model shares the angular velocity of the arm about a vertical axis.

Suppose that the model is set with its plane of symmetry vertical and at right angles to the perpendicular from the datum point* on to the axis of rotation. Then the angle of incidence can be set as desired and the motion consists of a forward velocity and an angular velocity in yaw. When the plane of symmetry is horizontal the motion will consist of a forward velocity and an angular velocity in pitch. By inclining the model suitably a velocity of sideslip and angular velocities in roll, pitch and yaw can be obtained. It is, however, not possible to combine a forward velocity with a purely rolling angular velocity.

The angular velocity of a whirling arm can, on account of its large inertia, be varied only slowly. This apparatus is therefore only suited to measurements made under steady conditions, i.e. with constant linear and angular velocities, and there is the restriction that the path in space is circular. One of the greatest troubles is the swirl set up by the arm itself and by the model. Since this varies with the setting of the model it is necessary to mount Pitot-static tubes on the arm to determine the effective air speed in the region of the model. It is also found that a whirling arm acts like a centrifugal pump so that, in the absence of correcting devices, the air will have an outward component of velocity at the model; this can be corrected by mounting suitable vertical aerofoils on the arm. The measurement of forces on the model by balances or strain gauges is greatly complicated by centrifugal effects while the accurate measurement of pressures is also difficult and troublesome.

8·6 Measurement of stiffness derivatives in oscillatory motion

Hitherto we have only considered the measurement of the quasi-static values of 'stiffness derivatives' such as N_v, but it is desirable to determine their dependence on the non-dimensional frequency parameter. For definiteness we shall now consider N_v in detail.

---

* Corresponding to the c.g. of the full-scale aircraft.

Let the model be mounted in the wind tunnel so that it can rotate about the axis OZ and let ψ be the angle of rotation. Then we have

$$\left.\begin{array}{l} r = \psi \\ v = -V\psi, \end{array}\right\} \qquad (8\cdot6, 1)$$

and

when ψ is small. Accordingly the equation of motion $(3\cdot12, 10)$ becomes*

$$C\ddot{\psi} - N_r\dot{\psi} + VN_v\psi = N(t). \qquad (8\cdot6, 2)$$

If we measure both the natural frequency and the rate of damping in a free oscillation $[N(t) = 0]$ we can determine the unknown derivatives N_r and N_v. In order to control the frequency we may add a spring constraint of stiffness σ (moment per radian). The equation of free motion then becomes

$$C\ddot{\psi} - N_r\dot{\psi} + (\sigma + VN_v)\psi = 0. \qquad (8\cdot6, 3)$$

Let the motion in a free oscillation be

$$\psi = \psi_0 e^{-\mu t}\sin(\omega_n t + \eta), \qquad (8\cdot6, 4)$$

and let μ and ω_n be obtained from records of the motion; further, let the moment of inertia C be determined by free oscillation tests in still air. †Then we have

$$-N_r = 2C\mu, \qquad (8\cdot6, 5)$$

$$VN_v = C(\omega_n^2 + \mu^2) - \sigma. \qquad (8\cdot6, 6)$$

Alternatively we may apply a periodic yawing moment, measure it by some suitable strain gauge and obtain simultaneous records of ψ and $N(t)$. The two derivatives can be obtained from an anlysis of the records and the mathematics is essentially the same as that given in § 8·4 in relation to L_p; the stiffness coefficient c, however, now contains the derivative N_v. This technique can be extended to the measurement of the cross derivatives L_v, L_r but the strain gauge is arranged to measure the rolling moment during a forced purely yawing oscillation.

* $N_{\dot{\psi}}$ is assumed zero.
† C contains a virtual moment of inertia and we assume here that the dependence of this on the frequency is negligibly small.

8·7   Measurement of acceleration derivatives or virtual inertias in still air

Measurements of the virtual inertias of a complete model are required in spinning tests and the virtual inertias of control surfaces are of importance in relation to mass-balancing. We shall now describe in outline how the values of such virtual inertias for 'still air' can be determined.

The simplest method in principle consists in measuring the sum of the corresponding structural and virtual inertias by a suitable experiment conducted in still air and then to repeat the experiment in a vacuum. The virtual inertia is obtained by difference since it is reduced to zero in vacuo. Alternatively, the second experiment may be made in a light gas, such as hydrogen or coal gas, whose density is known. Let I_a, I_g be the measured inertias in air and gas respectively, I_s the inertia of the structure and I_v the virtual inertia in air. Then we may assume that the virtual inertia in the gas is σI_v where σ is the density of the gas relative to that of air. Consequently

$$I_a = I_s + I_v,$$

$$I_g = I_s + \sigma I_v$$

and
$$I_v = \frac{I_a - I_g}{1 - \sigma}. \tag{8·7, 1}$$

Another technique, which is sometimes more convenient, is based on the comparison of tests in which the body is open to the atmosphere and enclosed in a closely fitting rigid envelope. Let the corresponding inertias of the envelope alone and of the envelope containing the body be measured. The difference gives the structural inertia of the body alone and then its virtual inertia can be obtained as the difference between the measured inertia in free air and the structural inertia. A variation of this method is sometimes applied to flat bodies, such as control surfaces. A uniform rigid sheet is cut to the same plan form as the body and its inertia measured in free still air. The structural inertia is calculated from the known form and uniform mass per unit area of the rigid sheet and the virtual inertia obtained by difference. It is then assumed that the virtual inertia of the sheet is the same as that of the flat body, but this assumption is open to question since the virtual inertia may depend on the rounding of the edges of the body and other geometrical minutiae. It is also to be noted that any selected virtual inertia of a control surface will not, in general, be the same when the control surface is isolated as when it is in situ on a main surface.

8·8   Measurement of derivatives in flight

In general the measurement of derivatives in flight is difficult and no great effort has been devoted to this subject. We shall content ourselves here with discussing a few relatively simple experiments.

Let us consider the measurement of the quasi-static derivatives which depend on v. We shall suppose that the aircraft is provided with instruments which are of such construction and so placed that they provide an accurate measure of a steady velocity of sideslip v and that there are means for applying known rolling and yawing moments. A known rolling moment can be produced by shifting a known mass laterally, as by moving fuel from one tank to another;* in a dive or climb this will also give a yawing moment. A yawing moment can be obtained by streaming a parachute or drogue from a wing tip and the tension in the towing cord must be measured in order to determine the moment. Since the towing cord will not, in general, lie parallel to OX, the yawing moment will, in general, be accompanied by a small rolling moment and lateral force.

Let us suppose that the aircraft is in steady horizontal flight with a steady velocity of sideslip and all angular velocities zero. Then the equations of motion $(3·12, 8)…(3·12, 10)$ can be written

$$-vY_v - mg\phi = Y + Y', \tag{8·8, 1}$$

$$-vL_v = L + L', \tag{8·8, 2}$$

$$-vN_v = N + N', \tag{8·8, 3}$$

where the constants Y, L, N represent the applied loads and the constants Y', L', N' represent the action of the controls.

---

* So long as we are concerned only with unaccelerated motions the resulting shift of the c.g. is of no consequence.

Suppose first that a purely rolling moment L is applied and that the ailerons are kept neutral. Then we shall have

$$Y = N = L' = 0,$$

and equation (8·8, 2) gives

$$-vL_v = L,$$

from which L_v can be found. Equation (8·8, 3) requires that a yawing moment N' be applied by the rudder to balance the yawing moment caused by the sideslip. We assume that the displacement of the rudder gives no rolling moment L', but it will give a lateral force Y'. If this can be measured or estimated and the angle of bank ϕ is known, equation (8·8, 1) enables us to find Y_v. Next, let a yawing moment N be applied by means of a streamed parachute while the rudder is kept neutral ($N' = 0$). Then equation (8·8, 3) gives N_v. Satisfaction of (8·8, 2) will require use of the ailerons or of a suitable equilibrating moment L obtained by shift of masses. It is preferable not to move the ailerons since this would, in general, introduce a yawing moment N' of unknown magnitude.

The values of derivatives in 'dynamic' conditions can be obtained by the analysis of records of the motion of the aircraft (in smooth air) which follow a sudden or periodic movement of a control but the discussion of this is beyond the scope of this book.*

---

* On this topic, see W. Milliken 'Dynamic Stability and Control Research' in *Proceedings, Third Anglo-American Aeronautical Conference, Brighton*, 1951.

Chapter 9

CONTROLS FOR ROLL, PITCH AND YAW

9·1 Introduction

In this chapter we consider in some detail the controls which give rise to rolling, pitching or yawing moments on aircraft, together with some aspects of aircraft controls in general. We have already pointed out in Chapter 1 that the attitude of the aircraft is altered at will by applying suitable moments which are of aerodynamic origin and caused by changes in the configuration of the surface exposed to the airstream. The usual and most convenient method of altering the surface configuration and so applying moments is to deflect a hinged flap and the characteristics of such flaps have been discussed in Chapter 7. The main problems of control design concern:

(*a*) The adequacy of the moment resulting from operation of the control.

(*b*) The avoidance of unduly large control forces which would be troublesome to the pilot.

(*c*) The avoidance of unduly light or sensitive controls.

Items (*b*) and (*c*) are intimately connected with the aerodynamic balance of the control flaps, a topic discussed in § 7·5, and with the mechanical gearing between the flap and the pilot's control lever or wheel. This question of gearing is treated in § 9·2 below and a table of permissible control forces is given in § 1·2.

9·2 The mechanical gearing of controls

For the present purpose we shall define the gear ratio G of a control flap to be the number of degrees of angular movement of the flap about its hinge per inch of linear movement of the pilot's hand or foot on the control and in the case of a 'stick' control shall reckon this movement at the mid-point of the hand-grip. Since, in the absence of friction, the work done by the pilot is equal to the work done on the flap, we have

$$\frac{F}{12} = \frac{\pi}{180}\left(\frac{GH}{e}\right)$$

or
$$F = \frac{\pi}{15}\left(\frac{GH}{e}\right) = 0 \cdot 209\left(\frac{GH}{e}\right), \qquad (9 \cdot 2, 1)$$

where F = pilot's force in lb., H = flap hinge moment in ft. lb., and e is the mechanical efficiency, as a fraction. The gear ratio need not be constant over the angular range of the flap. Since H is large when the flap angle is numerically large it will assist the pilot to arrange the mechanism so that G is a maximum for the neutral setting of the flap and relatively small at both extreme positions. The gear ratio is, however, subject to the restriction that the total travels of both the flap and of the pilot's control must be regarded as fixed. Thus the extreme positions of the flap are fixed by the required positive and negative moments on the aircraft while the travel of the pilot's control is regulated by his convenience. Corresponding movements are related by the equation

$$\eta^\circ = \int_0^x G(x)\,dx, \qquad (9 \cdot 2, 2)$$

where x is the displacement of the pilot's control from the neutral position, in inches.

When a geared tab is used to obtain some degree of aero-dynamic balance (see § 7·5) it may be of advantage to adopt a variable gear ratio for the tab, so arranged that the tab is most effective at the extremes of displacement of the flap. This may be combined with variable gearing between the pilot's control and the flap.

It should be noted that a low value of G, giving the pilot a 'mechanical advantage', necessarily reduces the rate at which the flap can be moved when the load is small. A relatively slow rate of movement, which would be unacceptable for a small aircraft, may be satisfactory for a large one where the loads are large.

9·3   Trimmers

A trimmer is a device which enables the force which the pilot must apply to his control to be adjusted, usually to zero, when the control flap is held fixed at any setting. The need for trimmers arises from the necessity to relieve the pilot of loads which would be fatiguing if long sustained. It is sometimes advantageous to have a 'fixed' trimmer for the purpose of adjusting the neutral setting of the control, but there must always be an adjustable trimmer under the pilot's control.

Trimmers are of two main types: (a) aerodynamic, and (b) mechanical. Most trimmers are aerodynamic and usually consist of a trimming tab mounted on the control flap at the trailing edge and so geared that the angle between the tab and the main surface is independent of the setting of the latter. It is, however, possible to arrange that one tab shall serve as a trimmer and as a balance or anti-balance tab. A mechanical trimmer consists of a spring or weight connected to the control or control operating system and so arranged that the applied moment can be varied at will. Such devices will, in general, influence the stick-free stability and perhaps the manœuvrability (see § 10·8). Adjustable trimmers of whatever type must provide smooth and fine control of the control force and they must be quite free from 'creep'.

Some degree of automatic trimming can be obtained by using a spring-mounted tab whose deflection depends on the equivalent air speed. Special attention must be given to such a tab to ensure that it will not flutter.

A simple and much used kind of fixed trimmer consists of a strip of sheet metal attached along the trailing edge of the flap. The neutral setting of the flap is adjusted by twisting the strip which is sufficiently stiff not to deflect appreciably under the aerodynamic load. A cord or ridge near the trailing edge and of suitable length will serve the same purpose (see the end of § 7·2).

A special application of the fixed trimmer to rudders serves to minimize the change of trim which occurs with change of throttle setting. The trimming tab or strip is so placed and deflected that the total moment caused by a change in the velocity of the slipstream is made small. It is usually necessary to 'joggle' the trimmer to get satisfactory results.

9·4 The spring tab

The spring tab is a special adaptation of the servo-tab or 'Flettner' designed to reduce the dependence of the stick force on the air speed. This is attained by so designing the mechanism that the tab is inoperative when the control hinge moment is small but fully effective when it is large.

Fig. 9·4, 1 shows a line diagram of the spring tab mechanism; the diagram is intended to make clear the mechanical relations and is not intended to show good practical proportions. In the figure A is the hinge of the main control surface and B that of the tab, while C is the trailing edge. AD is a lever pivoted at A to which the control operating rod is connected at D. The lever

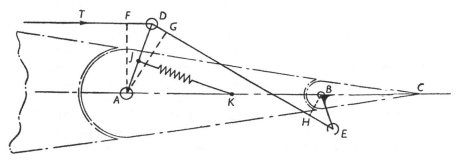

Fig. 9·4, 1. Schematic diagram of the spring tab.

BE is rigidly fixed to the tab BC and DE is a rigid rod pivoted freely at D and E. The link AD is connected to the main control surface through the spring unit JK; this may be arranged to remain rigid until the load in it reaches a predetermined value, after which the deflection is proportional to the excess of the load above this 'preload'.

In the diagram AF is perpendicular to the axis of the operating rod while AG and BH are perpendicular to DE, also $AF = r$, $AG = s$, $BH = t$. The aerodynamic hinge moment about A for the complete flap is H, the aerodynamic moment about B for the tab is H_t and the clockwise moment about A exerted by KJ on AD is M. Then the condition of equilibrium of the control as a whole is*

$$Tr + H = 0, \qquad (9·4, 1)$$

where T is the thrust in the control rod. Let T' be the thrust in the link DE. For equilibrium of the tab we have

$$T't = H_t, \qquad (9·4, 2)$$

* Moments due to gravity and acceleration are neglected.

and for equilibrium of AD

$$T's = Tr + M.$$ (9·4, 3)

By elimination of T' from (9·4, 2) and (9·4, 3) we derive

$$T = \frac{sH_t}{rt} - \frac{M}{r},$$ (9·4, 4)

and then (9·4, 1) yields

$$H = M - NH_t,$$ (9·4, 5)

where N is the 'follow-up ratio' given by

$$N = \frac{s}{t}.$$ (9·4, 6)

These equations exhaust the deductions from pure statics. It is now necessary to examine some kinematic relations.

The four links AD, DE, EB and BA constitute a four-bar linkage, so the angle DAB is fixed when the angle ABE is fixed. However, let the tab be deflected downwards through a *small* angle β, so ABE is reduced by β. It follows from elementary kinematics that the angle DAB is increased by

$$\beta\left(\frac{BH}{AG}\right) = \frac{\beta}{N}.$$

Hence the increase in the length of the spring unit JK is proportional to β and, provided β is not zero, the moment applied by it to AD is

$$M = M_0 + k\beta,$$ (9·4, 7)

where M_0 is the moment corresponding to the pre-load and k is a coefficient proportional to the stiffness of the spring. It is evident from the foregoing that a spring connection between the tab lever BE and the main control AB could be substituted for the spring unit JK. The pre-loading of the spring unit must be so arranged that for either extension or compression the initial movement occurs at a predetermined numerical value of the force transmitted.

The aerodynamic hinge moments are

$$H = qS_f\bar{c}_f(b_0 + b_1\alpha + b_2\xi + b_3\beta)$$
$$= q(\mathscr{B}_0 + \mathscr{B}_1\alpha + \mathscr{B}_2\xi + \mathscr{B}_3\beta)$$ (9·4, 8)

15

and
$$H_t = qS_t\bar{c}_t(c_0 + c_1\alpha + c_2\xi + c_3\beta)$$
$$= q(\mathscr{C}_0 + \mathscr{C}_1\alpha + \mathscr{C}_2\xi + \mathscr{C}_3\beta), \qquad (9\cdot4, 9)$$

where α is the incidence of the main surface, ξ is the flap deflection and q is the dynamic pressure. On substitution from $(9\cdot4, 7)$, $(9\cdot4, 8)$ and $(9\cdot4, 9)$ in $(9\cdot4, 5)$ and rearrangement we derive

$$\beta[k - q(\mathscr{B}_3 + N\mathscr{C}_3)]$$
$$= q\xi(\mathscr{B}_2 + N\mathscr{C}_2) + q[\mathscr{B}_0 + \mathscr{B}_1\alpha + N(\mathscr{C}_0 + \mathscr{C}_1\alpha)] - M_0 \quad (9\cdot4, 10)$$

but this relation only holds when the tab is operative, i.e. when the load in JK numerically equals or exceeds the pre-load. The last equation gives β as a function of ξ for any given value of the dynamic pressure q. Then H can be found from $(9\cdot4, 8)$ and finally T is given by $(9\cdot4, 1)$. Since $(\mathscr{B}_2 + N\mathscr{C}_2)$ will in practice be negative, an increase in ξ gives rise to an increase in $-\beta$, as required for the reduction of control force. The pre-load moment M_0 has the same sign as β.

The tab begins to come into operation when the main control angle is ξ_t. By putting β equal to zero in $(9\cdot4, 10)$ we obtain

$$\xi_t(\mathscr{B}_2 + N\mathscr{C}_2) = \frac{M_0}{q} - [\mathscr{B}_0 + \mathscr{B}_1\alpha + N(\mathscr{C}_0 + \mathscr{C}_1\alpha)]. \qquad (9\cdot4, 11)$$

Thus the larger the value of q the smaller numerically is the angle ξ_t and this is just the characteristic required to lighten the control at high speeds. However, even when pre-load is absent, equation $(9\cdot4, 10)$ shows that the ratio $-\beta/\xi$ increases with q, for the term proportional to q in the coefficient of β is relatively unimportant. Hence the spring tab, with or without pre-load, is effective in reducing the control force at high speeds of flight in comparison with that for a plain control whereas at low speeds the difference is zero (with pre-load) or small. Another valuable feature is that an increase of the degree of aerodynamic under balance of the control (increase of $-\mathscr{B}_2$) leads to an increase of $-\beta/\xi$ and this largely eliminates an increase in control force.

It is worthy of note that when the control rod is rigidly fixed the tab acts as an anti-balance tab since β/ξ is then positive.

9·5   Conventional ailerons

Usually rolling control of an aircraft is provided by mounting a hinged flap on each wing, in the outer part of the span, and so gearing these together that one moves up when the other moves down. These flaps are called *ailerons* and they differ from most other twin flaps by rotating in opposition instead of in-phase when in use. Thus, when the pilot moves the stick or wheel laterally the aileron on one wing moves up and the lift on that wing is reduced while the aileron on the other wing moves down so the lift on it is increased. The net result is to produce a rolling moment with little or no net change of lift; usually there is also a yawing moment but this is an undesired secondary effect.

The port and starboard ailerons are usually geared together directly so that, even if the pilot's control were disconnected, they would be constrained to move together in opposition. The result of the gearing is that the effective value of b_2 (see § 7·1) for the pair of ailerons is a mean of the values for the separate ailerons depending, however, on the relative gear ratios. At first suppose that the ailerons are connected together and to the control column by an inelastic mechanism. Let the upward movement of the starboard aileron and the downward movement of the port aileron, each per inch of movement of the pilot's control to the right, be G_1 and G_2 degrees respectively. Also let the corresponding hinge moments (both positive in the trailing edge down sense) be H_1 and H_2 respectively. Then the equation of work yields [cp. equation (9·2, 1)]

$$F = \frac{0·209}{e}(H_1 G_1 - H_2 G_2),\qquad (9·5, 1)$$

for a positive hinge moment on the port aileron assists the pilot to move the stick to the right. The linking of the ailerons and stick imposes a relation between the displacements which will be written

$$x = f_1(\xi_1) = f_2(\xi_2),\qquad (9·5, 2)$$

where x is measured in inches and ξ_1, ξ_2 in degrees. From a knowledge of this relation and of the dependence of the hinge moments on the respective deflections, F can be obtained for all settings of the aileron system. Clearly

$$G_1 = -\frac{d\xi_1}{dx} = -1/f_1'(\xi_1) \qquad (9\cdot5,\,3)$$

and

$$G_2 = \frac{d\xi_2}{dx} = 1/f_2'(\xi_2). \qquad (9\cdot5,\,4)$$

It is to be remarked that when x is positive ξ_1 is negative and ξ_2 positive. Also, when the gearing is symmetric, a reversal of sign of x has the effect of interchanging the angles ξ_1 and ξ_2. As a simple example take the case of a constant gear ratio and let ξ_1 and ξ_2 both be equal to ξ_0 when x is zero. Then we have

$$x = k(\xi_0 - \xi_1) = k(\xi_2 - \xi_0), \qquad (9\cdot5,\,5)$$

and equations $(9\cdot5, 3)$ and $(9\cdot5, 4)$ yield

$$G_1 = G_2 = \frac{1}{k}. \qquad (9\cdot5,\,6)$$

Hence $(9\cdot5, 1)$ becomes

$$F = \frac{0\cdot209}{ke}(H_1 - H_2), \qquad (9\cdot5,\,7)$$

while H_1 and H_2 refer to the angles ξ_1 and $(2\xi_0 - \xi_1)$ respectively.

We shall assume that the ailerons are of equal dimensions and have identical characteristics. Accordingly when the aerodynamic relations are linear we shall have for either aileron

$$H = q(\mathcal{B}_0 + K\mathcal{B}_2\xi) \qquad (9\cdot5,\,8)$$

where K is the response factor (see § 7·4) and the symbols have the same meanings as in $(9\cdot4, 8)$.* Equation $(9\cdot5, 7)$ now yields

$$F = \frac{0\cdot209\,qK\mathcal{B}_2(\xi_1 - \xi_2)}{ke}$$

$$= -\frac{0\cdot418\,qK\mathcal{B}_2 x}{k^2 e} \qquad (9\cdot5,\,9)$$

on account of $(9\cdot5, 5)$. Since $K\mathcal{B}_2$ is negative for an aileron which is not aerodynamically overbalanced we see that F is positive when x is positive, i.e. the pilot has to push the stick

---

* The coefficient $\mathcal{B}_2$ must now correspond to ξ measured in degrees. Further, $\mathcal{B}_0$ allows for that part of the wing incidence which is independent of ξ.

in the direction he wishes to move it. It is worthy of note that F varies inversely as the *square* of the gearing coefficient k.

When the gear ratio is constant and the hinge moment characteristic is linear the stick force is independent of the aileron droop ξ_0 as we have just seen. When, however, the hinge moment characteristic is non-linear this ceases to be true and with exaggerated non-linearities such as occur with Frise ailerons the stick force may be sensitive to the droop. Since, moreover, the droop may vary in flight on account of the lack of rigidity of the aileron system it is possible for the heaviness of the control to be largely affected. There are even cases on record where the ailerons have become over-balanced in flight on account of excessive up-float.

Some degree of aerodynamic balance can be secured by gearing the ailerons differentially, i.e. so that the gear ratios for the up-going and down-going ailerons are different. Suppose that a hinge moment H_0 is added (in the same sense) to both H_1 and H_2. Then we see from (9·5, 1) that the corresponding increment of F will be

$$\Delta F = \frac{0 \cdot 209 H_0}{e} (G_1 - G_2). \qquad (9 \cdot 5, 10)$$

Now suppose that the stick has been moved to the right so that, on the assumption that the ailerons are not over-balanced, the force F exerted by the pilot is positive (to the right). This force will be reduced, i.e. ΔF will be negative, when

(a) H_0 is positive and $G_2 > G_1$

or   (b) H_0 is negative and $G_1 > G_2$.

More generally, we can say that a positive hinge moment applied to both ailerons assists the pilot when the down-going aileron is more highly geared than the up-going aileron; conversely, a negative hinge moment applied to both ailerons assists the pilot when the up-going aileron is more highly geared than the down-going aileron. A hinge moment in either sense can be provided by a fixed tab suitably deflected. When the ailerons are 'convergent', i.e. when their b_1 is negative, the part of the hinge moment coefficient which depends on wing incidence will increase algebraically as the speed of flight increases and the incidence falls. Hence the balancing effect will be at a maximum at high speeds, when it is most needed, if the down-going aileron is more highly geared than the up-going one. The fixed tabs should then be deflected upwards to give a positive hinge moment. However, for the sake of reducing the tendency to tip stalling the up-going aileron should be more highly geared. On the whole it does not appear to be very profitable to seek aileron balance by differential gearing.

A secondary effect of deflecting the ailerons is the creation of a yawing moment. Thus the lift is increased on the wing where the aileron moves down and this is associated with an increase of induced drag as well as of profile drag; on the wing where the aileron moves up the induced drag is reduced and there may be little or no increase of profile drag. The net effect is a yawing moment tending to retard the up-going wing tip relative to the other wing tip. Now when the aircraft makes a banked turn the up-going wing tip is advanced by the rotation in yaw and the yawing moment brought into play by the ailerons opposes this rotation. Hence it is usual to speak of *the adverse yawing moment* of the ailerons. However, Frise ailerons are so arranged that the nose of the up-going aileron protrudes into the air stream below the wing whereas the nose does not protrude at all when the aileron is deflected downwards. The nose gives an increase of drag when it projects and so the adverse yawing moment is neutralized, wholly or in part. The adverse yawing moment can also be reduced by gearing the ailerons differentially so that the travel of the up-going aileron is always greater than that of the down-going aileron, for then the profile drag is increased on the wing where the lift is reduced.

The aspect ratio of ailerons varies much in practice. If flaps for take-off and landing could be dispensed with it would be advantageous to use long ailerons of small chord since the hinge moment per unit of rolling moment is then small. However, ailerons cannot overlap the flaps* and the effectiveness of the flaps falls off very rapidly as their span is reduced. Hence it is necessary to compromise and use shorter and wider ailerons. Experiments on very short and large chord ailerons

* There have been some not very successful attempts to mount ailerons on the flaps.

indicate that they appear sluggish to the pilot on account of their large moment of inertia but may be made satisfactory by fitting a spring tab.

9·6   Other rolling controls

If we imagine the chord of an aileron to become equal to that of the wing and adopt a suitably set-back hinge axis we arrive at the rotating wing tip which can provide adequate rolling control when its span is comparatively small. This was used with success on the Hill Pterodactyl tailless aircraft and has more recently been applied to other tailless aircraft with sweptback wings. The rotating tip will be aerodynamically balanced when the axis of rotation is somewhere near the quarter-chord axis and mass balancing will almost certainly be required as a preventive of flutter. It appears improbable that reversal of control can occur with rotating wing tips.

Elevons are essentially ailerons which operate in the usual way when the stick is moved laterally but which move in phase when the stick is moved fore-and-aft, so performing the function of elevators. They have been applied to tailless aircraft but suffer from the defect that their effective travel as ailerons is restricted at large 'elevator' deflections.

Spoilers on the port and starboard wings can provide a rolling moment by causing loss of circulation and so also of lift on the side where they are operated. Since a spoiler cannot add to the lift there is, in fact, always a loss of total lift when the spoiler is operated. There is an adverse yawing moment, as with ailerons, due to the change of the lift, but this is offset by the extra drag of the spoiler. At its simplest, a spoiler is a plate which can be made to protrude through a spanwise slit in the wing surface by a controlled amount and having a span comparable with that of an ordinary aileron. The spoiler is most effective in reducing the lift when it is far forward in the chord but there is then a rather serious lag between the operation of the spoiler and the fall in the lift. The lag can be reduced by placing the spoiler towards the rear but at the expense of some loss of effectiveness. It appears that the lag can also be reduced by 'venting' the spoiler, i.e. by allowing some air to pass between the top of the spoiler and the wing surface. The great advantage of spoilers is that they can be placed in front of flaps, which may then cover the entire span; moreover, they are probably not subject to reversal of their effect by wing twist. However, few entirely satisfactory systems of rolling control by spoilers alone have so far been produced although they have been combined with small ailerons with some success. Some of the troubles are the lag already mentioned, the difficulty of making the action smooth and continuous with stick movement, especially near the neutral setting, friction and inertia in the mechanism and a tendency to provoke tip stalling. Further, the absence of normal 'feel' will be disliked by many pilots though this could be simulated by springs. It is to be noted that when the stick is moved, say to starboard, the starboard spoiler is made to protrude but the port spoiler must not; to keep down drag the spoiler slit should be sealed when the spoiler is not in operation and it is difficult to arrange this without leaving a 'dead' region of stick travel where it is quite ineffective. In spite of all these difficulties, spoilers are well worth developing since they permit the use of full-span flaps for landing and take-off with a consequent large increase of available lift coefficient.

9·7   Elevators

The normal method of longitudinal control, at least for aircraft with tails, is to deflect in phase a pair of flaps hinged to the tailplane at the rear. These flaps are called elevators. The port and starboard elevators are usually connected by a tube having a high torsional stiffness* or they may be continuous.

The size of the elevators is fixed mainly by trimming requirements, i.e. it must be possible to set the elevators so as to secure balance of the pitching moments in all the circumstances of flight. Extreme pitching moments to be balanced will occur at the extremes of the c.g. range, at full throttle or gliding, flaps up or down, undercarriage up or down and at the extremes of flight speed. The size of the tailplane must be such that the tail volume ratio (see § 10·5) is adequate for stability but not excessive. Typically the elevator chord is of the order of 30 to 40 per cent of the tailplane chord, so elevators are typically of greater relative chord than ailerons.

* This is one of the measures taken to prevent antisymmetric flutter of the elevators.

Ideally there should be no change of longitudinal static stability when the elevators are released. This would require that b_1 for the elevators should be zero (see § 7·4 and § 10·4). The requirement that b_1 shall be zero or very small is most conveniently met by means of a horn balance (see § 7·5).

Elevators must not become aerodynamically overbalanced in any possible condition of flight. They must be provided with a trimmer (see § 9·3).

9·8 Adjustable and all-moving tailplanes

At one time it was common practice to make the tailplane angular setting relative to the body adjustable. With such an adjustable tailplane it becomes possible to fly at any speed or c.g. position within certain ranges with the elevators neutral. Unless, however, the tailplane setting can be changed very quickly, the adjustable tailplane is no substitute for elevators. It is not easy to arrange that the fillets at the roots of the tailplane shall in all settings be fair and properly sealed.

A power-operated tailplane without elevators has certain advantages for flight at very high speeds. It is relatively easy to make such a tailplane very stiff, so distortion can be minimized and in any event reversal of control cannot occur. Close aerodynamic balance is unnecessary and some aerodynamic overbalance not dangerous. The success of such an arrangement depends entirely on the quality of the power-operating unit.

9·9 Rudders

The function of the rudder or rudders is to provide a yawing moment as required by the pilot. A controllable yawing moment is required for the following purposes:

(a) To balance the yawing moment caused by the ailerons and other parts of the aircraft in a banked turn.

(b) To balance the yawing moment caused by the slipstream impinging on the fin and after body.

(c) To provide a controlling moment when taxiing.

(d) To balance the yawing moment caused by unsymmetric engine failure on an aircraft having engines carried on the wings.

On aircraft with one or more engines on each wing the most severe demand on the rudder is made in the worst case of unsymmetric engine failure which must be provided for. This is accordingly the 'design case' as regards rudder power. Twin rudders placed in the slipstreams are often adopted for aircraft of this kind in order to increase rudder power in taxiing. For single engined aircraft the power of the rudder may be fixed by the necessity to deal with 'swing at take-off'. The yawing moment here is caused by the rotating slipstream impinging on the fin, an effect which is specially great when the propeller thrust coefficient T_c is large, as at take-off and in the initial climb at full throttle.

Rudders must be provided with powerful trimmers, especially for aircraft with engines on the wings, and it must be possible to move the trimmers quickly. On large aircraft this may require the installation of power-operated trimmers.

It is undesirable that there should be a serious loss of 'weathercock stability' when the rudder is freed. Hence the coefficient b_1 for the rudder should be zero or small. There is a danger of 'snaking' when a positive b_1 is associated with a small negative b_2. Snaking is a lateral oscillation in which yawing and movement of the rudder on its hinge are important. The cure for snaking consists in mass-balancing the rudder (see § 7·7) and making b_1 zero or nearly so; these measures effectively destroy the coupling between the motions of the rudder and of the aircraft (see also § 15·1).

One of the worst dangers concerning the rudder is fin stalling associated with locking of the rudder. If for any reason the angle of sideslip becomes so large that the fin begins to stall there may be a change of rudder hinge moment tending to increase the rudder angle and so aggravate the stall; in bad cases the rudder may 'lock-on' and be uncontrollable by the pilot. Fin stall may be catastrophic since the aircraft is liable to be violently unstable when the fin loses its effectiveness. The danger of fin stalling can be greatly reduced by fitting a dorsal fin (see § 6·3) when there is a single central fin or by using twin fins of adequate area and *small* aspect ratio.

9·10   Mechanical and structural aspects of controls

It is most desirable that there should be, as nearly as possible, a definite relationship, irrespective of speed and air density, between the position of the pilot's control and the setting of the control surface. Since all materials lack complete rigidity it is never possible to reach perfection here, but it is essential that the stiffness of the gearing between the pilot's lever or wheel and the control should be high. It is important to note that the effective stiffness is greatly influenced by the stiffness of the supporting bearings, brackets, etc. which are used in the control circuit. The control flaps themselves must have adequate torsional stiffness.

Allied to the need for high stiffness in control circuits is the need to minimize backlash in all circumstances. This requires that the clearance in all bearings shall be no more than is strictly necessary and there must be means for taking up wear. Differences in the coefficients of thermal expansion of the materials used may cause either undesirable backlash or binding. Changes in the dimensions of a wooden structure, especially those caused by changes of humidity, may likewise cause backlash or binding.

It is of great importance to reduce friction in control systems as far as possible. Frictional resistances impose additional and unnecessary loads on the pilot and high friction makes accurate trimming impossible. In certain instances, e.g. snaking, friction in the control system may tend to cause instability but it is more usual for friction to have a stabilizing influence on motion with a control free. It is essential that lubricants should retain their lubricating properties adequately throughout the range of temperature and climate which the aircraft will encounter.

The surfaces of control flaps must be stiff enough to prevent the inevitable changes of form under varying aerodynamic loads from being large enough to influence their hinge moments appreciably (see § 7·5 and § 7·6). Reference has already been made to the need for adequate torsional stiffness of control flaps. It is relevant to remark that the need for high torsional stiffness will be less when the control moment is applied at the middle of the span of the flap than when it is applied at one end.

9·11 Sundry topics

Icing of controls is a matter of much importance. Apart from any danger of locking of the flap by a deposit of ice, which is probably small except for exposed horn balances, ice accretions may give rise to serious changes in the coefficients b_1 and b_2.

This imposes a lower numerical limit on b_2 on account of the danger of overbalance when icing occurs.

It is important that controls should be so designed and made that they add as little as possible to the drag of the aircraft. Thus the surfaces of flaps should be smooth while gaps should be sealed and be faired as far as possible. External balance masses and other excrescences should be avoided whenever possible.

Irreversible manual controls have been proposed as a means for preventing flutter and some attempts to produce such controls have been made, but with little success up to the time of writing. One trouble is a tendency for the control to 'judder' when it is moved towards neutral so that the hinge moment tends to accelerate the motion. This can be overcome, but, so far, only at the expense of making the control heavy to operate at all times. In order that the control flap shall be locked with sufficient rigidity to prevent flutter the irreversible unit must be close to the flap. Then it is found that the unavoidable elasticity of the gearing between the pilot's control and the irreversible unit gives rise to an unpleasant 'sponginess' of the control. It appears that there is a need for a new principle of irreversible operation.

The subject of power operation of controls lies beyond the scope of this book. However, it can be said that the main problem is to attain absolute reliability without excessive weight. In order that balance masses may be dispensed with it is necessary that the controls shall be irreversible.

Mass-balancing of control flaps is discussed in § 7·7.

GENERAL REFERENCE FOR CHAPTER 9

M. B. Morgan and H. H. B. M. Thomas, 'Control Surface Design in Theory and Practice', *J. Roy. Aero. Soc.*, vol. 49 (Aug., 1945), p. 431.

Chapter 10

STATIC STABILITY AND MANŒUVRABILITY

10·1 Introduction

We have already considered static longitudinal stability in § 5·5 where the discussion is based on the general dynamical equations of a rigid aircraft. In the present chapter we discuss the same problem in a more elementary manner ab initio and enter further into detail in several directions. We also treat the subject of longitudinal manœuvrability, following the method introduced by Gates.

Throughout this chapter the aircraft or other body considered is rigid with all controls locked unless the contrary is distinctly stated and the air density is assumed to be constant. The term body is used in a general sense, e.g. it is not restricted to mean the fuselage of an aircraft.

10·2 Equilibrium and stability of a pivoted body

Let us take a rigid body having a plane of symmetry, held fixed in a constant and uniform air stream whose velocity vector is parallel to the plane of symmetry. For simplicity we shall at present neglect the gravitational forces on the body. When the angle of incidence measured from some convenient datum line in the plane of symmetry is α the resultant aerodynamic force on the body will have a line of action AB lying in the plane of symmetry. Then if the body be provided with a fixed frictionless pivot whose axis is perpendicular to the plane of symmetry, but is otherwise free, it will be in equilibrium at the incidence α provided that the pivot lies on AB. Hence AB is the locus of pivot positions giving equilibrium or 'trim' at the incidence α; it is the 'trim line' corresponding to the incidence α.* Next, suppose that $A'B'$ is the trim line corresponding to the incidence α' and let AB and $A'B'$ meet at P.

---

* The trim line corresponding to α would depend on air speed, density and temperature, in general.

Then, if the body be pivoted at P, it will be in equilibrium for both the incidences α and α'. When $(\alpha' - \alpha)$ tends to zero with α fixed the point P will tend to a limiting position M which is the metacentre for the incidence α. If pivoted at M, the body will be in equilibrium for incidence α and for any incidence which differs from α by a very small angle; in other words, the equilibrium with pivot M and incidence α is neutral. In general M will describe a curve—the metacentric locus—when α is varied. This locus is the envelope in the body of the line of action of the resultant aerodynamic force.

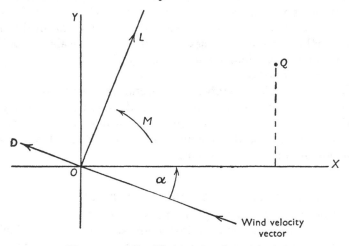

Fig. 10·2, 1. Equilibrium of a pivoted body.

The foregoing can be put in analytical form as follows (see Fig. 10·2, 1). OX and OY are rectangular axes fixed in the body and in the plane of symmetry. When the incidence measured from OX is α the resultant aerodynamic force on the body is equivalent to a lift force L, a drag force D, and a pitching moment M about O. The moment M' about an arbitrary point Q whose co-ordinates are x, y is evidently given by

$$M' = M - x(L\cos\alpha + D\sin\alpha) + y(L\sin\alpha - D\cos\alpha). \quad (10\cdot2, 1)$$

For any point on the trim line M' is zero so the equation of this line is

$$x(L\cos\alpha + D\sin\alpha) - y(L\sin\alpha - D\cos\alpha) - M = 0. \quad (10\cdot2, 2)$$

The *neutral line* corresponding to the incidence α is the locus

$$\frac{dM'}{d\alpha} = 0, \qquad (10\cdot2, 3)$$

or, from $(10\cdot2, 1)$,

$$x\left[\left(L - \frac{dD}{d\alpha}\right)\sin\alpha - \left(D + \frac{dL}{d\alpha}\right)\cos\alpha\right]$$
$$+ y\left[\left(L - \frac{dD}{d\alpha}\right)\cos\alpha + \left(D + \frac{dL}{d\alpha}\right)\sin\alpha\right] + \frac{dM}{d\alpha} = 0.$$
$$(10\cdot2, 4)$$

The metacentre corresponding to α is the point of intersection of the straight lines $(10\cdot2, 2)$ and $(10\cdot2, 4)$. When L, D and M are known functions of α the metacentric locus can be obtained by elimination of α from the equations.

Suppose next that the body is provided with an aerodynamic trimmer which consists *ideally* of a device by which the pitching moment can be adjusted without altering the values of L, D, $\frac{dL}{d\alpha}$, $\frac{dD}{d\alpha}$ and $\frac{dM}{d\alpha}$. Then if we take any point on the neutral line we can make the corresponding moment M' vanish by use of the trimmer and the point will accordingly be a metacentre since $\frac{dM'}{d\alpha}$ is zero. Moreover, $\frac{dM'}{d\alpha}$ will be positive for points on one side of the neutral line and negative on the other. Now, remembering that the wind velocity vector is, by hypothesis, fixed in space we see from Fig. $10\cdot2$, 1 that the condition for stability of the pivoted body is

$$\frac{dM'}{d\alpha} < 0. \qquad (10\cdot2, 5)$$

For, when this inequality is satisfied, an angular displacement of the body which increases α will give rise to a moment tending to reduce α, i.e. a restoring moment. When the body has the form of an aerofoil or conventional aircraft and the angle of incidence is not very large, the coefficient of x in $(10\cdot2, 4)$ is negative, for the dominant term is $-\frac{dL}{d\alpha}\cos\alpha$ and is negative.

Accordingly, the equilibrium will be stable for pivots situated to the right of the neutral line, i.e. upstream of this line.

It is evident from (10·2, 1) that the locus

$$\frac{d^2 M'}{d\alpha^2} = 0, \qquad\qquad (10\cdot2, 6)$$

is a straight line. Accordingly, for any given α, there is a point on the neutral line such that

$$\frac{dM'}{d\alpha} = \frac{d^2 M'}{d\alpha^2} = 0.$$

This is the point for which the pitching moment is independent of incidence to the highest possible order of small quantities; it is the aerodynamic centre *par excellence*. However, we are rarely, if ever, concerned with the second order changes in the pitching moment, and it is usual to apply the name *aerodynamic centre* to the point of intersection of the neutral line with OX. It is clear from considerations of symmetry that if the body is itself symmetric about OX and the angle of incidence zero, the aerodynamic centre *par excellence* as defined above must lie on OX. This centre will always lie near OX when the departure from symmetry is slight and the incidence small. By convention the term aerodynamic centre is usually applied to aerofoils and sometimes to wing-body combinations without tails, but seldom to complete aircraft or models with tails. In the last case the term neutral point is commonly used. It is worthy of emphasis that the whole of the foregoing analysis is valid for all rigid bodies and for all fluid media and velocities.

According to the theory of thin aerofoils in an incompressible medium, the point situated on the chord at one quarter of the chord aft of the leading edge is the aerodynamic centre for all small incidences. The aerodynamic forces for a given stream are equivalent to a constant couple, a lift force through the aerodynamic centre and a drag force. This theory accords well with facts, even for aerofoils that are not very thin, but the aerodynamic centre does not lie quite exactly at the quarter-chord point. However, the conception that the lift force acts always at a fixed aerodynamic centre is extremely useful in the theory of stability and control.

10·3   Neutral static stability of a body in flight

Before entering on the details of the argument it will be well for us to consider neutral stability in a general way. The basic characteristic of a body or system in neutral equilibrium in a certain datum state or configuration is that for certain other states or configurations in the immediate vicinity of the datum the body or system is also in equilibrium. There is a 'mode' of neutral displacement just as there is a mode of displacement for each free oscillation of a system. Another way of putting the matter is that no first order restoring forces are brought into play by certain infinitesimal deviations from the datum state. It follows from this that exceedingly slow deviations from the datum can occur in the absence of applied forces* and with all accelerations zero.

Now let us consider a symmetric rigid body in symmetric flight and with neutral static stability. In accordance with what has just been said it will be possible to deviate very slowly from the datum state of flight without finite acceleration. This implies that during any (very gradual) change of flight speed equilibrium is constantly preserved. Thus, during the change the resultant aerodynamic force will always balance the weight of the body so $\rho V^2 C_R$ is constant. Except for very steep dives or climbs this condition can be replaced by constancy of $\rho V^2 C_L$.

We shall now demonstrate that, subject to one proviso which will be explained, a rigid body in flight in equilibrium with neutral static stability will have its c.g. at a neutral pivot position corresponding to the actual incidence (see § 10·2). Suppose that the velocity of the c.g. relative to the surrounding medium has deviated slowly from its original value. Then we can imagine the deviation to occur in two steps:

(a) A change of relative speed at constant incidence.

(b) A change of incidence at constant relative speed.

Since the body is in equilibrium and since the gravitational forces have no moment about the c.g. the aerodynamic pitching moment about the c.g. is zero both before and after the deviation of velocity; moreover, since the angular velocity of the body and its accelerations are infinitesimal the pitching moment is determined by the air speed and incidence. Now let us assume that when the air speed varies at constant incidence the lift, drag and pitching moment about a standard origin all vary proportionally. Then, during any change of speed at

* Or under the action of vanishingly small applied forces.

constant incidence, the line of action of the resultant aero-dynamic force on the body will be invariable. Hence during stage (a) the pitching moment about the C.G. will remain zero. Now at the end of stages (a) and (b) the pitching moment is zero so that during the incidence change in stage (b) the pitching moment remains zero. Consequently the C.G. is at a neutral pivot position as discussed in § 10·2.

In the above argument we have made the assumption that the lift, drag and pitching moment for any constant incidence vary in strict proportion when the speed changes. This condition is satisfied, in particular, when the non-dimensional lift, drag and pitching moment coefficients are all independent of speed. However, when these coefficients alter with speed, say through their dependence on Mach number, it ceases in general to be true that the C.G. must be at a neutral pivot position for neutral stability in flight. The discussion of this case is beyond the scope of the elementary theory given in this chapter. See, however, § 5·5.

10·4   Neutral points of a rigid aircraft with tail

We shall now suppose that the axis OX is the 'wind axis' so that in the undisturbed state of the aircraft α is zero.* Then we have from (10·2, 4) that the abscissa of the neutral point, defined as the intersection of the neutral line with OX, is

$$x_n = \frac{\dfrac{dM}{d\alpha}}{D + \dfrac{dL}{d\alpha}}. \qquad (10\cdot4,\ 1)$$

Next, suppose that the body under consideration consists of two rigid parts, rigidly connected, and let the suffixes 1 and 2 be applied to the forces on these parts. Then the last equation becomes

$$x_n = \frac{\dfrac{dM_1}{d\alpha} + \dfrac{dM_2}{d\alpha}}{D_1 + D_2 + \dfrac{dL_1}{d\alpha} + \dfrac{dL_2}{d\alpha}}. \qquad (10\cdot4,\ 2)$$

* This does not imply that the incidence measured from the standard wing chord is zero.

Let us apply this to an aircraft and let the first part consist of the wings with the fuselage and the second part be the tail. Further, let us now take the origin O to be the aerodynamic centre of the wings cum fuselage, which implies*

$$\frac{dM_1}{d\alpha} = 0. \qquad (10\cdot4, 3)$$

Then we obtain
$$x_n = \frac{\dfrac{dM_2}{d\alpha}}{D + \dfrac{dL}{d\alpha}} \qquad (10\cdot4, 4)$$

where $$D = D_1 + D_2$$
and $$L = L_1 + L_2$$

are the total drag and lift respectively. Since $\dfrac{dM_2}{d\alpha}$ will in fact be negative and the positive sense of x is forward, the neutral point will lie aft of the aerodynamic centre. Let us now introduce the standard non-dimensional coefficients, thus

$$D = \tfrac{1}{2}\rho V^2 S C_D,$$

$$L = \tfrac{1}{2}\rho V^2 S C_L,$$

$$M_2 = \tfrac{1}{2}\rho V^2 S \bar{c} C_{m2}.$$

Then we obtain
$$\left(\frac{-x_n}{\bar{c}}\right) = \frac{-\dfrac{dC_{m2}}{d\alpha}}{C_D + \dfrac{dC_L}{d\alpha}}, \qquad (10\cdot4, 5)$$

which gives in non-dimensional form the distance of the neutral point of the complete aircraft behind the aerodynamic centre of the wings and fuselage. Since C_D is usually very much smaller than $\dfrac{dC_L}{d\alpha}$ there is little error in omitting C_D from the denominator of $(10\cdot4, 5)$ and it is customary to do this.

To complete the investigation we have to calculate $\dfrac{dC_{m2}}{d\alpha}$.

---

* We assume here that the aerodynamic interference due to the tailplane on the wings and body is negligible. Otherwise, the aerodynamic centre of the wings and fuselage could not be found by experiments on these bodies without the tail.

Now we have
$$\frac{dL_2}{d\alpha} = \tfrac{1}{2}\rho V^2 S' \frac{dC_{L2}}{d\alpha'}\frac{d\alpha'}{d\alpha}, \qquad (10\cdot4,\,6)$$

where S' and α' are the tailplane area and effective incidence respectively. For convenience put

$$\frac{dC_{L2}}{d\alpha'} = a_1', \qquad (10\cdot4,\,7)$$

which is thus the non-dimensional lift slope for the tailplane in situ behind the wings and body. We have seen in § 2·6 that

$$\frac{d\alpha'}{d\alpha} = \left(1 - \frac{d\epsilon}{d\alpha}\right), \qquad (10\cdot4,\,8)$$

where ϵ is the effective angle of downwash at the tailplane. Accordingly ($10\cdot4$, 6) becomes

$$\frac{dL_2}{d\alpha} = \tfrac{1}{2}\rho V^2 S' a_1'\left(1 - \frac{d\epsilon}{d\alpha}\right). \qquad (10\cdot4,\,9)$$

Let l be the distance of the aerodynamic centre of the tailplane aft of the aerodynamic centre of the wings and body. Then, if the pitching moment due to the drag on the tail be neglected,

$$\frac{dM_2}{d\alpha} = -l\frac{dL_2}{d\alpha} \qquad (10\cdot4,\,10)$$

and, by ($10\cdot4$, 9),

$$-\frac{dC_{m2}}{d\alpha} = \left(\frac{S'l}{S\bar{c}}\right)a_1'\left(1 - \frac{d\epsilon}{d\alpha}\right). \qquad (10\cdot4,\,11)$$

Put
$$\left(\frac{S'l}{S\bar{c}}\right) = \bar{V}', \qquad (10\cdot4,\,12)$$

a modified 'tail volume ratio',* and

$$\frac{dC_L}{d\alpha} = a_1.$$

Equation ($10\cdot4$, 5) accordingly becomes

$$\left(\frac{-x_n}{\bar{c}}\right) = \bar{V}'\left(\frac{a_1'}{a_1 + C_D}\right)\left(1 - \frac{d\epsilon}{d\alpha}\right), \qquad (10\cdot4,\,13)$$

or, approximately,

$$\left(\frac{-x_n}{\bar{c}}\right) = \bar{V}'\left(\frac{a_1'}{a_1}\right)\left(1 - \frac{d\epsilon}{d\alpha}\right). \qquad (10\cdot4,\,14)$$

* Compare equation ($10\cdot5$, 6). In practice the two tail volume ratios differ very little.

In the foregoing investigation we have postulated that the complete aircraft is rigid, which implies, in particular, that the elevators are rigidly fixed. Thus the neutral point whose position we have calculated is the elevator-fixed neutral point, usually known as the *stick-fixed neutral point*. We have also assumed that the proviso about the dependence of the aerodynamic forces on wind speed, mentioned in § 10·3, is satisfied.

It is not difficult to derive from the above results the position of the *stick-free neutral point*, provided that non-aerodynamic hinge moments are not applied to the elevator. Thus the elevator takes up its position of free equilibrium under the action of the aerodynamic forces while hinge moments due to gravity, springs and accelerations are absent;* any trimming tabs on the elevator are fixed rigidly. The equation of elevator hinge moments in its non-dimensional form gives (see § 7·1)

$$b_1' d\alpha' + b_2' d\eta = 0,$$

so the change in elevator angle is given by

$$d\eta = -\frac{b_1'}{b_2'} d\alpha'. \tag{10·4, 15}$$

The change in lift coefficient on the tailplane is

$$dC_{L2} = a_1' d\alpha' + a_2' d\eta$$

$$= a_1'\left(1 - \frac{a_2' b_1'}{a_1' b_2'}\right) d\alpha'$$

and

$$\frac{dC_{L2}}{d\alpha'} = a_1'\left(1 - \frac{a_2' b_1'}{a_1' b_2'}\right). \tag{10·4, 16}$$

Hence, by (10·4, 10), in order to derive the position of the stick-free neutral point it is only necessary to multiply a_1' in (10·4, 13) or (10·4, 14) by the factor

$$\left(1 - \frac{a_2' b_1'}{a_1' b_2'}\right).$$

Now a_1' and a_2' are positive and b_2' is negative when the elevator is not aerodynamically overbalanced, which we are entitled to assume. Hence the factor is greater than unity when b_1' is positive, but less than unity when b_1' is negative. An elevator

---

* We are entitled to postulate zero accelerations since we are considering a state of neutral stability.

without horn and appreciably underbalanced aerodynamically tends to align itself along the wind which implies that b_1' is negative [see equation (10·4, 15)]; such an elevator is called *convergent*. When b_1' is positive, as when a sufficiently large horn is fitted, the elevator is called *divergent*.* Thus, when the elevator is convergent the stick-free neutral point lies forward of the stick-fixed neutral point but the reverse is true when the elevator is divergent.

10·5   The C.G. margins, stick-fixed and stick-free

Let the origin O now be the c.g. of the complete aircraft and OX the wind axis, as explained in § 10·4, and at first suppose the elevator to be fixed. The c.g. is at a distance $h\bar{c}$ *aft* of some convenient datum point, say the leading edge of the mean wing chord, while the neutral point on OX is $h_n\bar{c}$ *aft* of the datum. Then the abscissa of the neutral point is

$$x_n = \bar{c}(h - h_n)$$

$$= \frac{\dfrac{dM}{d\alpha}}{D + \dfrac{dL}{d\alpha}} \quad \text{by (10·4, 1)}$$

$$= \frac{\bar{c}\,\dfrac{dC_m}{d\alpha}}{C_D + \dfrac{dC_L}{d\alpha}}.$$

Hence
$$\frac{dC_m}{d\alpha} = -(h_n - h)\left(C_D + \frac{dC_L}{d\alpha}\right), \qquad (10\text{·}5,\ 1)$$

which gives the rate of change with incidence of the coefficient of pitching moment about the c.g. in terms of the quantity $(h_n - h)$, known as the stick-fixed c.g. margin. If C_D can be neglected in comparison with $\dfrac{dC_L}{d\alpha}$ the last equation can be put in the simpler form

$$\frac{dC_m}{dC_L} = -(h_n - h). \qquad (10\text{·}5,\ 2)$$

* Note that this word does not here indicate instability.

When the elevator is free and not subject to gravitational or elastic moments, the neutral point is situated at a distance $h'_n \bar{c}$ aft of the datum. Hence, with stick free,

$$\frac{dC_m}{d\alpha} = -(h'_n - h)\left(C_D + \frac{dC_L}{d\alpha}\right), \qquad (10\cdot5, 3)$$

and, when C_D is neglected,

$$\frac{dC_m}{dC_L} = -(h'_n - h), \qquad (10\cdot5, 4)$$

where $(h'_n - h)$ is called the stick-free C.G. margin. For positive static stability $\dfrac{dC_m}{d\alpha}$ is negative and the C.G. margin must therefore be positive.

The stick-fixed C.G. margin is directly proportional to the rate of change of elevator angle with lift coefficient. When the elevator is moved the lift on the tail is changed and with it the pitching moment about the C.G. A new state of equilibrium is reached when the incidence has adjusted itself so that the total pitching moment is zero. Suppose that the elevator angle is increased by $d\eta$. Then the lift coefficient on the tailplane is increased by $a'_2 d\eta$ and the pitching moment coefficient is increased by

$$-\frac{S' l_T a'_2 d\eta}{S\bar{c}},$$

where l_T is the distance of the aerodynamic centre of the tail-plane aft of the C.G. Thus

$$\frac{\partial C_m}{\partial \eta} = -\bar{V} a'_2, \qquad (10\cdot5, 5)$$

where

$$\bar{V} = \frac{S' l_T}{S\bar{c}}, \qquad (10\cdot5, 6)$$

is the tail volume ratio. But, by (10·5, 1)

$$\frac{\partial C_m}{\partial \alpha} = -(h_n - h)\left(C_D + \frac{dC_L}{d\alpha}\right),$$

where we now adopt the notation for a partial differential coefficient since C_m depends on two variables. The condition that the total change of C_m shall be zero is

$$\frac{\partial C_m}{\partial \eta}\,d\eta + \frac{\partial C_m}{\partial \alpha}\,d\alpha = 0,$$

which yields

$$\frac{d\eta}{d\alpha} = -\frac{(h_n - h)\left(C_D + \dfrac{dC_L}{d\alpha}\right)}{\overline{V}a_2'}. \tag{10.5, 7}$$

When allowance is made for the additional lift on the tailplane we have

$$dC_L = \frac{dC_L}{d\alpha}\,d\alpha + \frac{S'}{S}\,a_2'\,d\eta$$

$$= d\eta\left(\frac{dC_L}{d\alpha}\frac{d\alpha}{d\eta} + \frac{S'\,a_2'}{S}\right).$$

Hence

$$\frac{d\eta}{dC_L} = \frac{\dfrac{d\eta}{d\alpha}}{\dfrac{dC_L}{d\alpha} + \dfrac{S'}{S}\dfrac{a_2'}{d\alpha}\dfrac{d\eta}{d\alpha}}$$

$$= \frac{-(h_n - h)\left(C_D + \dfrac{dC_L}{d\alpha}\right)}{\overline{V}a_2'\left[\dfrac{dC_L}{d\alpha} - \dfrac{\bar{c}}{l_T}(h_n - h)\left(C_D + \dfrac{dC_L}{d\alpha}\right)\right]}, \tag{10.5, 8}$$

by equation (10·5, 7). Now the factor

$$\frac{\bar{c}}{l_T}(h_n - h)$$

is typically small. If we treat it as zero and neglect C_D in comparison with $\dfrac{dC_L}{d\alpha}$ equation (10·5, 8) becomes

$$\frac{d\eta}{dC_L} = -\frac{(h_n - h)}{\overline{V}a_2'}, \tag{10.5, 9}$$

which is the commonly used form of the relation. We can now derive the rate of change of elevator angle with speed of flight. Since $V^2 C_L$ is constant we obtain by logarithmic differentiation

$$\frac{dC_L}{dV} = -\frac{-2C_L}{V}, \tag{10.5, 10}$$

and then

$$\frac{d\eta}{dV} = -\frac{2C_L}{V}\frac{d\eta}{dC_L} \tag{10.5, 11}$$

$$= \frac{2C_L(h_n - h)}{V\bar{V}a_2'}, \tag{10.5, 12}$$

on substitution from (10·5, 9). Thus, subject to the conditions for the validity of the approximate relation (10·5, 9), the elevator adjustment or stick movement corresponding to unit change of speed is proportional to the stick-fixed c.g. margin.

Next, let us consider the state of affairs when the elevator is free and not subject to any non-aerodynamic moments. When the angle β which the tab chord makes with the elevator chord is varied, the elevator will take up such an angle that its total hinge moment is zero and there will be concomitant changes of incidence and flight speed. For any given aircraft and condition of flight there will therefore be a definite value of $\frac{d\beta}{dC_L}$, but this will depend on the stick-free c.g. margin. When this margin is zero the aircraft has neutral static stability with the elevator free, so an exceedingly small tab movement will give a finite change of C_L (see § 10·3). Hence $\frac{d\beta}{dC_L}$ vanishes when $(h_n' - h)$ is zero.

The condition of balance of elevator hinge moments yields

$$b_1'\,d\alpha' + b_2'\,d\eta + b_3'\,d\beta = 0$$

or

$$b_1'\left(1 - \frac{d\epsilon}{d\alpha}\right)d\alpha + b_2'\,d\eta + b_3'\,d\beta = 0. \tag{10.5, 13}$$

Also the equation of balance of pitching moments gives

$$\frac{\partial C_m}{\partial\alpha}\,d\alpha - \bar{V}(a_2'\,d\eta + a_3'\,d\beta) = 0$$

or

$$(h_n - h)\left(C_D + \frac{dC_L}{d\alpha}\right)d\alpha + \bar{V}(a_2'\,d\eta + a_3'\,d\beta) = 0, \tag{10.5, 14}$$

by (10·5, 1). When allowance is made for the contribution of the tail to the lift on the aircraft we obtain

$$dC_L = \frac{dC_L}{d\alpha}\,d\alpha + \frac{S'}{S}(a_2'\,d\eta + a_3'\,d\beta). \tag{10.5, 15}$$

Equations (10·5, 13) and (10·5, 14) can be solved for $\dfrac{d\alpha}{d\beta}$ and $\dfrac{d\eta}{d\beta}$ and then (10·5, 15) yields

$$\frac{dC_L}{d\beta} = \frac{dC_L}{d\alpha}\frac{d\alpha}{d\beta} + \frac{S'}{S}\left(a_2'\frac{d\eta}{d\beta} + a_3'\right). \qquad (10\cdot5, 16)$$

The explicit general solution is cumbrous and will not be quoted, but when the numerical values of the coefficients are known the value of $\dfrac{dC_L}{d\beta}$ can easily be calculated. We can, however, derive a simple and useful explicit formula by neglecting some small terms. When a_3' and C_D are neglected in (10·5, 14) we obtain

$$\frac{d\eta}{d\beta} = -\frac{(h_n - h)\dfrac{dC_L}{d\alpha}}{\bar{V}a_2'}\frac{d\alpha}{d\beta},$$

and (10·5, 13) then yields

$$\frac{d\alpha}{d\beta}\left[b_1'\left(1 - \frac{d\epsilon}{d\alpha}\right) - \frac{b_2'(h_n - h)\dfrac{dC_L}{d\alpha}}{\bar{V}a_2'}\right] = -b_3'. \qquad (10\cdot5, 17)$$

Now if we neglect a_2' and a_3' in (10·5, 15), i.e. neglect the lift on the tailplane in comparison with that on the wings, we obtain

$$\frac{d\beta}{dC_L} = \frac{d\beta}{d\alpha}\bigg/\frac{dC_L}{d\alpha}$$

$$= \frac{b_2'(h_n - h)\dfrac{dC_L}{d\alpha} - \bar{V}a_2'b_1'\left(1 - \dfrac{d\epsilon}{d\alpha}\right)}{\bar{V}a_2'b_3'\dfrac{dC_L}{d\alpha}}. \qquad (10\cdot5, 18)$$

This shows that $\dfrac{d\beta}{dC_L}$ is a linear function of h and the coefficient of h is

$$-\frac{b_2'}{\bar{V}a_2'b_3'}.$$

But we have seen above that $\dfrac{d\beta}{dC_L}$ vanishes when $h = h_n'$. Accordingly

$$\frac{d\beta}{dC_L} = \frac{b_2'(h_n' - h)}{\bar{V}a_2'b_3'}. \qquad (10\cdot5, 19)$$

By (10·5, 10) it follows that

$$\frac{d\beta}{dV} = -\frac{2C_L b_2'(h_n' - h)}{V\overline{V}a_2'b_3'}. \qquad (10\cdot5, 20)$$

Thus the tab movement per unit change of speed is proportional to the stick-free c.g. margin. If we had kept the tab fixed we could have produced the same displacement of the elevator by applying a stick force proportional to $\rho V^2\beta$. Hence the stick force per unit change of speed is proportional to the stick-free c.g. margin.

10·6 Trim curves

Trim curves have been briefly discussed in § 2·10. We now consider the theory of such curves in some detail.

When allowance is made for the influence of the lift on the tail the pitching moment coefficient about the c.g. for the complete aircraft is

$$C_m = C_{MO} + \frac{dC_m}{d\alpha}\alpha - \overline{V}(a_2'\eta + a_3'\beta) = 0, \qquad (10\cdot6, 1)$$

for equilibrium, where C_{MO} is the pitching moment coefficient at the no-lift incidence and with the elevator and tab in their datum positions. The incidence α is measured from the no-lift incidence and the contribution to the pitching moment made by changes of drag on the tail are neglected. In general, the lift coefficient is given by

$$C_L = \frac{dC_L}{d\alpha}\alpha + \frac{S'}{S}(a_2'\eta + a_3'\beta). \qquad (10\cdot6, 2)$$

Case A. Tab absent or fixed, elevator controlled

The terms in β are absent from equations (10·6, 1) and (10·6, 2) so (10·6, 1) yields

$$\alpha = -\frac{C_{MO} - \overline{V}a_2'\eta}{\dfrac{dC_m}{d\alpha}}.$$

On substitution in (10·6, 2) we obtain after reduction

$$C_L = -C_{MO}\frac{dC_L}{dC_m} + a_2'\eta\left(\overline{V}\frac{dC_L}{dC_m} + \frac{S'}{S}\right). \qquad (10\cdot6, 3)$$

Thus, provided that the aerodynamic coefficients are constant throughout the range of C_L considered, η will plot linearly against C_L. It follows from (10·6, 3) that the slope of this 'trim curve' is

$$\frac{d\eta}{dC_L} = \frac{1}{a_2'\left(\overline{V}\dfrac{dC_L}{dC_m} + \dfrac{S'}{S}\right)}$$

$$= \frac{\dfrac{dC_m}{dC_L}}{\overline{V}a_2'\left(1 + \dfrac{\overline{c}}{l_T}\dfrac{dC_m}{dC_L}\right)}. \qquad (10·6, 4)$$

On substitution from equation (10·5, 1) we recover equation (10·5, 8). The ordinate corresponding to zero value of the lift coefficient is

$$\eta_0 = \frac{C_{MO}}{\overline{V}a_2'\left(1 + \dfrac{\overline{c}}{l_T}\dfrac{dC_m}{dC_L}\right)}. \qquad (10·6, 5)$$

When the small term $\quad\dfrac{\overline{c}}{l_T}\dfrac{dC_m}{dC_L}$

is neglected in the denominator of (10·6, 5), which amounts to neglect of the contribution of the tailplane to the overall lift, we derive the simple result

$$\eta_0 = \frac{C_{MO}}{\overline{V}a_2'} \qquad (10·6, 6)$$

With the same simplification and on substitution from (10·5, 2) the slope of the trim curve becomes

$$\frac{d\eta}{dC_L} = -\frac{(h_n - h)}{\overline{V}a_2'}. \qquad (10·6, 7)$$

Thus, the trim curves for various values of the c.g. margin are concurrent at the ordinate η_0 when C_L is zero and the negative slope of any trim curve is proportional to the c.g. margin. If we regard the tailplane cum elevator merely as providing pitching moment, it is obvious that when the lift is zero there will be balance of pitching moments at a fixed elevator setting such that the pitching moment coefficient due to the tail just balances

C_{MO}. It is to be remarked that the ordinate for C_L zero cannot be determined by direct measurement since the corresponding flight speed would be infinite (see § 10·3).

It is possible, on the basis of equation (10·6, 7), to determine the position of the stick-fixed neutral point if the value of the slope

$$s = \frac{d\eta}{dC_L}$$

has been determined for two positions of the c.g. Corresponding to h_1 and h_2 let the slopes be s_1 and s_2 respectively. Then we have, on the assumption that a'_2 is constant,

$$\frac{s_1}{s_2} = \frac{h_n - h_1}{h_n - h_2}$$

and
$$h_n = \frac{s_1 h_2 - s_2 h_1}{s_1 - s_2}. \tag{10·6, 8}$$

If need be the more accurate formula (10·6, 4) can be used with the value of $\dfrac{dC_m}{dC_L}$ in the denominator deduced from the above approximate value of h_n.

Case B. Elevator free, tab controlled

We now have to introduce the condition of balance of the elevator hinge moments, namely

$$b'_0 + b'_1\left(1 - \frac{d\epsilon}{d\alpha}\right)\alpha + b'_2\eta + b'_3\beta = 0. \tag{10·6, 9}$$

If β be prescribed equations (10·6, 1) and (10·6, 9) determine α and η as linear functions of β, provided that all the aerodynamic coefficients are constant. Then (10·6, 2) gives C_L as a linear function of β, and if we plot a trim curve with β as ordinate and C_L as base we shall obtain a straight line. As pointed out in § 10·5 the slope of this trim curve vanishes when $(h'_n - h)$ is zero. Subject to the simplifications (a) that the term in β in (10·6, 1) is neglected, and (b) that the terms in η and β in (10·6, 2) are neglected, it follows, as in equation (10·5, 19), that the slope of the trim curve is

$$\frac{d\beta}{dC_L} = \frac{b'_2(h'_n - h)}{\overline{V}a'_2 b'_3}. \tag{10·6, 10}$$

On the foregoing assumptions the elevator angle for zero lift coefficient is still given by (10·6, 5) and the corresponding value of α is zero. Then if we take b'_0 to be zero, which merely implies a special datum for the elevator angle, equation (10·6, 9) yields

$$\beta_0 = -\frac{b'_2}{b'_3}\eta_0 = -\frac{b'_2 C_{MO}}{b'_3 \overline{V} a'_2} \qquad (10\cdot6, 11)$$

by (10·6, 6). Hence, provided that the aerodynamic coefficients are constant throughout the range of C_L considered, the trim curves will be concurrent at the ordinate β_0 for C_L zero. If the slope of the trim curve be determined for two values of h, the value of h'_n can be deduced from (10·6, 10) by means of a formula similar to (10·6, 8).

10·7   Manœuvrability and the manœuvre points

In § 10·6 we have seen that the stick-fixed neutral point has the property that the stick movement per unit change of speed is proportional to the distance of the c.g. forward of this point. Likewise the change of stick force per unit change of speed is proportional to the distance of the c.g. forward of the stick-free neutral point. These two quantities must be neither too small nor too large for the aircraft to have satisfactory handling qualities; if they are too small the controls are too 'sensitive' while if they are too large the controls are 'heavy'. Experience indicates that the stick force is more important to the pilot than the stick movement, so the stick-free c.g. margin is of special importance. In pull-outs from dives the quantities *stick movement per unit increment of normal acceleration* and *change of stick force per unit of normal acceleration* are of great significance. They are usually known respectively as the *stick movement per g* and the *stick force per g* and the latter is regarded as being particularly important.

The stick movement per *g* vanishes when the c.g. coincides with a point called by Gates the *stick-fixed manœuvre point* while the stick force per *g* vanishes when the c.g. coincides with the *stick-free manœuvre point*. In calculating the positions of the manœuvre points it is usual to make the simplifying assumptions already mentioned in § 2·13. Thus it is assumed that the stick position and stick force are measured in a steady motion in a vertical circle, i.e. the forward speed and the angular velocity in pitch are taken to be constant. Also the resultant of the reversed mass-acceleration and weight is taken to be constant in magnitude and always radial.

Let the normal acceleration be ng. Then the angular velocity in pitch q is given by

$$q = \frac{ng}{V}. \qquad (10\cdot7, 1)$$

Also, since the total lift is $(n+1)\,W$,

$$C_L = \frac{2(n+1)\,W}{\rho V^2 S}. \qquad (10\cdot7, 2)$$

The presence of the angular velocity gives rise to an additional pitching moment qM_q (see § 5·3). Since in any steady state the total pitching moment is zero

$$\frac{dC_m}{dn} = \frac{dC_m}{dC_L}\frac{dC_L}{dn} + \frac{dC_m}{d\eta}\frac{d\eta}{dn} + \frac{M_q}{\frac{1}{2}\rho V^2 S\bar{c}}\frac{dq}{dn} = 0.$$

On substitution from (10·5, 2) and (10·5, 5) we obtain*

$$\bar{V}a_2'\frac{d\eta}{dn} = -(h_n - h)\frac{dC_L}{dn} + \frac{2M_q\dfrac{dq}{dn}}{\rho V^2 S\bar{c}} = C_{LO}\left(h - h_n + \frac{gM_q}{WV\bar{c}}\right) \qquad (10\cdot7, 3)$$

on substitution from (10·7, 1) and (10·7, 2), where C_{LO} is the lift coefficient in rectilinear horizontal flight at the speed V. But if the stick-fixed manœuvre point is situated at a distance $\bar{c}h_m$ aft of the c.g. we have by the definition that $\dfrac{d\eta}{dn}$ vanishes when $h = h_m$. Hence we derive from (10·7, 3)

$$h_m = h_n - \frac{gM_q}{WV\bar{c}}. \qquad (10\cdot7, 4)$$

Since M_q is in fact negative the manœuvre point lies aft of the neutral point. Now the major part of M_q is contributed by the tailplane and if we neglect the remainder we get from (5·2, 8) and (5·3, 14)

$$M_q = -\tfrac{1}{2}\rho VS' l_T^2 a_1', \qquad (10\cdot7, 5)$$

where a_1' is the lift slope for the tailplane and the drag co-efficient for the tailplane has been neglected in comparison

---

* In the above equation we treat C_L as independent of η and q which implies neglect of the lift on the tail.

with a_1'. Accordingly

$$-\frac{gM_q}{WV\bar{c}} = \frac{\bar{V}a_1'}{2\mu_1},$$ (10·7, 6)

where μ_1 is the conventional density ratio given by (5·2, 3). Finally we obtain from (10·7, 4) the approximate relation

$$h_m = h_n + \frac{\bar{V}a_1'}{2\mu_1},$$ (10·7, 7)

which gives a slight underestimate of h_m.

We have seen in § 10·4 that the effect of freeing the elevator (when no non-aerodynamic moments are applied to the elevator) is to multiply the effective tail lift slope by the factor

$$\left(1 - \frac{a_2'b_1'}{a_1'b_2'}\right).$$

Accordingly we can deduce from (10·7, 7) that the position of the stick-free manoeuvre point is given approximately by

$$h_m' = h_n' + \frac{\bar{V}a_1'}{2\mu_1}\left(1 - \frac{a_2'b_1'}{a_1'b_2'}\right).$$ (10·7, 8)

The stick-free manoeuvre point will lie aft of the corresponding neutral point.

For a rigid aircraft and with the non-dimensional derivatives independent of V the stick force per g is constant. This follows from the fact that the increments of lift and of hinge moment corresponding to any small movement of the elevator are then both proportional to ρV^2. Since $\eta\rho V^2/g$ is constant, it follows that η/g is proportional to C_L, i.e. the stick movement per g is proportional to the lift coefficient in steady rectilinear flight at the same equivalent airspeed.

Gates and Lyon have pointed out that the stick-fixed manoeuvre margin is approximately proportional to the co-efficient **C** of the longitudinal stability quartic [see equation (5·2, 17)]. For a normal aeroplane the dominant terms in the expression for this coefficient are

$$\frac{1}{i_B}(m_q z_w - \mu_1 m_w) = -\frac{\mu_1 z_w}{i_B}\left(\frac{m_w}{z_w} - \frac{m_q}{\mu_1}\right)$$

$$= \frac{\mu_1 \dfrac{dC_L}{d\alpha}}{2i_B}\left(-\frac{\bar{c}}{l_T}\frac{dC_m}{dC_L} + \tfrac{1}{2}\frac{S'a_1'}{S\mu_1}\right),$$

approximately, from (5·3, 5), (5·3, 7) and (5·3, 14),

$$= \frac{\mu_1 \bar{c} \dfrac{dC_L}{d\alpha}}{2l_T i_B} \left(h_n - h + \frac{\overline{V} a_1'}{2\mu_1} \right)$$

$$= \frac{\mu_1 \bar{c} \dfrac{dC_L}{d\alpha}}{2l_T i_B} (h_m - h) \qquad (10·7, 9)$$

by (10·7, 7). On the basis of this approximate relation $\mathbf{C}$ is positive so long as the C.G. is forward of the stick-fixed manœuvre point whereas $\mathbf{E}$ becomes negative as soon as the C.G. moves aft of the stick-fixed neutral point. According to calculations made by Gates and Lyon the divergence which occurs when the C.G. passes aft of the neutral point remains fairly gentle until the C.G. reaches the manœuvre point but then soon becomes violent as the C.G. continues to move aft.

We have noted in § 5·8 that the values of λ for the 'rapid incidence adjustment' are approximately the roots of

$$\lambda^2 + \mathbf{B}\lambda + \mathbf{C} = 0.$$

If this were exactly true the vanishing of $\mathbf{C}$ would indicate the state of neutral static stability so far as this particular mode of motion is concerned. Since we have just seen that, with close approximation, $\mathbf{C}$ vanishes when the C.G. coincides with the stick-fixed manœuvre point, it follows that in this condition of the aircraft the 'rapid incidence adjustment' is nearly neutral. This was pointed out by Gates and Lyon and goes far to explain the special importance of this point in relation to the general handling qualities of the aircraft.

10·8 The influence of weights and springs attached to the elevator

So long as the elevator is fixed the attachment to it of springs or weights giving rise to hinge moments obviously does not affect the stability but it is otherwise when the elevator is free. Consider first the influence on the stick-free static stability of attaching a spring to the elevator or its control system in such a manner as to give a constant hinge moment in the trailing edge down sense. If for any reason the speed of flight becomes reduced the aerodynamic hinge stiffness of the elevator, which is proportional to the dynamic pressure, is reduced. Hence the downward elevator deflection caused by the spring is increased and the elevator automatically makes a correcting movement; the elevator likewise makes a correcting movement when the speed increases. Thus the spring increases the stick-free static stability and moves aft the neutral point. A constant moment in the same sense due to an attached weight gives the same result. Other things being equal, the static stability will be greater with an under mass-balanced elevator than with a mass-balanced one

Next consider the stick-free manoeuvre point. Since this is a neutral point for flight in a circle at constant speed it is clear that the attachment of a spring has no influence on its position for the hinge moment is independent of 'g'. However, the attachment of a mass giving an elevator hinge moment in the T.E. down sense for positive g adds a restoring moment proportional to g and thus adds to the stick-free manoeuvre margin.

Reversal of the sense of the moments due to springs or weights will, of course, reverse the effects on the stability.

10·9 Influence of the propulsive system and other factors

The foregoing discussion of static stability is based on the assumption (see § 10·3) that any change of speed without change of incidence causes strictly proportionate changes of lift, drag, and pitching moment. This condition is not, in general, fulfilled when the propulsive system of the aircraft is in operation, so the theory strictly applies only to motion in the glide.

The influence of the propulsive system is considered in § 5·6 where it is pointed out that the main effect is a reduction of static stability associated with increased downwash at the tail. This is most severe for twin tractor screws and may be slight or absent with jet propulsion.

The effect of varying the vertical position of the C.G. is discussed in §§ 5·4 and § 5·6 while the influence of distortion of the structure is considered in Chapter 12. The joint effects of compressibility of the air and of distortion on static stability are discussed in § 13·8.

Chapter 11

STALLING AND THE SPIN

By PROFESSOR A. D. YOUNG

Part I:—THE STALL

11·1 Introductory remarks

When we refer to the stalling behaviour of an aeroplane we generally mean its behaviour at incidences in the neighbourhood of that corresponding to its maximum lift coefficient. At such incidences separation of the air flow from parts of the wings has begun and consequently more or less rapid changes in lateral and longitudinal stability and trim result. The lateral instability is usually accompanied by the dropping of a wing tip and it is this which provides the most dangerous feature of the stall. The lateral instability generally appears at some incidence near that corresponding to the maximum lift coefficient, or stalling incidence, and persists over a considerable range of higher incidences.

The reason for this lateral instability is best seen by considering first a wing alone, without fuselage, nacelles, etc. Suppose for simplicity that the variation of lift coefficient with incidence is the same for all sections of the wing and is as illustrated in Fig. 11·1, 1. If a small rate of roll is now initiated by some slight asymmetric disturbance then the incidence of the down-going wing will be increased by the roll, whilst that of the up-going wing will be decreased by it. Consequently, if the initial incidence is below that of the stall S the lift of the down-going wing will be increased and that of the up-going wing will be decreased and the roll will therefore be heavily damped. If, on the other hand, the initial incidence is above that of the stall, the lift of the down-going wing will be further decreased, whilst that of the up-going wing will be increased, and the rate of roll will therefore increase. The wing is then laterally unstable. This phenomenon is readily demonstrated in a wind tunnel by supporting a wing so that it is free to roll at a fixed incidence. At incidences above the stall, the wing can be readily induced to roll and its rate of roll will increase until it reaches a value for which the effective incidence of the two half wings are such that the aerodynamic rolling moment just balances the frictional moment at the pivot, and the rate of roll of the wing will then remain constant. The wing is then said to auto-rotate.*

The fall in lift of a wing section with further increase of incidence beyond the stall is associated with a separation of flow

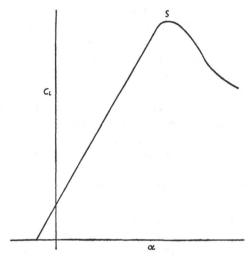

Fig. 11.1, 1. Curve of lift coefficients showing the stall at S.

from part of the wing upper surface. It will be clear therefore that the magnitude of the rolling moment that results when a wing has reached its stall will depend on the distance from the rolling axis of the parts of the wings from which the flow has begun to separate and the rate at which those parts lose lift with increase of incidence. These factors are themselves dependent on the wing sections used and the spanwise distribution of lift and aerodynamic incidence near the stall, and the latter in turn depend on the sections, taper, plan form and wash out of the wings. We can conclude, therefore, that the rolling instability of a plain wing at the stall will be a function of the

* The significance of this phenomenon of auto-rotation in relation to the spin is discussed in § 11·5.

geometry of the wing and the sections used, and the sections over the outer parts of the wing will be the most important.

When we consider a complete aeroplane we find additional factors which influence the spread of breakaway of flow and the consequent stalling behaviour. Wing-nacelle and wing-fuselage junctions are often ready sources of early separation and are of equal importance with the geometry of the wing in determining the origin and rate of spread of the flow breakaway. Further, we find that the longitudinal stability, control and trim at stalling incidences have an important bearing on the lateral stability characteristics. It can be stated that in general any aeroplane will be laterally unstable if its incidence can be raised to a high enough value, but this brings into question the ability of the elevators to raise the aeroplane to the required incidence. Thus, the thickened wake due to an early breakaway of flow from the wing-fuselage junction may influence the elevator and reduce its efficiency. This may result in a fairly sudden nose-down pitching moment which would be reinforced by a change in the pitching moment of the wing in the same direction due to the stalling of the centre section. This change of trim may in some cases be large and sudden enough for the pilot to interpret as the stall, although in point of fact the main parts of the wing may be unstalled. If the pilot pulls the stick further back he may succeed in stalling the wings completely so that the aeroplane becomes laterally unstable. On the other hand the elevator range or efficiency may be such as to make it normally impossible for him to bring the outer parts of the wings to their stalling incidence.* Slip-stream, flaps and engine gills may be expected to modify the intensity and direction of the centre section wake as well as the spanwise lift distribution and hence will affect the stalling behaviour. Wing tip slots will affect it by increasing the stalling incidence over the outer parts of the wings. Thus, it appears that, in addition to being influenced by wing section, plan-form and washout, the stalling behaviour of an aeroplane will be affected by the cleanness of the wing-fuselage and wing-nacelle junctions, the relative positions of wing and tailplane,

* In some cases these aircraft can be made to stall completely in a rapid manoeuvre such as the pull-out from a dive; such stalls are sometimes referred to as dynamic stalls.

the elevator efficiency and range, slipstream, flap and gill settings and the presence and efficiency of wing tip slots. In § 11·3 the separate effects of these factors will be discussed in more detail.

11·2   Assessment of the stall of an aircraft. Stall warning

From the point of view of the pilot the features of main interest to consider in assessing the stalling behaviour of an aeroplane are:

(a) Is there an adequate warning of the stall?

(b) Does a wing or nose drop first and how far and fast does it drop?

(c) How effective are the controls at and beyond the stall in preventing a wing from dropping or in raising a wing once it has fallen?

(d) Is there any tendency to enter a spin after the stall?

Since we are here considering the matter from the pilot's viewpoint, the stall in this context is said to occur when there is some sudden change of longitudinal or lateral trim or some loss of control as the incidence is slowly increased. This does not necessarily correspond to a complete separation of flow from either wing.

It is usual in testing the stall of an aircraft to follow a procedure designed to provide the answers to the above queries. A valuable aid in assessing the onset and spread of flow breakaway is obtained by observing wool tufts, which can freely align themselves with the local direction of the wind, and which are pivoted at various heights to light posts distributed over the wing upper surface. These wool tufts readily indicate the violent eddying and reversed flow characteristic of regions in which the flow is separated from the surface.

Considerable variations in the stalling behaviour of different aircraft are found, depending on the relative influence of the different factors discussed above. Thus, we may find that for aircraft of one type the stall may be marked by a slight gentle change in longitudinal trim with the controls remaining reasonably effective beyond the stall, whilst with another type of aircraft a wing may drop viciously through more than ninety degrees with no possibility of being checked by ailerons or rudder; in some cases the aircraft may enter a spin. Between these extremes all varieties of stalling behaviour are met, as for example, an aircraft may exhibit an initial partial stall marked by the nose dropping a few degrees with no lateral instability, but a further rearward movement of the stick may provoke a complete stall with a wing dropping more or less violently.

In every case the question as to how much warning the pilot gets of the impending stall is of paramount importance. A pilot will generally tolerate an aeroplane with bad stalling characteristics if the aeroplane's behaviour just prior to the stall provides him with adequate warning. The warning usually takes the form of vibration, pitching, tail buffeting (see § 15·6), rapid change of longitudinal trim, or reduction in effectiveness of the control. Unfortunately, the conditions associated with a good warning are usually those resulting from an early partial root stall which is in most cases gentle as it is not necessarily accompanied by lateral instability. On the other hand, the vicious wing drop resulting from initial separation of flow near a wing tip is frequently unaccompanied by any adequate warning. For this reason, the installation of artificial stall warning devices in certain types of aircraft has sometimes been considered. Such devices usually involve a pressure hole near the nose of a wing, placed so that at a few degrees below the stalling incidence the stagnation line moves across the hole over a small range of incidence and the pressure there changes rapidly in consequence. This change of pressure can be arranged to initiate automatically various forms of stall warning, e.g. lights, horns or even stick vibration, but there are well-based objections to all these forms of warning and to date the problem of providing a completely satisfactory form of warning has not been solved.

11·3   Effect of aircraft geometry and design on the stall

(A) *Wing section*

Systematic flight tests have demonstrated that the shape and thickness of the wing section towards the outer parts of the wings play a most important part in determining the nature of the stall. Thus it is found that increase in the camber and thickness* of these sections improves the stall, and a forward position of the maximum camber position (as with the NACA 230 series of sections) is associated with a sharper stall than with wing sections having the maximum camber at a more normal position.

To appreciate why the stall is so sensitive to the wing section used towards the tips, we must first note that the stall can only become dangerous when, with a small change of incidence, the flow breakaway quickly becomes widespread over the outer parts of the wings with a consequent rapid loss of lift there. If the sections towards the tips are such that the spread of flow breakaway is relatively slow with increase of incidence near the stall then we may expect the aircraft to have a gentle stall. We must therefore consider the factors that determine the rate of spread of flow breakaway with incidence on the upper surface of a wing section near the stall.

The pressure distribution around a wing section is determined by its shape, incidence and to a small extent by the Reynolds number. At incidences approaching the stall the pressure on the upper surface generally rises rapidly from a high suction peak near the leading edge to a small positive pressure at the trailing edge, and separation of flow at the trailing edge begins when the positive pressure gradient there becomes so large that the thickened boundary layer can no longer flow against it. If the positive pressure gradient at such an incidence increases in magnitude from the trailing edge forward (as with the NACA 0012 section, see Fig. 11·3, 1a) then it may be expected that the breakaway of flow will spread rapidly forwards and become practically complete with a relatively small increase of incidence. On the other hand, if the positive pressure gradient decreases from the trailing edge forward (as with the NACA 6512 section, see Fig. 11·3, 1d) then we may expect the rate of forward spread of breakaway to be slow. It follows that a section with the former type of pressure distribution will show a large and rapid fall of lift with a small increase of incidence at the stall, but a section with the latter type of pressure distribution will lose lift gradually with increase of incidence at the stall.

* Provided the thickness is greater than about 8 per cent for conventional sections, having the maximum thickness position at about 0·3c behind the L.E., and greater than about 10 per cent for low-drag or high-speed sections having the maximum thickness position between 0·4c and 0·5c behind the L.E.

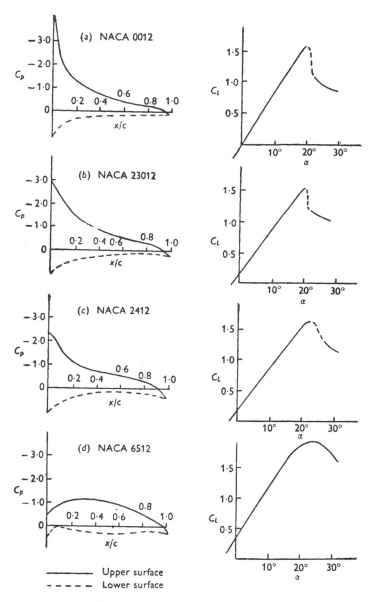

Fig. 11·8, 1. Calculated pressure distributions at $C_L = 1·0$, and measured lift curves (aspect ratio = 6) for a number of NACA sections of thickness-chord ratio = 0·12.

It is found that the characteristic shape of the pressure distribution of a wing section is well in evidence at lift coefficients of the order of 1·0. Hence by obtaining the pressure distribution for a given section at this lift coefficient, either theoretically or experimentally, one can rapidly assess the probable properties of that section. These remarks are illustrated in Fig. 11·3, 1 where the calculated pressure distributions at a C_L of 1·0 and the measured lift coefficient-incidence curves are shown for the NACA 0012, 23012, 2412, and 6512 sections.

It follows from this argument that the stalling characteristics of a wing section can be gauged by the degree of concavity downwards presented by the measured or calculated upper surface pressure distribution when plotted as in Fig. 11·3, 1 with negative pressure coefficients upwards against distance from the leading edge at a C_L of 1·0, say. The greater the concavity the more sudden the stall, and a simple if rough guide to this concavity is the magnitude of the suction peak or minimum pressure coefficient. The following table lists a number of aerofoil sections in order of magnitude of the suction peak:

TABLE 11·3, 1. MAGNITUDE OF SUCTION PEAKS
FOR VARIOUS AEROFOILS

Aerofoil	t/c	Camber (per cent chord)	Suction peak $(p-p_0)/\frac{1}{2}\rho V^2$(min.) for $C_L = 1\cdot0$
NACA 6512	0·12	6·0	−1·06
NACA 4412	0·12	4·0	−1·30
CLARK Y	0·117	3·9	−1·52
NACA 2415	0·15	2·0	−2·00
NACA 2412	0·12	2·0	−2·16
NACA 2212	0·12	2·0	−2·27
RAF 28	0·12	2·0	−2·58
NACA 2409	0·09	2·0	−2·75
NACA 23012	0·12	2·0	−2·83
NACA 0012	0·12	0	−3·79

The above order of listing of these sections corresponds very closely to the order in which one would list them according to the rapidity with which they lose lift with a small change of incidence at the stall.

It may be assumed that the stalling properties of a wing section are a good guide to the stalling properties of a wing having that section towards its tips. This assumption appears to be well borne out in practice and can be supported by the argument that separation of flow may be expected to be contagious; a rapid forward spread of flow breakaway will result in a rapid spanwise spread of breakaway. Thus we find, as indicated by the above table, that increase in camber and thickness of the wing sections near the tips of a wing result in an improvement in the stalling behaviour. The poor stalling properties of the NACA 230 sections are noteworthy.

It is implied in the above discussion that stalling always begins with separation of the turbulent boundary layer from the trailing edge. This is true of wing sections of thickness above a certain critical value depending on the Reynolds number. At normal flight Reynolds numbers this value is about 8 per cent for conventional sections and about 10 per cent for low drag or high speed sections. For sections of thickness below the critical value the positive pressure gradients immediately aft of the leading edge are so intense at even moderate incidences that an early front separation of the laminar boundary layer occurs near the leading edge, followed by the reattachment of a somewhat weakened and turbulent boundary layer which separates in its turn at a higher incidence. Once this process begins the growth of circulation with incidence is effectively stunted and the lift coefficient-incidence curve is flat topped, as is illustrated by the curve for the NACA 2306 section shown in Fig. 11·3, 2. The lateral stability of aeroplanes with wing sections of this order of thinness over an appreciable outer part of their span may therefore be expected to be good at the stall although the maximum lift coefficient attained may be low.

As already remarked, the critical thickness of low drag or high speed sections, below which we find the characteristic stunted lift curve of thin sections, is about 2 per cent greater than that for conventional sections. The more rearward position of the maximum thickness for the former sections is associated with a smaller nose radius of curvature for a given thickness than for conventional sections, and this results in a greater peak suction near the nose at high lift coefficients. For low drag sections of moderate thickness greater than the critical and with some measure of concavity over the rear a typical pressure distribution at a C_L of 1·0 is shown in Fig. 11·3, 3.

Its characteristics are a high peak suction at the nose, a rapid decrease of pressure gradient (convex downwards) from the

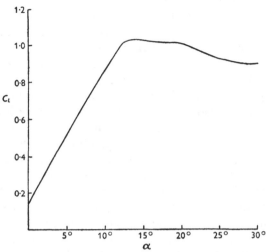

Fig. 11·3, 2. Measured lift curve of aerofoil with NACA 2306 section (A = 6).

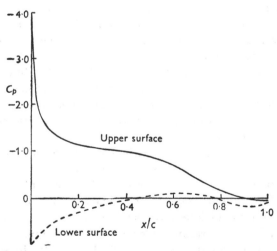

Fig. 11·3, 3. Pressure distribution on low drag aerofoil section (NACA 66-2-216) at $C_L = 1·0$.

nose to about 0·4c, then an increase of pressure gradient (convex upwards) to about 0·7c followed by a decrease of pressure gradient to the trailing edge. One would anticipate, therefore, that for such sections near the stall separation from the trailing edge would start early and spread forward rapidly with a small increase of incidence over the rear third of the section. The separation would then slow up considerably over the middle third of the section and would subsequently spread forward and become complete very rapidly. This leads us to expect that whilst the final stall of a low drag section of moderate thickness may be fairly sharp, the early but limited separation of flow over the rear part of the section and the consequent thickened wake and change in trim characteristics should help to provide some stall warning. The somewhat scanty available experimental data tend to confirm this prediction.

(B) Wing plan form and washout

It can be readily demonstrated that at a given overall lift coefficient the aerodynamic incidence over the outer parts of a wing increases relative to the aerodynamic incidence over the inner parts with increase in taper ratio. For example, for an unswept, untwisted wing the maximum aerodynamic incidence is at the root if the taper ratio is 1 : 1, at about 0·5 of the semi-span if the taper ratio is 2 : 1, and is at about 0·8 of the semi-span for a taper ratio of 5 : 1. It is not surprising therefore to find an increase in taper accompanied by a worsening of stalling behaviour since it encourages an earlier breakaway of flow from the outer parts of the wings.

One might expect that for the same reason washout* should improve the stalling behaviour. This is found to be the case, but for the effect to be at all marked a considerable amount of washout is required. Large washout is generally undesirable since it is accompanied by an appreciable increase in induced drag.

A sweptback wing is more likely to have a stall starting near the tips than a non-swept wing, whilst a swept forward wing is more likely to have a root stall. This is partly due to the changes in spanwise loading produced by sweep back or forward, partly due to the boundary layer drift that occurs from root to tip with sweptback wings and from tip to root with

* A wing is said to have washout when the incidence of the no-lift line at the tip is less than that at the root.

swept forward wings, and partly due to the induced curvature of the flow at the tips. This last effect gives the tip sections an effective negative camber if the wing is sweptback and an effective positive camber if the wing is swept forward (see also § 15·5).

It must be emphasized that the spanwise position where flow breakaway first begins may be considerably influenced by the fuselage, nacelles, slipstream, etc. In any case its importance should not be exaggerated since the rate of spread of flow breakaway when it reaches the outer parts of the wings is more important than where it starts.

(C) Flaps

The lowering of flaps may affect the stalling behaviour of an aeroplane in a number of different and conflicting ways, and the net result may in some cases be a worse stall with flaps down than with flaps up whilst in others the stall may be more gentle.

The following summarizes the main contributions to the total effect of lowering the flaps on the stall of an aircraft:

(*a*) The stall of a wing section with flaps deflected is usually sharper than that of the section with flaps up.

(*b*) The upwash over the outer unflapped parts of the wing is increased and hence the local aerodynamic incidence there is increased.

(*c*) The flaps tend to clean up any incipient breakaway of flow or 'dirtiness' over the wing centre section.

(*d*) The downwash at the tailplane is generally increased.

(*e*) The flap wake may envelop the tailplane at incidences in the neighbourhood of the stall, reducing its efficiency (including that of the elevator) and causing a change of trim. This is generally accompanied by tail buffeting (see § 15·6) and pitching which increase the stall warning.

Effects (*b*), (*c*) and (*d*) tend to worsen the stall, effect (*a*) will also worsen it if the flap is full span or nearly full span, whilst effect (*e*) tends to improve it. To assess the importance of the latter effect for any given design, charts and data are available* from which it is possible to make an estimate of the position,

* Silverstein, Katzoff and Bullivant, 'Downwash and Wake behind Plain and Flapped Airfoils', *NACA T.R. No.* 651 (1939); and Silverstein, Katzoff, 'Design Charts for providing Downwash Angles and Wake Characteristics behind Plain and Flapped Wings', *NACA T.R. No.* 648 (1939).

intensity and width of the wake and the downwash at the tail-plane. It should be noted that buffeting of the tailplane may be felt when the tailplane is outside the wake proper, which is normally defined as the region in which the total head is less than the main stream value. Experiments indicate that the 'buffeting wake' is about twice as wide as the actual wake.*

(D) Wing tip slots, automatic and fixed

If automatic slots of the Handley Page type are fitted to the outer parts of the wing of an aircraft and are efficiently designed and of adequate span they are found to be very effective in producing good stalling behaviour. They increase the stalling incidence over the parts of the wings to which they are attached to a value which is generally outside the range which can be obtained locally by movement of the elevators. In consequence, the stall is then confined to the root section and is mild. In addition, they improve the effectiveness of the ailerons at high incidences. It must be emphasized, however, that with air-craft for which the elevators can bring the slotted parts of the wings to their stalling incidence and so completely stall the wings, a vicious wing drop may readily result. It is probable that in such cases there would be ample warning due to the buffeting and vibration induced by the wake of the earlier stalled centre section. It is clear that full span automatic slots will not necessarily be accompanied by a mild stall, unless the slots are divided into inner and outer portions so designed that the outer portions have a higher final stalling incidence than the inner portions.

Fixed slots or slits cut through the outer parts of the wing also improve the stalling behaviour if properly designed, by increasing the local stalling incidence. However, they are rarely so effective as automatic slots and they cause an appreciable increase in drag at high speeds which is detrimental to per-formance.

(E) Nacelles and slipstream

As already remarked, nacelles frequently induce local centres of early separation which, being usually located over the inner parts of the wings, help to improve the stalling behaviour.

* W. J. Duncan, D. L. Ellis and C. Scruton, 'First Report on the General Investigation of Tail Buffeting', *R. & M.* 1457 (1932).

Thus, the wake from the disturbed regions may reduce the efficiency of the elevator and introduce buffeting as a stall warning, whilst the flow separation may produce a nose down change of trim. These effects are, however, likely to be less marked with the small nacelles of modern jet engines than with the larger nacelles of piston engines.

Slipstream has an opposite effect since it 'cleans' up the flow round the nacelles and over the root section, and in addition may increase the efficiency and downwash over the tail plane. The importance of including in routine stalling tests some tests with throttles open will be realized from these remarks; an aeroplane which may have a gentle stall with engines idling may have a dangerously vicious stall with the throttles open. With jet engines, however, the throttle opening may be expected to have a much less significant effect on the stall.

Part II:—THE SPIN

11·4   Introductory remarks

When an aeroplane has been brought to the stall with the consequent development of lateral and longitudinal instability, and if no attempt is then made by the pilot to unstall the aeroplane, it will execute a complicated manoeuvre involving in general rolling, pitching, yawing and sideslipping and leading in some cases to a spin. The latter when fully developed is a more or less steady manoeuvre which in its simplest terms can be described as a rapid descent following a steep helical path about a vertical axis. The attitude of the aircraft to the vertical may vary from about 30° to 80°, and since the attitude is approximately the mean incidence of the wings, the latter are partly or completely stalled. The radius of the helix is usually a fraction of the span of the aircraft. The steady spin being a state of equilibrium requires a balance between the aerodynamic forces, inertia forces and the weight, and this balance can only occur if the resultant aerodynamic force passes through the vertical axis of spin. At the large incidences of a spin the resultant aerodynamic force always acts very close to the normal to the centre section wing chord through the centre of gravity and in the plane of symmetry; therefore it is usually accepted that this normal passes through the axis of spin.

It is important to note, however, that the plane of symmetry of the aircraft does not necessarily contain the axis of spin, in other words the span of the wing need not be horizontal. Any departure of the wing span from the horizontal may be regarded as obtained by a rotation of the aircraft about the normal to the centre section wing chord through the centre of gravity. A sketch illustrating an aircraft in a steady spin and some of the standard notation adopted in the literature of the subject is shown in Fig. 11·4, 1.

To give some idea of the orders of magnitude of the quantities describing a spin we may note that for aircraft of conventional layout the rate of descent (V) may vary between about 60 f.p.s. and 350 f.p.s., the angle of the flight path to the vertical ($\Omega R/V$) is small, less than 10°, the rate of spin about the vertical axis (Ω) may be between 2 and 8 radians per second, and the radius of the helix (R) may be anything up to about 0·5 wing span. The particular spin which an aeroplane may take up will depend on its geometry, mass distribution, and control settings, but aeroplanes have been known to have more than one steady type of spin for given control settings. In general, with increase of incidence or attitude (α) the rate of spin gets faster and the radius of spin decreases, and we distinguish between the slow steep spin (α less than about 45°) and the fast flat spin (α greater than about 45°). The latter type of spin has been generally regarded as the more dangerous as it tends to be more stable and recovery from it is more difficult. This is in large measure due to the fact that the attitude is such that the fin and rudder, which normally play an important part in the recovery from the spin, may be heavily shielded by the wake from the tail plane. Depending on the rotation and the setting of the wing to the horizontal there will in general be a sideslip of the wing either inwards or outwards. As will be seen later this sideslip plays a vital part in determining the nature of the spin and its stability.

The above presents a simplified picture of a steady spin. In practice an aeroplane may rotate for several turns before its motion reaches a stage that can be described as steady and it may pass with little modification of the control settings from one steady spin to another. In addition, aircraft have been known to oscillate in pitch and yaw about a mean spinning

18

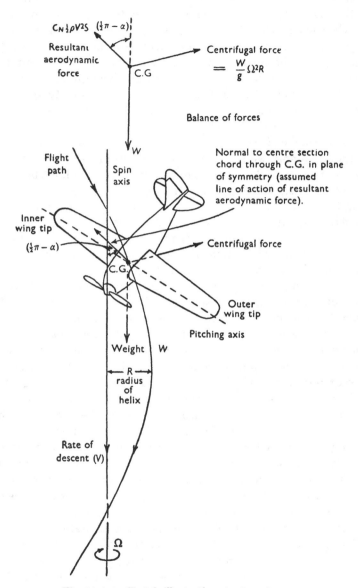

Fig. 11·4, 1. Sketch illustrating steady spin.

condition. The entry to the spin may be made either inadvertently or deliberately, and most aircraft can be made to spin by appropriate movement of the controls at the stall. It is a general principle, therefore, that when an aircraft can be made to spin its design should be such that recovery from the spin is readily possible and without undue loss of height.

It is clear that in a steady spin there must be a balance of the inertia forces, aerodynamic forces and weight as well as a balance of the inertia moments, and aerodynamic moments. These requirements readily yield on analysis certain important facts and data about the steady spin; nevertheless the subject is in detail a complicated one because of the dependence of both types of moments and forces on the motion and vice versa. Thus, it is very difficult in certain cases to unravel the full consequences of say a change in the geometry of an aircraft, or its mass distribution, or in its control settings. In the following only the essentials of the motion will be discussed; for the less basic but by no means insignificant details required by the specialist reference should be made to the extensive literature of the subject (see the list at the end of the chapter).

11·5   Auto-rotation

Reference has already been made (see § 11·1) to the fact that a wing supported so that it is free to roll will auto-rotate at incidences above the stall. Neglecting the frictional torque at the pivot the rate of auto-rotation is determined by the condition that the aerodynamic rolling moment due to roll is zero. A little thought will soon show that the rate of auto-rotation must be a function of the incidence of the wing, and must also depend on the lift and drag characteristics of the wing over the whole incidence range. In the simplest arrangement with the axis of rotation in the wind direction through the centre of gravity there is no sideslip. It is then found that the non-dimensional rate of rotation $\left(\dfrac{\Omega b}{2V}\right)$ for a monoplane increases with incidence to a maximum (of the order of 0·3) at an incidence of about 30° and then falls rapidly, so that above about 40° it will not auto-rotate. A biplane with little stagger, on the other hand, will auto-rotate up to a very large incidence. If the wing is offset from the spinning axis and if it is mounted so that its plane of symmetry does not necessarily include the axis then it may sideslip during the motion. The sideslip has a profound effect on the rate of auto-rotation and the range of incidences over which it occurs: thus both rate and range are increased if the sideslip is in the direction of the outer or rising wing, and conversely both are reduced if the sideslip is in the opposite direction. This is due to the fact that the sign of the derivative, l_v, is always negative and its magnitude is always large above the stall. The same phenomenon of auto-rotation and the effect on it of sideslip are to be observed if the wing is replaced by a complete model, and it is clear from the experimental data that any aeroplane if left free to roll (in the simple manner of our wind tunnel test) can be made to auto-rotate at any incidence from the stall up to, say, 80° by having a suitable amount of sideslip. Hence, although the characteristics of a steady spin are largely determined by the conditions of equilibrium of the forces and of the pitching and yawing moments, the property of auto-rotation of lifting surfaces when stalled is a primary cause of the spin.

11·6   Balance of forces

Referring to Fig. 11·4, 1 we see that the weight of the aircraft must equal the component of the resultant aerodynamic force along the vertical (which we might conveniently call the drag), whilst the horizontal component of the aerodynamic force (or lift) must balance the centrifugal force. Since the resultant aerodynamic force acts along the normal to the root wing chord in the plane of symmetry, we have

$$W = \tfrac{1}{2}\rho V^2 S . C_N \sin \alpha \qquad (11\cdot6, 1)$$

where C_N is the resultant aerodynamic force expressed in the usual way as a coefficient. Hence

$$V^2 = \frac{W}{\tfrac{1}{2}\rho S . C_N \sin \alpha}.$$

After an initial fall with increase of incidence immediately above the stall the resultant force coefficient, C_N, in general increases slowly with further increase of incidence. Hence we find that the rate of descent, V, increases with wing loading and decreases with increase of incidence.

From the balance of the horizontal forces

$$\frac{W}{g}\,\Omega^2 R = C_N \cdot \tfrac{1}{2}\rho V^2 S \cdot \cos\alpha. \qquad (11\cdot6,\ 2)$$

Hence

$$\left.\begin{aligned}
R &= C_N \cdot \cos\alpha\,\frac{g\rho b^2 S}{8W}\Big/\Big(\frac{\Omega b}{2V}\Big)^2 \\
&= g\cot\alpha/\Omega^2,
\end{aligned}\right\} \qquad (11\cdot6,\ 3)$$

so the helix radius decreases with increase of incidence for constant rate of spin or with increase of rate of spin for constant incidence.

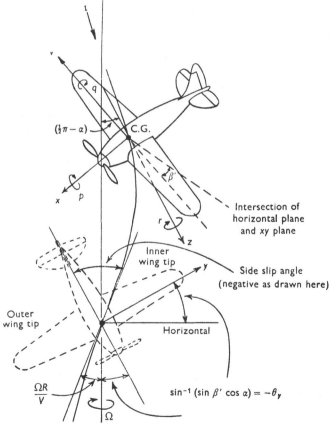

Fig. 11·7, 1. Body axes and sideslip in spin.

11·7   Balance of moments. Sideslip

We will consider body axes with the centre of gravity as origin, the x-axis parallel to the wing chord line in the plane of symmetry, the y-axis in the spanwise direction normal to the plane of symmetry, and the z-axis along the normal to the wing chord line in the plane of symmetry. The angle of rotation of the aircraft about the z-axis (for the general asymmetrical attitude of the spin) will be denoted by β' (see Fig. 11·7, 1). The axes are right-handed and the usual conventions as to the signs of rotations, moments, etc. are observed. If we denote the components of the vector Ω along the three axes by p, q and r, then we readily see that

$$\left. \begin{aligned} p &= \Omega \cos \alpha \cos \beta', \\ q &= -\Omega \cos \alpha \sin \beta', \\ r &= \Omega \sin \alpha. \end{aligned} \right\} \qquad (11\cdot7,\ 1)$$

Ω is positive when right-handed about the downwards vertical axis of spin.

We will assume that the axes taken are also principal axes so that products of inertia about them are zero. The reversed centrifugal or inertia moments are shown in § 3·4 to be $(C-B)qr$, $(A-C)rp$, $(B-A)pq$ about the x, y and z axes respectively. Here A, B and C are the corresponding moments of inertia about the three axes. Consequently, if the aerodynamic moments about the axes are denoted by L, M, N, then for balance of the moments we must have

$$\left. \begin{aligned} L &= -\tfrac{1}{2}\Omega^2 (C-B) \sin 2\alpha \sin \beta', \\ M &= \tfrac{1}{2}\Omega^2 (A-C) \sin 2\alpha \cos \beta', \\ N &= -\tfrac{1}{2}\Omega^2 (B-A) \cos^2\alpha \sin 2\beta'. \end{aligned} \right\} \qquad (11\cdot7,\ 2)$$

We may note that the sideslip angle, β, defined as the angle which the relative velocity makes with the plane of symmetry (positive when the sideslip is towards the starboard wing) is related to the helix angle $(\Omega R/V)$, and α and β' by the equation (see Fig. 11·7, 1).

$$\beta = -\frac{\Omega R}{V} - \sin^{-1}(\sin \beta' \cos \alpha).$$

This is sometimes written as

$$\beta = -\Omega R/V + \theta_y,$$

where

$$\theta_y = -\sin^{-1}(\sin\beta'\cos\alpha)$$

= angle between y-axis and horizontal plane (positive with starboard wing down). $\left.\right\}$ (11·7, 3)

Hence, in a right-handed spin dropping the inner or starboard tip makes the sideslip more positive, and if θ_y is sufficiently great the net sideslip will be towards the inner wing. Conversely raising the inner tip increases the sideslip towards the outer tip.

We can write equations (11·7, 2) in alternative non-dimensional forms as follows:

$$\left.\begin{aligned}
\overline{l\alpha} &= -\tfrac{1}{2}\lambda^2\mu(i_C - i_B)\sin 2\alpha\sin\beta' \\
&= \tfrac{1}{2}\lambda^2 k_A\sin 2\alpha\sin\beta', \\
\overline{m\alpha} &= \tfrac{1}{2}\lambda^2\mu(i_A - i_C)\sin 2\alpha\cos\beta' \\
&= -\tfrac{1}{2}\lambda^2 k_B\sin 2\alpha\cos\beta', \\
\overline{n\alpha} &= -\tfrac{1}{2}\lambda^2\mu(i_B - i_A)\cos^2\alpha\sin 2\beta' \\
&= \tfrac{1}{2}\lambda^2 k_C.\cos^2\alpha\sin 2\beta',
\end{aligned}\right\} \quad (11·7, 4)$$

where*

$$\overline{l\alpha} = L/\tfrac{1}{2}\rho V^2 Sb, \quad \overline{m\alpha} = M/\tfrac{1}{2}\rho V^2 Sb, \quad \overline{n\alpha} = N/\tfrac{1}{2}\rho V^2 Sb,$$

$$\lambda = \Omega b/2V, \quad \mu = 2W/g\rho Sb, \quad i_A = 4Ag/Wb^2,$$

$$i_B = 4Bg/Wb^2, \quad i_C = 4Cg/Wb^2, \quad k_A = 8(B-C)/\rho Sb^3,$$

$$k_B = 8(C-A)/\rho Sb^3, \quad k_C = 8(A-B)/\rho Sb^3.$$

Considering first the balance of the rolling moments we may note that for all aircraft $C > B$ and hence the inertia rolling moment has the same sign as β', i.e. positive with inner wing up and negative with inner wing down in a right-handed spin. Therefore the required aerodynamic rolling moment for balance must be negative when the inner wing is up and positive when it is down. In general, however, the inertia rolling moment is small and the condition for balance is not very different from

* In the literature of spinning it is usual to find the additional symbol α to denote aerodynamic moment coefficients and the symbol i to denote inertia coefficients.

that for auto-rotation. We have already noted that at any given incidence auto-rotation is always possible by an adjustment of sideslip, the balance of rolling moments therefore implies a requirement for the sideslip which is generally easily met.

Turning now to the balance of pitching moments we note that A is always less than C, but the difference is less marked for multi-engined aircraft where much of the weight is in the wings than for single-engined aircraft where most of the weight is concentrated in the fuselage. Hence the inertia pitching moment is always positive tending to increase the incidence and for a given value of $\lambda = \Omega b/2V$ it reaches a maximum when $\alpha = 45°$. The aerodynamic pitching moment to balance it must therefore be negative tending to pitch the nose down. This moment is in the main provided by the normal force on the tail plane, and to a lesser extent by that on the wing, and it is found to increase steadily with incidence and with downward deflection of the elevator. Small increments to the aerodynamic pitching moment are also provided by the rates of roll and pitch and the sideslip, but a good approximation to the aerodynamic pitching moment as a function of incidence can be obtained by static tests of a model. Since the inertia pitching moment for a given value of λ reaches a maximum for $\alpha = 45°$ when plotted against α, and the aerodynamic moment is very roughly linear with α, it follows that for a given value of λ and a given elevator setting there may be two, one or no values of the incidence where a balance is possible (see Fig. 11·7, 2). From the figure it will be clear that where there are two such values the balance is unstable at the lower incidence and stable at the higher.

It follows from the above that at a given incidence the balance of pitching moments determines the rate of rotation, and the particular relation between incidence and rotation rate for any aircraft can be obtained from (11·7, 2) or (11·7, 4) if experimental data for the aerodynamic moments are available and the moments of inertia are known. In general it is found that the larger the incidence above about 45° the higher is the rate of rotation. It is easy to deduce from (11·7, 2) or (11·7, 4) that at a given incidence the following will all result in a reduction of the rotation rate:

(a) Increase in $C - A$ (or k_B), i.e. increase in the loading of the fuselage as compared with the loading of the wings.

(b) Rearward movement of the stick.

(c) Rearward movement of the centre of gravity.

(d) Increase in μ (as, for example, due to an increase of altitude).

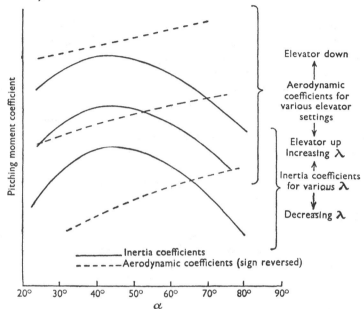

Fig. 11·7, 2. Balance of pitching moments in spin.

It now remains for us to consider the balance of yawing moments. The sign of the inertia yawing moment depends on the sign of $(B - A)$ and of β'. The sign of $(B - A)$ is positive in those aircraft where the fuselage inertia is large relative to the inertia of the wings, as for a single-engined aircraft, whilst it may have a negative sign where the converse holds, as for multi-engined aircraft. Thus, in the former case if the inner wing is high the inertia moment will be pro-spin in the sense that it will tend to speed up and flatten the spin by yawing the aircraft so as to increase the sideslip towards the outer wing and hence increase the auto-rotation rate. With the inner wing low, however, the effect of the inertia moment will be the reverse. Likewise, when the wing inertia is dominant over the fuselage inertia and $B < A$, then the inertia moment will be anti-spin if the inner wing is up and pro-spin if it is down. However, we may note that the inertia yawing moment is usually small compared with the possible values of the aerodynamic yawing moments. This does not mean that the former can be neglected, since in the steady spin the two must balance, but it does mean that by careful design the aerodynamic yawing moment within the pilot's control should be adequate to ensure recovery from the spin and it should be possible to design against a flat spin developing.

The aerodynamic yawing moments derive from the spinning and sideslipping motion of the aircraft as well as the rudder control. These moments can be measured on a model in a wind tunnel in a manner to be described later. All moments are usually quoted in units, a unit being a non-dimensional moment coefficient (e.g. $N/\frac{1}{2}\rho V^2 Sb$) of 0·001. Thus, the full effectiveness of a rudder is generally regarded as being of the order of 10 units. It will be clear that in a right-handed spin a positive yawing moment is pro-spin, and a negative yawing moment is anti-spin. Considering the contributions due to the spinning motion first, these can be regarded as provided by the wing, body and fin. The wing contribution is difficult to predict, it may be of either sign depending on the incidence and wing section, and contributions up to ± 20 units are possible at large incidences and rates of spin. The body contribution is generally anti-spin and can play a vital part in making a dangerous spin impossible. It is not surprising to find that fuselages of round cross-section provide less anti-spin moment than do fuselages of rectangular cross-section, or that increase in length of the fuselage or side area towards the rear (as when tapered to a vertical knife-edge) all help to increase the anti-spin moment. The body contribution generally increases with incidence and with rate of spin; in the latter case the relation is approximately a linear one. Experiments* indicate that a poor-shaped body from a spinning point of view may have a body yawing moment contribution as little as -5 units at 60° incidence and $\lambda = 0·7$, whilst for a good-shaped body the contribution may be of the order of -50 units. Likewise, under the same conditions the

* H. B. Irving, 'Simplified Presentation of the Subject of Spinning of Aeroplanes', *R. & M.* 1535 (1933).

fin and rudder should contribute a moment of the order of
— 30 units, but contributions considerably smaller in magnitude
and even positive in sign have been known at large incidences
where the wake from the tail plane has effectively shielded the
fin and rudder and even reversed their damping effectiveness.
Again, full movement of the rudder should readily provide
moments of the order of ± 10 units or more, but where it is
badly shielded by the tail plane wake its full effectiveness may
be reduced to as low as ± 2 units. It will therefore be apparent
that it is a matter of vital importance in any design to ensure
that as much as possible of the fin and rudder is unshielded by
the tailplane wake. This provides the basis of design criteria
that have been laid down for avoiding the flat spin and ensuring
a ready recovery from the spin. The main methods of achieving
it are either to raise the tail plane as high as possible, or to
move the fin and rudder as far forward of the tail plane as
possible.

The contribution to the yawing moment due to sideslip are
usually small and may be pro-spin or anti-spin depending on
whether the sideslip is towards the inner or outer wing of the
spin. In general aircraft have small but positive weathercock
stability at the incidences of a spin, and the moments due to
sideslip therefore tend to turn the aircraft into wind. Thus,
with sideslip towards the inner wing the yawing moment is
pro-spin tending to tighten and speed up the spin, and the
converse holds with sideslip towards the outer wing. The
former case is likely to occur with biplanes which auto-rotate
readily at high incidences and therefore require inward sideslip
to balance the rolling moments; in such cases a reduction of
weathercock stability is an advantage. Sideslip towards the
outer wing is likely to occur with monoplanes or highly stag-
gered biplanes where outward sideslip is required to preserve
auto-rotation at high incidences; in such a case an increase of
weathercock stability is an advantage.

By way of recapitulation, we note that if we wish to determine the possible steady spin, or spins, which an aeroplane
may take up with given control settings, we require data,
either estimated or experimental, of C_N and $\overline{m\alpha}$ as functions of
α, and of $\overline{l\alpha}$ and $\overline{n\alpha}$ as functions of α and λ and sideslip angle. In
addition, of course, we require to know the moments of inertia.

From the balance of vertical forces we can then deduce for a given α the rate of descent [see (11·6, 1)]. From the balance of pitching moments (11·7, 4) we then deduce λ as a function of α (if in the first instance we treat $\cos\beta' \approx 1·0$) and the balance of rolling moments then gives us β' and therefore the sideslip angle β (11·7, 3) as a function of α, use being made of the balance of horizontal forces (11·6, 3) to determine R. Now having λ and β as functions of α we then seek for the combinations of λ, β and α for which the yawing moments balance. Such combinations determine possible steady spins.

From this point of view, it may be seen that it is the balance of yawing moments that is decisive as to whether a spin can occur. The balance of rolling and pitching moments are generally readily met by adjustments of sideslip and rate of spin, but the yawing moments are as a rule fairly insensitive to the resulting changes of incidence and it is only the rudder that can provide major changes of yawing moment. If for some reason the rudder is relatively ineffective in a steady spin recovery is bound to be very difficult. The importance of ensuring effective rudder control under the conditions of a spin will thus be apparent.

11·8   Recovery from a spin

Let us suppose that an aircraft has been deliberately manœuvred into a spin, the pilot having applied full rudder whilst keeping the stick well back at the stall. There will at first be a stage of incipient spin, as it is sometimes called, during which the aircraft will rotate for a few turns before it finally takes up the attitude and speed of rotation of the steady spin. Recovery from the incipient spin is usually much easier than recovery from the steady spin, as it is generally steeper, and aircraft, including fighters, for which the spin is not regarded as a useful manœuvre are at the most required to recover with appropriate control movements after two turns of the incipient spin. For aircraft such as trainers, however, for which the fully developed spin is a requisite training manœuvre, a ready recovery is required from spins of up to eight turns' duration. The action required of the pilot to bring an aircraft out of the spin may vary a little from aircraft to aircraft but in general for normal types of aircraft it involves applying opposite rudder and about half a turn later downward elevator. It will be readily appreciated from the foregoing discussion how these control movements help to bring the aircraft out of the spin. The advantage of a time lag between the rudder and elevator movement follows from the fact that the unshielded part of the fin and rudder below the tail plane plays a vital part in the early stages of the recovery in providing damping and control effectiveness. Downward movement of the elevator reduces this unshielded part and if made too early may hamper rather than help the recovery.

The ailerons are sometimes used to assist recovery from the spin, but it is found that the optimum direction of movement of the ailerons varies from aircraft to aircraft. In general, for aircraft for which $B < A$, the wing inertia dominates the fuselage inertia, so the recovery is better when the ailerons are set to give an anti-spin rolling moment, whilst for aircraft for which $B > A$, and the fuselage inertia dominates the wing inertia, the recovery is better when the ailerons are set to give a pro-spin rolling moment. An explanation for this can be offered on the following lines. In the steady spin an increase in pro-spin rolling moment will result in an increase in inward sideslip and therefore in a dropping of the inner wing. We have seen that if $B < A$ this in turn results in a pro-spin inertia yawing moment increment, whilst if $B > A$ the resulting increment in inertia yawing moment is anti-spin. As already remarked, it is these final changes of yawing moment that are all-important in deciding whether or not a spin persists. The picture is to some extent complicated by the yawing moments that result directly from the movements of the ailerons. Pro-spin movement will produce an anti-spin yawing moment and vice versa.

In extreme cases the thrust from one or more of the engines of a multi-engined aircraft is sometimes used to provide the yawing moment needed to recover from a spin. The effectiveness of this method may be augmented or reduced by the gyroscopic couples that develop on the rotating parts of the engines concerned (including any propellers), depending on the sense of rotation of these parts and the sense of the spin.

For certain experimental aircraft and also prototype aircraft first undergoing spinning or stalling trials it is customary to fit an anti-spin parachute, anchored to a convenient strong point towards the rear of the fuselage. The parachute can be opened in a spin from which recovery by all the more normal methods has proved impossible. It provides a very powerful anti-spin yawing moment and pitching moment which make recovery certain, but the loads on the aircraft may be severe. The parachute is, of course, discarded as soon as the aircraft has left the spin.

Finally it should be noted that the inertia loads on an aircraft, both during the spin and the recovery, may be large, particularly on those parts furthest from the axis of spin. In general the problem of stressing for these loads becomes more serious with increase in the size of the aircraft. This consideration, alone, is sufficient to exclude the spin as an acceptable manœuvre for large aircraft. The stick forces required in the recovery also present a problem as they may in certain cases be very heavy, possibly beyond the strength of the pilot, for even small aircraft.

11·9   Experimental methods

The methods of test available for experimental research on the general problem of spinning as well as on the spinning characteristics of particular aircraft include full-scale flight tests, model tests on a spinning balance in a wind tunnel, and free flight tests of spinning models in a vertical jet (or spinning tunnel). All three methods have their uses and limitations. Full-scale flight tests are clearly limited on the grounds of expense, danger, the narrow range of variables under control and general inconvenience. Nevertheless, a great deal of valuable knowledge has been obtained from such tests, and, as long as scale effect is an unknown factor in spinning, flight tests will be required to provide checks on the results obtained with models.

Model tests on a spinning balance in a wind tunnel are generally conducted with the axis of spin parallel to the wind direction, and the rig permits the model to be rotated steadily in an attitude simulating a steady spin. Quantities that can be varied from test to test are incidence, angle of sideslip, rate of rotation, radius of helix, and wind speed (or rate of descent) and the rig may be designed to measure some or all of the six

components of the resultant force and moment on the model.*†
By completely enclosing the model in a shield which rotates
with it, the inertia contributions to the force and moment
components can be derived. Thus, such a rig can be used to
provide extremely valuable and systematic data for fundamental research on the steady spin.

In the spinning tunnel,‡ as it is called, a model can be made
to reproduce in free flight the latter stages of the entry to the
spin as well as the steady spin. The axis of the wind tunnel jet
is vertical with the air coming from below. The model is
launched in a spinning attitude either from a spindle or by
hand into the jet, with the controls usually set in pro-spin
positions. When the spin has become steady the wind speed
is adjusted so as to keep the model conveniently at the level
of the observation room and then, either by remote control or
by the automatic operation of a mechanism (usually clock-
work) inside the model, the controls can be made to take up
settings designed to bring the model out of the spin. Usually
the mechanism permits a variety of recovery procedures to be
tried, including the time-lag generally required between rudder
and elevator movement. At the end of the test, or if the model
recovers and enters a dive, it is caught in a safety net below
the working section.

The spinning tunnel that is proposed for the National Aeronautical Establishment at Bedford is illustrated in Fig. 11·9, 1.
This tunnel is planned, by current standards, on ambitious lines;
the power available is 3000 B.H.P. and the tunnel will be capable
of being pressurized up to four atmospheres, giving a welcome
range of Reynolds numbers. The guide vanes and screens,
after the annular return, at the bottom of the tunnel are
designed to produce slightly faster moving air towards the
outside of the working jet than in the centre. This helps to
keep the model in the centre of the jet; with slower moving air
towards the outside, as with a thick boundary layer, the model

* M. J. Bamber and C. H. Zimmerman, 'The Aerodynamic Forces and
Moments exerted on a Spinning Model of the $NY-1$ Airplane as measured
by the Spinning Balance', *NACA Report No.* 456 (1933).

† P. H. Allwork, 'A Continuous Rotation Balance for the Measurement of
Yawing and Rolling Moments in a Completely Represented Spin', *R. & M.*
1579 (1933).

‡ A. V. Stephens, 'Free Flight Spinning Experiments with Several Models',
R. & M. 1404 (1931).

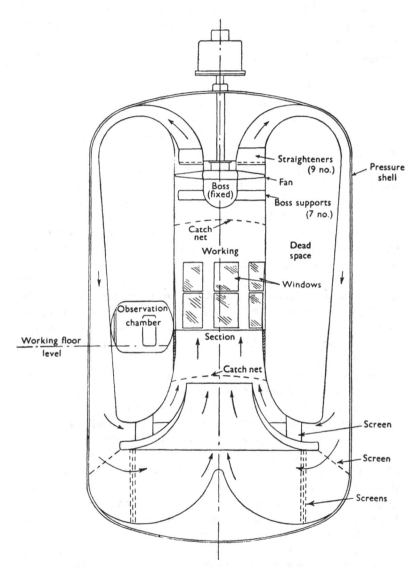

Fig. 11·9, 1. N.A.E. spinning tunnel. Max. power—8000 H.P. Pressure—1 atmos.–4 atmos. (*Abs.*). Corresponding max. speed—140 f.p.s.–90 f.p.s. Working section—15 ft. diameter. Model span—8 ft. 6 in.

tends to move to the outer parts of the jet and damages itself
on the tunnel walls.

It will be clear that the model must be made as far as is
possible both geometrically and dynamically similar to the full-
scale aircraft that it is intended to represent, and allowance
must be made for the altitudes at which the full-scale aircraft
may be expected to spin. The model is usually made of balsa
wood and the correct weight and moments of inertia are
obtained by inserting lead weights. It has been generally found,

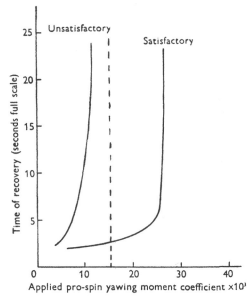

Fig. 11·9, 2. Influence of pro-spin yawing moment on time of
recovery.

however, that models spin steeper and recover more readily
than do the full-scale aircraft they represent. This difference
has been ascribed to scale effect; for example, there is evidence
to indicate that the damping in yaw due to the body decreases
rapidly in certain cases with increase of Reynolds number. To
meet any differences due to scale effect, models are usually
tested with small vanes on arms attached to the inner wing tip
giving known pro-spin yawing moments. If the time to recover
is plotted against the applied yawing moment coefficient curves
such as those illustrated in Fig. 11·9, 2 are obtained. It will

be seen that in each case there is a value of the applied yawing moment coefficient for which the time to recover becomes very large with a further small increase of applied yawing moment, and this value is sometimes referred to as the threshold value. If this threshold value is of the order of seventeen units or more then the recovery of the full-scale aircraft from a fully developed spin is expected to be satisfactory, but if the threshold value is markedly less than seventeen units then the aircraft may not be expected to recover readily. This is a rough and ready rule which has been found to apply in most cases, but cases have arisen where it has failed.* A factor which has only recently been appreciated is the effect of the rolling moment introduced by the yaw vane on the nature and stability of the spin. This rolling moment will modify the sideslip in the manner discussed in § 11·7, and its resultant effects on the spin will change sign as $(B-A)$ changes sign. There is evidently room for a considerable amount of research on scale effects in spinning.

* G. E. Pringle, 'The Difference between the Spinning of Model and Full-scale Aircraft', *R. & M.* 1967 (1943).

GENERAL REFERENCES FOR CHAPTER 11

General reviews as well as comprehensive reference lists will be found in the following:—

STALLING

A. D. Young, 'A Review of Some Stalling Research', *R.A.E. Report No. Aero.* 1718, *R. & M.* 2609 (1942).

THE SPIN

H. B. Irving, 'Simplified Presentation of the Subject of Spinning of Aeroplanes', *R. & M.* 1535 (1933).

B. M. Jones, *Aerodynamic Theory* (ed. Durand), Berlin, 1935, vol. 5, p. 204 *et seq.*

J. H. Crowe, *Aircraft Engineering*, vol. XI (1939), pp. 39 *et seq.*, 111 *et seq.*, 150 *et seq.*, 203 *et seq.*, 273 *et seq.*

Chapter 12

THE INFLUENCE OF
DISTORTION OF THE STRUCTURE

12·1 Introduction

In the theory we have till now assumed that the structure of the aircraft and of each control flap is rigid. Provided that the stiffnesses are high and the speed of flight low the assumption of rigidity is justified, but in practice the effects of the flexibility of the structure are often not negligible, especially in flight at high speeds. The first phenomenon depending on structural flexibility to receive much attention was flutter, but this subject lies beyond the scope of this book. At present the subject of 'aero-elasticity' covers a wide field and is recognized as being of great importance. It includes, besides the very extensive and intricate subject of flutter, such matters as wing divergence (see § 2·11), reversal of control (see § 2·12) and the influence of structural flexibility on general stability and manœuvrability.

Structural distortions are of two main types:

(a) Those involving the main structure and thus altering the configuration of the aircraft as a whole. Such distortions can be treated adequately as elastic, as a rule.

(b) Local distortions, especially those affecting the surfaces exposed to the air stream. Such distortions may be inelastic in the sense that they do not obey Hooke's law.

Distortions of type (a) are of fundamental importance as regards flutter, divergence, reversal of control and general stability and manœuvrability. Those of type (b) are chiefly of importance in relation to control forces since the 'heaviness' of a control flap may be sensitive to small distortions of its surface.

In this chapter our principal aim is to explain the methods which can be applied to problems of aircraft dynamics involving distortion of the structure and we shall not attempt to give an exhaustive catalogue of the results of distortion. The joint influence of distortion and of the compressibility of the air on longitudinal stability is considered in § 13·8.

12·2   Methods for investigating the effects of structural flexibility

In all treatments of problems of aircraft stability it is necessary to allow for the general freedoms of the aircraft but, in addition, we have now to take account of the distortional freedoms. Provided that the aircraft is completely symmetric about the plane OXZ (this includes the condition of symmetry as regards the flexibilities) and that the undisturbed steady state of flight is symmetric (see § 3·9), it will remain true that small motions of deviation of the symmetric and of the anti-symmetric types are independent. At present we shall suppose that the structure obeys Hooke's law, so all distortions are linear homogeneous functions of the applied loads.

The principal methods for investigating the effects of structural flexibility are:

(a) Exact solution of the differential equations governing the distortions.

(b) Iterative determination of the distortions.

(c) The method of distortional co-ordinates or representation of the flexible structure by a 'semi-rigid' counterpart.

(d) The method of the modification of derivatives.

A few remarks on each of these will now be given before entry into any detailed discussions.

Method (a) is not of general applicability for the reason that aircraft structures are so complicated that the representation of their elastic properties by differential equations is, in general, impracticable. However, the method can be applied to specially simplified structures which may be supposed to be substituted for the real structures. For example, exact and tractable differential equations governing the distortions of wings or tailplanes can be established when the surface is supported by parallel uniform cantilever spars and the load on each perpendicular strip of the surface is supposed to be directly transmitted to the spars.* This method will not be further discussed here.

In method (b) we require to know the coefficients of flexibility (sometimes known as influence coefficients) of the elastic

* See H. Bolas, 'Tail Flutter—A New Theory', *Aircraft Engineering* (March 1929). This paper is misnamed, since the subject discussed is divergence, not flutter. Wing flutter is discussed on the same basis by Frazer and Duncan in 'Conditions for the Prevention of Flexural-torsional Flutter of an Elastic Wing', *R. & M.* 1217 (1928).

structure for the kinds of load which occur and for the important kinds of distortion, either at all points affected by distortion or (in a less exact treatment) at a number of representative stations. We begin by finding the distortions under an assumed load system, corresponding perhaps to some guessed state of distortion or to the state of zero distortion. The loads are then re-calculated with these distortions and new distortions calculated. The process is continued and, if the calculated distortions converge, the required solution is obtained. This method is much too laborious for general use, but it is briefly discussed in § 12·3.

In method (c) we allow one or more modes of distortion, in proportion to suitably chosen known function of position in the structure, for each important kind of distortion and then derive the corresponding dynamical equations by the Lagrangian method of virtual work. This is the only method hitherto devised which is both generally applicable and workable. It is discussed in § 12·4 and applied later to particular problems (see also § 15·10).

In method (d), which is due to Gates and Lyon,* the effects of distortion are allowed for by modifying the values of the ordinary aerodynamic derivatives. This procedure can only be accurate when the distortions occur very slowly in relation to the lowest relevant natural frequency of structural oscillation and it is inapplicable when the forces and moments proportional to the distortional velocities and accelerations cease to be negligible. However, the method has the great merit of simplicity and, within its limited field of applicability, it is very valuable. This method is considered in § 12·6.

12·3 Iterative determination of the distortions

This method has been described in general outline in § 12·2. We shall now illustrate the method by applying it to determine the twist and the loads on an elastic aerofoil with fixed root, placed in a given uniform air stream. The aerofoil considered may be sweptback but we shall suppose it divided up, for the purpose of the calculation, into strips lying fore-and-aft and we shall suppose that these strips move as rigid bodies when the aerofoil distorts. This last assumption is not quite accurate

* 'A Continuation of Longitudinal Stability and Control Analysis, Part I, General Theory', *R. & M.* 2027 (1944).

since there will, in general, be some distortion of the profile shape of a fore-and-aft strip when the aerofoil twists or bends. We shall neglect the influence of any fore-and-aft force in distorting the structure.

The elastic specification of the wing is provided by the flexibility or influence functions, which are supposed known. Take a convenient reference point P in the fore-and-aft strip situated at a perpendicular distance y' from the fixed root and let the normal loads on the strip of width dy' be equivalent to a normal force $N(y')dy'$ at P and a couple $M(y')dy'$ whose axis is perpendicular to the strip and in the plane of the aerofoil. Let the twist measured in the fore-and-aft plane at section y due to unit normal load at P be $n(y, y')$ and let the twist due to unit couple at P be $m(y, y')$. Then the total twist at section y is

$$\theta(y) = \int_0^s n(y, y')N(y')dy' + \int_0^s m(y, y')M(y')dy'. \quad (12\cdot3, 1)$$

Since the twist is added to the local angle of incidence it will be possible to express the local aerodynamic loads by the equations

$$N(y') = N_0(y') + \theta(y')N_1(y'), \quad\quad (12\cdot3, 2)$$

$$M(y') = M_0(y') + \theta(y')M_1(y'). \quad\quad (12\cdot3, 3)$$

We begin the process by calculating the load system from the last equations with $\theta(y')$ taken zero or assigned some value estimated to be near the truth. The value of $\theta(y)$ is then found from $(12\cdot3, 1)$ and this in turn is used to find the loads from $(12\cdot3, 2)$ and $(12\cdot3, 3)$. The process is continued until successive values of θ agree sufficiently closely or it is clear that the process fails to converge. The latter occurrence would indicate that the air speed equalled or exceeded the divergence speed of the aerofoil.

Strictly speaking the functions $N_1(y')$ and $M_1(y')$ depend on the distribution of incidence along the span and therefore on $\theta(y)$. To allow for this it would be necessary to begin with the functions N_1 and M_1 appropriate to an assumed function $\theta(y)$, as given by lifting line theory, say. Then the function $\theta(y)$ finally obtained from the iteration would be compared with the originally assumed function and, if the discrepancy were serious, the functions M_1 and N_1 could be redetermined. It is to be noted that the originally assumed and finally obtained functions must be compared after their values at say the tip had been made to agree by multiplying one of them by a constant. The whole process could be continued until a final convergence was obtained but this is of little practical concern.

In the practical application of this method it will usually be necessary to make an approximation by supposing the wing to be divided into a finite number of strips of equal width and then to use Simpson's rule or other formula of approximate integration in evaluating the integrals in (12·3, 1). It is now only necessary to know the values of a finite number of flexibility coefficients evaluated for loads applied and twist measured at mid-span of each strip.

For aerofoils whose elastic properties are simple it may be possible to simplify the procedure. For example, let the aerofoil have a locus of flexural centres such that a normal load applied anywhere on this locus causes no twist *anywhere*. Then if we take our reference point P on the locus of flexural centres the first integral in (12·3, 1) will disappear and equation (12·3, 2) need not be used.

The iterative method may be used to determine the divergence speed of the aerofoil by the following device.* Find for a range of air speeds the twist at some chosen section and plot the reciprocal of this against the reciprocal of V^2. The intersection of the curve, which will differ little from a straight line, with the axis of abscissae determines the divergence speed and it can be found with sufficient accuracy without calculating the twist at a speed very near the divergence speed (cp. § 2·11).

12·4 The method of distortional coordinates or of semi-rigid representation

We have already explained in § 12·2 that in this method we assume that each principal kind of distortion is linearly compounded from suitably chosen known functions,† and then derive a complete set of dynamical equations by the Lagrangian principle of virtual work. In its simplest form there is just one

---

* See 'The Divergence Speed of an Elastic Wing', by A. G. Pugsley and G. A. Naylor, *R. & M.* 1815 (1937).

† The functions may in practice be defined by graphs or tables.

distortional function and co-ordinate for each kind of distortion (e.g. wing twist or bending). For complete accuracy it would, in general, be necessary to have an infinite set of co-ordinates for each type of distortion but in practice a very small number suffices, provided that the functions representing the distortional displacements are well chosen. This point is further discussed below.

There is another way of looking at this method which may be helpful. When we assume, for instance, that the wing twist θ is expressible by the equation

$$\theta = \theta_1 F_1(y) + \theta_2 F_2(y), \qquad (12\cdot4, 1)$$

we are, in effect, substituting a mechanism having just two degrees of freedom for the elastic wing; in other words, we are replacing the elastic wing by a 'semi-rigid' structure. The word 'semi-rigid' is used to indicate that the structure has only a finite (usually small) number of degrees of freedom. When the displacement functions are of simple type it may be possible to imagine simple mechanisms consisting of rigid elements by which they could be realized. For example, if we assume that the wing has a pair of rigid and parallel spars with parallel hinges at the root the flexural and torsional displacements will both be proportional to the distance from the root. It is, however, entirely superfluous to consider the details of the mechanism since all that is needed for the formulation of the dynamical equations is the expressions for the displacements in terms of the distortional co-ordinates, such as θ_1 and θ_2 in equation (12·4, 1).

We must now examine the adequacy of semi-rigid representation of an elastic component of an aircraft. For definiteness let us think of the bending deflections of a wing or fuselage. If the loads were always distributed in the same way, i.e. if the loads at all points varied in proportion, the mode of bending would be fixed, and we could write with complete accuracy

$$z = z_1 f_1(y), \qquad (12\cdot4, 2)$$

where z is the deflection, y the distance from the root or other reference point, z_1 the distortional co-ordinate and $f_1(y)$ gives the deflection for one particular value of the load intensity, corresponding to the unit value for z_1. Thus, so long as the loads are distributed in this particular manner we shall represent the distortion accurately by the formula (12·4, 2). In practice it is common for the load distribution on a given structural component to vary only slightly; for example, the bending loads on a fuselage are almost wholly attributable to the tail and are therefore of a fixed type to a considerable degree of accuracy. Representation of one type of distortion by two well-chosen functions will nearly always be amply accurate for practical purposes and often a single function will suffice. It can be accepted that a single well-chosen function will always be adequate in investigations of stability and control unless very accurate quantitative results are needed.*

We shall suppose now that the distortional co-ordinates have been selected and shall pass on to consider the construction of the dynamical equations. The complete set of these will consist of the following: (a) The dynamical equations for the 'rigid body freedoms' of the aircraft as a whole, for the type of motion under consideration (i.e. longitudinal-symmetric or lateral-antisymmetric). These will be the standard equations as obtained in Chapter 3 but with additional terms in the distortional co-ordinates representing the coupling between these co-ordinates and the particular dynamical variable which corresponds to the equation. For example, in the dynamical equation corresponding to u the coupling terms will represent the influence of the distortions on the force corresponding to u, namely, the longitudinal force X. (b) Additional dynamical equations, one corresponding to each of the distortional co-ordinates. The equation corresponding to a distortional co-ordinate q_r will, in general, contain terms representing the couplings between this co-ordinate and the dynamical variables for the aircraft as a whole as well as the other distortional co-ordinates.

The first step towards constructing the dynamical equations consists in obtaining the complete expressions for the displacements, velocities and accelerations of all points of the aircraft.†

---

* For further discussion of the justification of semi-rigid representation the reader may consult 'The Representation of Aircraft Wings, Tails and Fuselages by Semi-rigid Structures in Dynamic and Static Problems', by W. J. Duncan, *R. & M.* 1904 (1943).

† We shall not be concerned with linear displacements referred to fixed axes; angular displacements are important.

When the point in question is moved when a distortional co-ordinate changes, the expressions for the velocity and acceleration will contain respectively terms in the first and second differential coefficients of the distortional co-ordinate with respect to the time. From a knowledge of the displacements, velocities and accelerations we can derive the local forces, including the inertia forces (reversed mass-accelerations). Finally, the dynamical equation corresponding to any dynamical variable represents mathematically the condition that the work done by all the forces (including the inertia forces) is zero in an imaginary or 'virtual' displacement corresponding to a small increment of this dynamical variable alone (the other dynamical variables being taken as fixed). This procedure will be exemplified below by one specific example. It should be noted that the compact forms of Lagrange's dynamical equations (containing derivatives of the kinetic and potential energies) are only applicable when *all* the dynamical variables are displacement co-ordinates. Since *u, v, w,* for example, are not displacement co-ordinates but velocity deviations referred to body axes, the compact forms of the equations cannot, in general, be used in the present field of inquiry.

The distortions are assumed to be so small that squares and products of these and of their time rates of change can be neglected. Thus the dynamical equations are linear with co-efficients which can be treated as constant in investigating the nature of the small motions of deviation from a steady state.

12·5   Influence of the flexibility of the fuselage

As an example of the use of distortional co-ordinates or 'semi-rigid representation' we shall consider the influence of the vertical flexibility of the fuselage in longitudinal-symmetric motion. We shall suppose that there is a single distortional co-ordinate Φ and that the bending of the fuselage gives a displacement in the direction of the body axis OZ equal to

$$z = \Phi f(x), \qquad (12.5, 1)$$

where the function $f(x)$ defines the mode of bending.* Since Φ is supposed to be a small quantity whose square can be neglected

---

* Note that z is positive when Φ is positive. Hence $f(x)$ is positive with x negative.

we derive from § 3·3 the following expressions for the components of velocity and acceleration in the direction OZ for any point of the rear fuselage or tail

$$w' = w - qx + \Phi f(x), \qquad (12\cdot5, 2)$$

$$\gamma = \dot{w} - qV - \dot{q}x + \ddot{\Phi}f(x), \qquad (12\cdot5, 3)$$

where we have assumed that OX is the wind axis ($W = 0$, $U = V$). The inertia force in the direction OZ of a particle of mass δm is $-\gamma\delta m$. Hence the whole inertia force in the direction of OZ is

$$-\Sigma\gamma\delta m = -\Sigma[\dot{w} - qV - \dot{q}x + \ddot{\Phi}f(x)]\,\delta m, \qquad (12\cdot5, 4)$$

and the inertial pitching moment is

$$\Sigma\gamma x\delta m = \Sigma x[\dot{w} - qV - \dot{q}x + \ddot{\Phi}f(x)]\,\delta m, \qquad (12\cdot5, 5)$$

where it is to be remembered that x is positive for points forward of the c.g. The expression in $(12\cdot5, 4)$ will appear on the right-hand side of the equation of normal forces. Consequently the coupling term $\ddot{\Phi}\Sigma f(x)\,\delta m$ will appear on the left, where the summation covers all points of the fuselage and tail which partake of the flexural motion. The complete equation of normal forces will accordingly be [cp. equation $(3\cdot11, 12)$]

$$-uZ_u + \dot{u}m - wZ_w + \theta mg\sin\Theta - q(mV + Z_q)$$
$$+ \ddot{\Phi}\Sigma f(x)\,\delta m - \Phi Z_{\dot{\Phi}} - \Phi Z_{\Phi} = Z(t), \qquad (12\cdot5, 6)$$

where $Z_{\dot{\Phi}}$ and Z_{Φ} are aerodynamic derivatives. The latter are almost wholly attributable to the tailplane whose effective angle of incidence depends on both Φ and $\dot{\Phi}$. Thus, when Φ is positive (downward bending) the angle of incidence of the tailplane is increased by the slope of the deflection at the tailplane, i.e. by the value of $-\Phi f'(x_t)$ since the positive sense of x is forward, where x_t is the value of x at, say, the quarter-chord of the tailplane. The downward velocity of the tailplane due to bending is $\dot{\Phi}f(x_t)$ and this gives rise to an increase of tailplane incidence amounting to $\dot{\Phi}f(x_t)/V$. The increments of normal force, of longitudinal force and of pitching moment corresponding to these incidence changes can easily be calculated and so the derivatives

$$Z_{\Phi},\ Z_{\dot{\Phi}},\ X_{\Phi},\ X_{\dot{\Phi}},\ M_{\Phi}\ \text{and}\ M_{\dot{\Phi}}$$

can be found.

The longitudinal force is unaffected by the bending except through the derivatives X_Φ and $X_{\dot\Phi}$. Accordingly equation (3·11, 11) becomes

$$\dot u m - u X_u - w X_w + \theta mg \cos\Theta - q X_q - \dot\Phi X_{\dot\Phi} - \Phi X_\Phi = X(t).$$
(12·5, 7)

It readily follows from (3·11, 13) and (12·5, 5) that the equation of pitching moments is

$$- u M_u - \dot w M_{\dot w} - w M_w + \dot q B - q M_q - \ddot\Phi\Sigma x f(x)\,\delta m$$
$$- \dot\Phi M_{\dot\Phi} - \Phi M_\Phi + \Phi g\sin\Theta\Sigma f(x)\,\delta m = M(t), \quad (12\cdot5, 8)$$

where the last term on the left allows for the change in the gravitational pitching moment due to the bending. In order to find the flexural equation we must calculate the work done in the virtual normal displacement $\delta z = f(x)\,\delta\Phi$ and with the associated increment of slope $-f'(x)\,\delta\Phi$. In this manner we derive an equation which may be written

$$- \dot u F_u - u F_u + \dot w\Sigma f(x)\,\delta m - \dot w F_w - w F_w - \dot q\Sigma x f(x)\,\delta m$$
$$- q V\Sigma f(x)\,\delta m - \dot q F_{\dot q} - q F_q + \theta g\sin\Theta\Sigma f(x)\,\delta m + \ddot\Phi\Sigma f(x)^2\,\delta m$$
$$- \dot\Phi F_{\dot\Phi} + \Phi(-F_\Phi + f_\Phi) = F(t).$$
(12·5, 9)

Here we use F to denote the generalized flexural force with a suffix to indicate the particular derivative, as usual. $F(t)$ represents the generalized flexural force caused by operation of the controls or the impact of gusts, while f_Φ represents the elastic and gravitational flexural stiffness. Some further remarks on the coefficients in this equation are given at the end of this section; several of them can be neglected without serious error. The complete set of dynamical equations corresponding to forward forces, normal downward forces, pitching moment and generalized flexural force are (12·5, 7), (12·5, 6), (12·5, 8) and (12·5, 9) respectively. The corresponding determinantal equation for free motion is a sextic in λ.

We shall content ourselves here with considering the influence of the flexibility of the fuselage on the longitudinal static stability. The static stability will be positive so long as the constant term p_0 in the determinantal sextic is positive (see § 4·10). This coefficient is the determinant of the terms independent of λ in the equations of motion. Accordingly

$$
p_0 = \begin{vmatrix} -X_u & -X_w & mg\cos\Theta & -X_\Phi \\ -Z_u & -Z_w & mg\sin\Theta & -Z_\Phi \\ -M_u & -M_w & 0 & -M_\Phi + g\sin\Theta\Sigma f(x)\,\delta m \\ -F_u & -F_w & g\sin\Theta\Sigma f(x)\,\delta m & -F_\Phi + f_\Phi \end{vmatrix}.
$$
$$(12\cdot5, 10)$$

In the further discussion we shall take Θ to be zero (horizontal flight, since we have adopted wind axes). We then derive immediately

$$
\frac{p_0}{mg} = \begin{vmatrix} -Z_u & -Z_w & -Z_\Phi \\ -M_u & -M_w & -M_\Phi \\ -F_u & -F_w & -F_\Phi + f_\Phi \end{vmatrix}.
$$

It is convenient to expand this in terms of the elements in the third column. Then

$$
\frac{p_0}{mg} = (-F_\Phi + f_\Phi)(Z_u M_w - Z_w M_u)
$$
$$
+ M_\Phi(Z_u F_w - Z_w F_u) + Z_\Phi(M_w F_u - M_u F_w). \quad (12\cdot5, 11)
$$

Now $(-F_\Phi + f_\Phi)$ is the total effective flexural stiffness and is certainly positive. When we divide by this we obtain

$$
\frac{p_0}{mg(-F_\Phi + f_\Phi)} = (Z_u M_w - Z_w M_u)
$$
$$
+ \frac{M_\Phi}{(-F_\Phi + f_\Phi)}(Z_u F_w - Z_w F_u)
$$
$$
+ \frac{Z_\Phi}{(-F_\Phi + f_\Phi)}(M_w F_u - M_u F_w). \quad (12\cdot5, 12)
$$

Now the first bracket on the right of this equation is the value of p_0 in the absence of fuselage flexure. Hence the remaining terms give the effect of the flexibility in modifying the static stability.

The derivative Z_Φ is small and will now be neglected.* Moreover, M_Φ is negative since the tailplane incidence is increased when the bending is downward (Φ positive). *Hence*

* The multiplier of this quantity is also small.

the stability will be increased when $(Z_w F_u - Z_u F_w)$ *is positive.* Now
we have (see § 5·3)

$$z_w = -\tfrac{1}{2}\left(C_D + \frac{dC_L}{d\alpha}\right) \quad \text{and} \quad z_u = -C_L,$$

so Z_w is large and negative and Z_u is also negative. F_u gives the
rate of increase of the flexural generalized force with forward
speed; consequently F_u is positive when the tail load is downward and negative when the tail load is upward. Thus when
the tail load is downward the whole expression $(Z_w F_u - Z_u F_w)$ is
negative, for F_w is obviously negative, so the influence of the
flexibility is destabilizing. However, when the tail load is
upward, so that F_u is negative, the destabilizing effect will be
reduced and when the tail load in this sense is sufficiently large
the effect may be stabilizing. This will happen most easily
when C_L is very small for then Z_u is relatively small. Thus
there tends to be a stabilizing effect at high speeds of flight
when the tail load is upward. Over a large range of C_L the tail
load for trim may vary largely, with a corresponding influence
on the stability. The trim curves (see § 10·6) will not be straight
lines.

A few words must be said about the fuselage stiffness coefficient f_Φ, which is of elastic origin with a small gravitational
component which will not be further considered. If for simplicity we assume that the Bernoulli-Euler theory of bending
can be applied to the fuselage we can derive an expression for
f_Φ by considering the flexural elastic energy. Since the generalized flexural restoring force corresponding to the displacement
Φ is Φf_Φ, it follows that the elastic energy is $\tfrac{1}{2} f_\Phi \Phi^2$. Next, since
the deflexion at x is $\Phi f(x)$, the curvature is $\Phi f''(x)$ and the
energy stored per unit length is $\tfrac{1}{2} EI[\Phi f''(x)]^2$ where EI is the
local value of the flexural rigidity. Hence we obtain on equating
the two expressions for the stored energy

$$f_\Phi = \int EI f''(x)^2 dx, \qquad (12\cdot 5, 13)$$

where the integral covers the fuselage. We may also obtain f_Φ
from a static loading test if the aircraft is available. Let a
normal load W be applied at the tail so that the actual deflexion
is z and let the corresponding value of the flexural co-ordinate

as derived from equation (12·5, 1) be Φ. Then the energy stored is

$$\tfrac{1}{2}Wz = \tfrac{1}{2}\Phi^2 f_\Phi \qquad (12\cdot5, 14)$$

and from this f_Φ can be found. Particular attention must be given to the method of supporting the fuselage in such a test since it is the deflexion of the tail relative to the centre section of the fuselage that is required.

12·6   The method of the modification of derivatives

We shall begin by considering the problem of the flexible fuselage already treated in § 12·5. Let us consider merely static deviations from the datum condition and suppose that the undisturbed flight path is horizontal. The equation of flexural forces (12·5, 9) then becomes when $F(t)$ is zero

$$-uF_u - wF_w + \Phi(-F_\Phi + f_\Phi) = 0$$

or

$$\Phi = \frac{uF_u + wF_w}{-F_\Phi + f_\Phi}. \qquad (12\cdot6, 1)$$

We may now use this relation to eliminate Φ from the remaining dynamical equations. For the case considered equation (12·5, 7) merely serves to determine θ, which does not occur in the other equations, and this equation need not be treated further. Equation (12·5, 6) becomes

$$-u\left(Z_u + \frac{Z_\Phi F_u}{-F_\Phi + f_\Phi}\right) - w\left(Z_w + \frac{Z_\Phi F_w}{-F_\Phi + f_\Phi}\right) = 0, \quad (12\cdot6, 2)$$

while (12·5, 8) becomes

$$-u\left(M_u + \frac{M_\Phi F_u}{-F_\Phi + f_\Phi}\right) - w\left(M_w + \frac{M_\Phi F_w}{-F_\Phi + f_\Phi}\right) = 0, \quad (12\cdot6, 3)$$

when the externally applied forces are zero. Now these equations are of exactly the same form as when the fuselage is rigid but the effective values of the aerodynamic derivatives are altered. Let us denote the effective value of the derivative by adding an accent. Then we have

$$Z'_u = Z_u + \frac{Z_\Phi F_u}{-F_\Phi + f_\Phi}, \qquad (12\cdot6, 4)$$

$$M'_u = M_u + \frac{M_\Phi F_u}{-F_\Phi + f_\Phi}, \qquad (12\cdot6, 5)$$

$$Z'_w = Z_w + \frac{Z_\Phi F_w}{-F_\Phi + f_\Phi}, \tag{12·6, 6}$$

$$M'_w = M_w + \frac{M_\Phi F_w}{-F_\Phi + f_\Phi}. \tag{12·6, 7}$$

We may note, in particular, that M'_u will not in general vanish even when M_u is zero. The value of

$$Z'_u M'_w - Z'_w M'_u,$$

which must be positive for stability, is

$$Z_u M_w - Z_w M_u + \frac{M_\Phi(Z_u F_w - Z_w F_u) + Z_\Phi(F_u M_w - F_w M_u)}{-F_\Phi + f_\Phi},$$

$$\tag{12·6, 8}$$

where the second term gives the influence of the flexibility of the fuselage. This is in complete agreement with the results obtained in § 12·5.

The modified derivatives, as given by equations (12·6, 4)... (12·6, 7), have been obtained on the assumption that the deviations from the datum state occur extremely slowly and at present we have no justification for giving them a wider application. The essential part of the theory is the use of the dynamical equation corresponding to the flexural co-ordinate to express this co-ordinate explicitly in terms of the other deviations. Equation (12·5, 9) shows at once that, in free motion, we can always express the flexural co-ordinate Φ linearly in terms of u, $\dot{u}$, w, $\dot{w}$, θ, q and $\dot{q}$ provided that

$$\ddot{\Phi} \Sigma f(x)^2 \delta m - \Phi F_\Phi$$

is negligible in comparison with $\Phi(-F_\Phi + f_\Phi)$. This will be true when λ is numerically small, whether it be real or complex. The term in $\ddot{\Phi}$ will be negligible in an oscillatory motion whose frequency is decidedly less than that of the fundamental frequency of the fuselage for bending in the plane OXZ. For instance, when the frequency is less than 1/10 of the fundamental frequency the error in neglecting the term in $\ddot{\Phi}$ will be less than 1 per cent. In order to judge when the term in Φ can be neglected it will be necessary to estimate the derivative F_Φ, which arises almost wholly from incidence changes at the tail-plane.

We may sum up by saying that the method of the modification of derivatives is applicable when the rates of change of the distortions are sufficiently small and that the precise conditions require investigation in each case. In oscillatory motion the inertia terms will be negligible when the frequency is small in relation to the lowest natural frequency of the structure involving distorting of the type considered.* The method of the modification of derivatives can be applied, subject to the above reservations, when both distortion of the structure and the compressibility of the air are influential. This is discussed in § 13·8.

12·7 Influence of the torsional flexibility of the tailplane

We shall content ourselves with a qualitative discussion of this question. Torsional flexibility of the tailplane will lead to reversal of elevator control at a value of the equivalent air speed which, in the absence of compressibility effects, will be independent of the air density (see § 2·12). At lower speeds the flexibility will result in a reduction of the effectiveness of the elevator. It is clear that flexibility of the fuselage increases the effective torsional flexibility of the tailplane since the angle of pitch at the root of the tailplane is increased by the slope of the fuselage deflection. However, as pointed out by Collar and Grinsted, the critical reversal speed is hardly influenced by the flexibility of the fuselage since, at the critical speed, the lift caused by operating the elevator is zero or nearly so and the corresponding flexural load on the fuselage is trifling. At speeds below the critical the flexibility of the fuselage contributes to a reduction of the effectiveness of the elevator.

When the tailplane is torsionally flexible the stability will be influenced by the setting of the tailplane. If we neglect the small aerodynamic pitching couple on the tailplane the condition of balance of the pitching moments on the whole aircraft requires that the total tail lift shall be fixed for a given speed of flight in air of given density and temperature. The constancy of this lift requires that $(a'_1 \alpha' + a'_2 \eta)$ shall be constant. Thus any change in tail setting must be compensated by a change in

---

* See further 'Ignoration of Distortional Co-ordinates in the Theory of Stability and Control', by W. J. Duncan, *Report No.* 1 of the College of Aeronautics (1946).

elevator angle. The important point is that the aerodynamic twisting moment is not fixed although the lift is fixed and the twisting moment may in fact have either sign.* Suppose now that the tail setting is such that the aerodynamic twisting moment is in the nose-up sense† and that the speed of flight increases as the result of some disturbance. The aerodynamic twisting moment will increase and the tailplane incidence and lift will increase. Hence the flexibility has a destabilizing influence. But if the tail setting were such that the twisting moment was in the nose-down sense the flexibility would have a stabilizing effect.

We may draw the general conclusion that when the structure is flexible, as it must be, the stability of whatever type will, in general, depend on the initial angular settings of surfaces. For instance, a change of fin setting will influence N_u, a derivative which is zero for a rigid symmetric aircraft in symmetric flight.

12·8   The influence of local surface distortions

Local surface distortions are chiefly of importance in relation to control forces. It has already been pointed out in § 7·5 that the hinge moment on a flap control depends on the convexity of its surfaces and any change of camber will likewise influence the hinge moment. Now when the surfaces of the flap are flexible they will deflect to some extent when the flap is moved or the angle of incidence is changed. Even when the control setting and incidence are constant the surface may deflect when the speed of flight varies on account of the pressure difference across the surface changing; this pressure difference is largely affected by the venting arrangements. It is evident that a flap control with a relatively flexible surface will show complex non-linearities in its hinge moment characteristics and will almost certainly be unsatisfactory in practice. Hence it is now recognized as essential to make the surfaces of control flaps amply stiff. Metal covering is almost essential for high-speed aircraft and whenever fabric is used the spacing of the ribs and stringers must be close enough to provide adequate stiffness.

---

* The centre of pressure of the load caused by a change of incidence is near the quarter chord point but that caused by a change of elevator angle is much further back. (See § 7·2)

† This implies that for trim at the speed considered the T.E. of the elevator is raised.

REFERENCES FOR FURTHER READING

A. R. Collar and F. Grinsted, 'The Effects of Structural Flexibility of Tail-plane, Elevator and Fuselage on Longitudinal Control and Stability', *R. & M.* 2010 (1942).

S. B. Gates and H. M. Lyon, 'A Continuation of Longitudinal Stability and Control Analysis, Part II, Interpretation of Flight Tests', *R. & M.* 2028 (1944).

H. M. Lyon and J. Ripley, 'A General Survey of the Effects of Flexibility of the Fuselage, Tail Unit and Control System on Longitudinal Stability and Control', *R. & M.* 2415 (1950).

Chapter 13

THE INFLUENCE OF THE COMPRESSIBILITY OF THE AIR

13·1 Introductory survey

It is beyond the scope of this book to discuss the fundamental theory of the flow of compressible fluids. The aim of this survey is merely to give such explanations and quote such results from theory and experiment as will help the reader to gain a general understanding of the influence of the compressibility of the air on the dynamical behaviour of aircraft.

The compressibility of the air begins to modify the forces on a body moving through it when the velocity of the air relative to the body is an appreciable fraction of the velocity of sound in the undisturbed air. It is thus convenient to use the ratio (velocity of flight): (velocity of sound in the undisturbed air in the vicinity of the aircraft) as an indication of the importance of compressibility. This non-dimensional quantity is called the Mach number, in honour of Ernst Mach, and denoted by the symbol $\mathbf{M}$. The value of $\mathbf{M}$ at which compressibility effects become appreciable depends on the nature of the moving body. For example, the onset of an appreciable effect occurs at a lower Mach number for a thick aerofoil at zero incidence than for a thin one of the same basic form. The reason for this is that the excess of the relative velocity of flow near the aerofoil above the velocity of flight reaches higher values for the thick aerofoil than for the thin one. Let a be the velocity of sound in the undisturbed air in the vicinity of the body and V the speed of flight. Then the Mach number of flight or general Mach number is

$$\mathbf{M} = \frac{V}{a}, \tag{13·1, 1}$$

while the velocity of sound is given by

$$a = \sqrt{(\gamma R T)} \tag{13·1, 2}$$

where γ is the ratio of the specific heats of air, R the gas constant for air and T the absolute temperature (degrees K).

Thus for air the velocity of sound is a function of the temperature alone. At 15° C. the velocity of sound in air is 1117 ft. per sec. or 340·5 metres per sec. or 761·6 miles per hour or 661·4 knots. At the standard temperature of the stratosphere (−56·46° C.) the velocity of sound is 968 ft. per sec. or 295 metres per sec. or 660·3 miles per hour or 573·4 knots.

The local Mach number at any point in the neighbourhood of a body in flight is the velocity of the air at the point measured relative to the body, divided by the *local* velocity of sound at the point. Wherever the local velocity of flow is high the temperature will be low, as follows from the conservation of energy. On account of the dependence of the velocity of sound on temperature, the local velocity of sound will be low where the local velocity of flow is high; hence the local Mach number is increased more than in proportion to the local velocity. It is possible for the local Mach number to become equal to or greater than unity while the general Mach number is below unity.

When the general Mach number is less than unity the flight is said to be *subsonic* and when it is greater than unity the flight is called *supersonic*. The term *transonic* is applied somewhat loosely to the region of flight speeds near unit Mach number, usually with the implication that there is at least local supersonic flow (see above).

The critical Mach number for a body in steady motion in a given attitude is defined as the general Mach number when the greatest local Mach number just reaches unity. Below the critical Mach number the flow is everywhere subsonic, whereas above it there is a region or regions of supersonic flow. Experiment shows that there is no sudden or catastrophic change in the flow pattern or in the aerodynamic forces exactly at the critical Mach number. However, when the Mach number is somewhat above the critical there is a rather sudden drop in the lift coefficient accompanied by a large increase in the drag coefficient and change in the pitching moment coefficient; in the same region of Mach number more or less violent buffeting (see § 15·6) may occur. The similarity of these phenomena to those of the ordinary low-speed stall (see Chapter 11) has led to the use of the phrase 'shock stall'. At the shock stall shock waves have already developed (see below) and there may be breakaway of the flow. It is clear that the shock stalling Mach number, not the critical Mach number, is of practical importance.

The separation of the shock stalling and critical Mach numbers may vary from almost zero to about 0·2 and cannot be predicted with certainty on the basis of present knowledge. It has been observed, however, that for two-dimensional flow the separation is great when the critical Mach number is low and vice versa. The separation will depend considerably on the attitude of the body.

When the flow is supersonic, at least locally, *shock waves* may occur. A shock wave is a front where there is a sudden increase of pressure, density and temperature and fall of velocity as the air moves through it. In steady flight shock waves, if present, are either fixed relative to the aircraft or oscillate about a fixed mean position. The process of compression of the air in the shock wave involves increase of entropy or degradation of energy. When shock waves develop there is an increase of drag, partly attributable directly to the dissipation of energy in the shock wave and partly caused by the influence of the shock wave on the boundary layer. The strength of a shock wave as measured by the pressure difference across it is very small when the flow is only just supersonic and increases with the speed of flow in front of and normal to the wave. When shock waves are established there is a considerable change in the regime of flow associated with changes of pressure distribution, lift, drag, pitching and hinge moment and downwash. These effects make themselves felt as changes of trim, stability and control characteristics.

In supersonic flight shock waves spring from or forward of the most forward points or edges of the aircraft so the aircraft in general is immersed in an airstream which has already been modified by the shock waves. The whole regime of flow in supersonic flight is very different from that in definitely subsonic flight and there are correspondingly large differences in pressure distribution, position of aerodynamic centre, trim, stability and control characteristics.

A point source of disturbance moving through air at rest with supersonic speed V gives rise to a weak conical shock wave of which it is the vertex (see Fig. 13·1, 1). The axis of the cone lies along the direction of motion and the semi-vertical angle, called the Mach angle, is denoted by μ and given by

$$\sin \mu = \frac{1}{\mathbf{M}}. \qquad (13\cdot1, 3)$$

This is a consequence of the fact that a wave front is propagated normal to itself through the air with a speed equal to the speed of sound when the wave is weak. The triangle of velocities shown in the figure leads at once to the relation ($13\cdot1$, 3). For two-dimensional motion we have to suppose a line perpendicular to the plane of motion to be substituted for the point disturbance and the complete wave consists of two half planes meeting

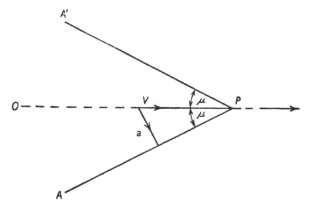

Fig. $13\cdot1$, 1. Weak conical shock wave or Mach cone.

on the line and making the angle μ with the direction of motion. Fig. $13\cdot1$, 1 can be regarded as showing the traces of these half planes. The pair of half planes is called a Mach wedge.

Any weak disturbance is propagated with the speed of sound in the undisturbed air. In Fig. $13\cdot1$, 1 no point lying to the right of the Mach cone (or wedge, in the two-dimensional case) APA' can be influenced by the disturbance at P at the instant considered.

The leading point or edge of a finite body in supersonic flight will give rise to a shock wave which is strong in the neighbourhood of the body and which can therefore propagate at a speed higher than that of sound in the undisturbed air. In front of the nose the air will be compressed, its temperature raised and the local velocity of sound increased. For a blunt-nosed body the 'bow wave' will always be detached from the body, i.e. there will be a space between the wave and the nose. For a body with a sharp nose of finite angle the bow wave will be detached for a range of low supersonic Mach numbers but will be attached to the nose for higher Mach numbers. The thinner the body the lower the Mach number at which attachment occurs. In the limit when the thickness tends to zero the wave is attached for unit Mach number and upwards.

When the component of the velocity of flight normal to the leading edge of a wing is subsonic the leading edge itself is said to be subsonic; similarly when the normal component of velocity is supersonic the leading edge is said to be supersonic. It is shown in § 13·3 that there are important differences between the characteristics of wings with subsonic and supersonic leading edges. With some reservations, it may be said that a wing with a subsonic leading edge behaves like a wing in subsonic flow.*

It is evident from what has been said that we have the following conditions of flight to consider:

(*a*) Low-speed flight where the influence of the compressibility of the air is negligible.

(*b*) Definitely subsonic flight where the influence of compressibility is felt but shock waves are absent.

(*c*) High subsonic and transonic flight where shock waves are present.

(*d*) Supersonic flight with subsonic leading edges.

(*e*) Supersonic flight with supersonic leading edges.

13·2   Linearized theories of subsonic and supersonic flows

Linearized theories of both subsonic and supersonic flows have been developed, based on the assumption that the deviation of the local velocity from the velocity of flight (general velocity of the airstream) is everywhere a small fraction of this velocity. The assumption of the smallness of the velocity perturbations is obviously violated whenever a stagnation point

---

* The sub- or supersonic nature of the component of velocity perpendicular to the trailing edge is also of importance. For the supersonic case no disturbance can be passed between the upper and lower wing surfaces round the trailing edge.

exists, but the regions where the assumption is invalid are often very small, especially when the bodies considered are thin and set at small angles of attack. Experiments show fair agreement with the linearized theories for slender bodies (e.g. thin aerofoils) at small angles of incidence provided that the flow is either definitely subsonic or definitely supersonic.* In developing the theories the squares and products of the components of the velocity deviations (velocity perturbations) are neglected. In consequence, the deviations of density and of pressure are also linear functions of the perturbations.

The oldest and simplest form of the linearized theory of subsonic flow is due to Prandtl and Glauert; the theories of von Kármán and Temple, which are based on the hodograph method, show, on the whole, rather better agreement with experiment but unfortunately they do not lend themselves to the development of simple formulae and will not be further discussed here. Let us define the pressure coefficient as

$$C_p = \frac{p - p_0}{\frac{1}{2}\rho_0 V^2}, \qquad (13\cdot2,\,1)$$

where V is the velocity of flight and p_0, ρ_0 are the pressure and density in the undisturbed air. Then, if we use the suffices i and c for incompressible and compressible flows respectively, the Prandtl-Glauert theory yields for two-dimensional subsonic flow

$$\frac{C_{pc}}{C_{pi}} = \frac{1}{\sqrt{(1 - \mathbf{M}^2)}}, \qquad (13\cdot2,\,2)$$

where $\mathbf{M}$ is the Mach number of flight as defined in § 13·1. Since the pressure coefficients at all points are in a constant ratio it follows that

$$\frac{C_{Lc}}{C_{Li}} = \frac{C_{mc}}{C_{mi}} = \frac{1}{\sqrt{(1 - \mathbf{M}^2)}}, \qquad (13\cdot2,\,3)$$

while the slope of the curve of lift coefficients is changed in the same ratio. It also follows that the positions of centres of pressure and aerodynamic centres are unaffected by the compressibility. For Mach numbers below the critical the compressibility has a negligible effect on the profile drag coefficient.

* Somewhat similar approximate theories of transonic flow have also been proposed but will not be discussed here.

For finite aspect ratios and again for definitely subsonic flow the linearized lifting line theory yields

$$\frac{C_{Lc}}{C_{Li}} = \frac{1 + \dfrac{a_0}{\pi A}}{\sqrt{(1 - \mathbf{M}^2)} + \dfrac{a_0}{\pi A}}, \tag{13.2, 4}$$

where a_0 is the slope of the curve of lift coefficients for two-dimensional incompressible flow. This formula shows that the influence of compressibility falls as the aspect ratio is reduced.

The linearized theory of two-dimensional supersonic flow is due to Ackeret and the results are particularly simple. Consider an element of surface making a small angle ϵ with the direction of flow.* Then the pressure coefficient at the element is

$$C_p = \frac{2\epsilon}{\sqrt{(\mathbf{M}^2 - 1)}}. \tag{13.2, 5}$$

This formula can be applied to the upper and lower surfaces of an aerofoil independently and leads to the following value of the lift coefficient

$$C_L = \frac{4\alpha}{\sqrt{(\mathbf{M}^2 - 1)}}, \tag{13.2, 6}$$

where the angle of incidence is measured from the line joining the leading and trailing edges. The pitching moment coefficient referred to the leading edge is

$$C_M = \frac{2}{\sqrt{(\mathbf{M}^2 - 1)}} \left[-\alpha + \int_0^1 (\phi_2 - \phi_1)\, \xi d\xi \right], \tag{13.2, 7}$$

where ξ is distance aft of the leading edge expressed as a fraction of the chord and ϕ_1, ϕ_2 are the inclinations of the tangents to the upper and lower surfaces, respectively, to the chord line. These angles are to be taken positive when the distance of the surface from the chord line increases towards the rear, so both are positive towards the leading edge of a symmetric aerofoil. The wave drag coefficient is

$$C_{DW} = \frac{2(2\alpha^2 + \overline{\phi_1^2} + \overline{\phi_2^2})}{\sqrt{(\mathbf{M}^2 - 1)}}, \tag{13.2, 8}$$

---

* The sign of ϵ is positive when the normal to the element drawn from it *into the fluid* is inclined upstream. Thus, for a thin aerofoil at positive incidence ϵ is positive on the lower surface and negative on the upper.

where the bar indicates that the mean value over the chord is to be taken. The whole drag coefficient is to be obtained by adding the contribution of the skin friction. Another deduction from the theory is that the aerodynamic centre of a wing is situated at the centroid of its area.

Ackeret's theory applies only to fully established supersonic flow and when the shock waves present are weak. Thus it is inapplicable near unit Mach number (as is indicated by the theoretical pressure coefficient tending to infinity) and near the leading edge of an aerofoil with a blunt nose.

For an unyawed rectangular wing of finite span in supersonic flight the flow remains two-dimensional except within the Mach cones springing from the wing tips. The pressure difference between the lower and upper surfaces is reduced within the cones and the lift coefficient is given by

$$\frac{C_L}{C_{L\infty}} = 1 - \frac{1}{2A\sqrt{(M^2-1)}},\qquad (13{\cdot}2,\,9)$$

where $C_{L\infty}$ is the lift coefficient for infinite aspect ratio and the same incidence while A is the aspect ratio. This formula holds only when $A\sqrt{(M^2-1)} > 1$; no simple formula has yet been obtained when the inequality is not satisfied. The aerodynamic centre is theoretically situated at

$$\frac{1}{6[2A\sqrt{(M^2-1)}-1]}\qquad (13{\cdot}2,\,10)$$

chords in front of the mid-chord position and experiments show that it lies rather further forward.

The wave drag coefficient due to the lift is

$$C_{DL} = \alpha C_L \qquad (13{\cdot}2,\,11)$$

and the total wave drag coefficient is

$$C_{DW} = C_{DO\infty} + \alpha C_{L\infty}\left[1 - \frac{1}{2A\sqrt{(M^2-1)}}\right],\qquad (13{\cdot}2,\,12)$$

where $C_{DO\infty}$ is the two-dimensional wave drag coefficient at zero lift. This formula is only valid when the inequality given above is satisfied.

13·3   The influence of sweepback

Sweeping back the wings of an aircraft is recognized as a valuable means for raising the Mach number at the shock stall. For a wing of infinite span it is strictly demonstrable that, apart from possible boundary layer effects, the component of the velocity of the airstream parallel to the span is without influence upon the aerodynamic forces. Thus the effective relative speed is $V \cos \Lambda$, where Λ is the angle of sweepback. It is possible for the effective velocity to be subsonic and below that at the shock stall when V is supersonic.

The simple conclusion stated above cannot remain strictly valid for an aircraft with sweptback wings. It is clear that for a symmetric aircraft in symmetric flight the flow at the plane of symmetry is always in that plane, so in this region the full velocity of flight is effective. Likewise there is a large departure from the state of flow for infinite aspect ratio near the wing tips. It is found that the departures from the conditions for infinite aspect ratio are of the following nature:

The suction peak on a wing of symmetric section at zero incidence occurs further aft near the root and further forward near the tip. At positive incidence this is accentuated on the upper surface but there is an opposite tendency on the lower surface. This implies that at high subsonic speeds the shock waves on the upper surface will lie aft of those on the lower surface near the roots but forward of them near the tips. Now the lift at high supercritical speeds depends largely on the difference between the high suctions in front of the shock waves on the upper and lower surfaces. Hence there is a tendency at high Mach numbers for sweepback to cause a gain of lift near the roots and a loss of lift near the tips. When the loss of lift near the tips becomes great we have, in effect, a tip stall with all its consequences as regards trim and stability.* On the other hand, there are usually other agencies tending to cause premature stalling near the roots and only experiment can show which tendency will prevail in any instance.

Forward sweep of the wings can be used in the same way as backward sweep to raise the shock stalling Mach number.

* This argument is due to Professor A. D. Young (private communication).

However, forward sweep has been seldom or never adopted and little is known of its aerodynamic consequences. It is likely to lead to root stalling at high speeds of flight.

Attention may be drawn here to the contribution to the derivative L_v proportional to C_L which occurs with swept wings (see § 6·3).

13·4   The dynamical equations, derivatives and stability

The dynamical equations developed in §§ 3·11 and 3·12 are valid for a rigid aircraft for all speeds of flight and, in order to apply them, it is only necessary to substitute the appropriate values of the aerodynamic derivatives and other coefficients. Likewise the methods given in Chapter 4 for the solution of the dynamical equations and for the investigation of stability and response are universally valid. The outstanding problem is to obtain the values of the derivatives for flight at high subsonic and at supersonic Mach numbers.

When the influence of the compressibility of the air becomes appreciable, the non-dimensional force and moment coefficients are no longer independent of the speed of flight. As already pointed out in § 2·5 this implies that new terms appear in the expressions for some of the derivatives. The derivatives are also modified by the influence of compressibility on the co-efficients themselves and on their rates of change with angle of incidence at constant Mach number.

For convenience we shall, as usual, adopt wind axes (see § 3·11) and shall begin by considering the quasi-static derivatives of longitudinal-symmetric motion. We then have the simplification that the Mach number is to be taken as constant in the calculation of the derivatives with respect to the normal velocity deviation w since, to the first order of small quantities, the speed of flight is independent of w. The same is true of the derivatives with respect to the angular velocity of pitch q. Accordingly, the expressions for the derivatives with respect to w and q given in § 5·3 remain valid but the values of C_L, C_m, C_D and their rates of change with incidence will of course depend on the Mach number. The derivatives with respect to α in the formulae are now to be interpreted as partial differential coefficients with respect to α at constant M.

The general expression for the quasi-static derivative of a force F with respect to u as given in equation (2·5, 28) is

$$\frac{F_u}{\rho V S} = C_F \cos \alpha - \tfrac{1}{2}\frac{\partial C_F}{\partial \alpha}\sin \alpha + \tfrac{1}{2}\mathbf{M}\frac{\partial C_F}{\partial \mathbf{M}}\cos \alpha$$

$$= C_F + \tfrac{1}{2}\mathbf{M}\frac{\partial C_F}{\partial \mathbf{M}}, \qquad (13\cdot4,\,1)$$

for wind axes. Also equation (2·5, 14) becomes

$$C_F = C_L \cos A + C_D \cos B.$$

Hence we obtain

$$\frac{F_u}{\rho V S} = C_L \cos A + C_D \cos B + \tfrac{1}{2}\mathbf{M}\left(\frac{\partial C_L}{\partial \mathbf{M}}\cos A + \frac{\partial C_D}{\partial \mathbf{M}}\cos B\right)$$

$$(13\cdot4,\,2)$$

By Table 2·5, 1 we have, for the longitudinal force, $A = -\dfrac{\pi}{2}$ and $B = \pi$. Hence

$$x_u = \frac{X_u}{\rho V S} = -C_D - \tfrac{1}{2}\mathbf{M}\frac{\partial C_D}{\partial \mathbf{M}} \qquad (13\cdot4,\,3)$$

or

$$x_u = -C_D - \tfrac{1}{2}\mathbf{M}\frac{\partial C_D}{\partial \mathbf{M}} - C_{AS}, \qquad (13\cdot4,\,4)$$

when allowance is made for the influence of the propulsive system (see § 5·6). For the normal force we have $A = \pi$ and $B = \dfrac{\pi}{2}$. Hence

$$z_u = \frac{Z_u}{\rho V S} = -C_L - \tfrac{1}{2}\mathbf{M}\frac{\partial C_L}{\partial \mathbf{M}}. \qquad (13\cdot4,\,5)$$

For wind axes equation (2·5, 31) yields, when allowance is made for the influence of compressibility and l is identified with $\bar{c}$,

$$\frac{M_u}{\rho V S \bar{c}} = C_M + \tfrac{1}{2}\mathbf{M}\frac{\partial C_M}{\partial \mathbf{M}} = \tfrac{1}{2}\mathbf{M}\frac{\partial C_M}{\partial \mathbf{M}}, \qquad (13\cdot4,\,6)$$

since the aircraft is trimmed in the datum state. Hence

$$m_u = \frac{M_u}{\rho V S l_T} = \frac{\bar{c}}{2 l_T}\mathbf{M}\frac{\partial C_M}{\partial \mathbf{M}}, \qquad (13\cdot4,\,7)$$

In estimating the values of the derivatives, reliable experimental values of the force and moment coefficients over a
range of Mach numbers, if available, should be used. Failing
these, values based on theory may be adopted. It is clear from
equations (13·4, 4)...(13·4, 7) that the additional terms in the
derivatives depending directly on $\mathbf{M}$ will be important when the
non-dimensional force and moment coefficients change rapidly
with Mach number, and this occurs usually in the transonic
region. Theoretical estimates of the rates of change of the
coefficients with Mach number in the region where the rates
are large are quite unreliable.

For lateral-antisymmetric motion we have to consider the
derivatives with respect to v, p and r (see § 6·3). To the first
order of small quantities the presence of the deviations v and p
does not alter the resultant speed relative to the air, so the
rates of change of the force and moment coefficients should be
taken for constant Mach number and there are no additional
terms in the derivatives depending on $\mathbf{M}$. However, the presence of the deviation r causes differences of relative speed over
the wings so additional terms depending on $\mathbf{M}$ are here present.
For example, we have from equation (6·3, 37)

$$l_r = -\frac{1}{Ss^2}\int_{-s}^{s} cz_u y^2 \, dy,$$

and on account of (13·4, 5) this becomes

$$l_r = \frac{1}{Ss^2}\int_{-s}^{s} c\left(C_L + \tfrac{1}{2}\mathbf{M}\frac{\partial C_L}{\partial \mathbf{M}}\right) y^2 \, dy, \qquad (13\cdot4, 8)$$

where C_L and its derivative with respect to $\mathbf{M}$ should have their
local values. Similarly, we derive from (6·3, 40) and (13·4, 3)
that

$$n_r = \frac{-1}{Ss^2}\int_{-s}^{s} c\left(C_D + \tfrac{1}{2}\mathbf{M}\frac{\partial C_D}{\partial \mathbf{M}}\right) y^2 \, dy. \qquad (13\cdot4, 9)$$

It is to be understood that these are the quasi-static values of
the derivatives contributed by the wings, as given by 'strip
theory'.

The whole theory of stability as given in this book is based
on the assumption that the aircraft flies in a homogeneous
atmosphere. The real atmosphere is not homogeneous and in

the standard atmosphere the density and temperature are decreasing functions of the altitude. Neumark has examined mathematically the influence of this atmospheric stratification on the longitudinal stability of an aircraft in horizontal flight. It is necessary to bring in altitude as a new variable and to allow for the dependence of the aerodynamic forces upon it. The rate of change of the deviation in altitude is itself expressed linearly in terms of the deviations w and θ and it is found that the determinantal equation becomes a quintic instead of a quartic. For details the reader must refer to the original paper,* but we shall quote the conclusion that the stability of the phugoid motion may be considerably influenced by the altitude effect, especially in supersonic flight.

Some general comments on static and dynamic stability in flight at high speeds are given in § 13·7 while the extended theory of static stability is given in § 13·8.

13·5 Control characteristics

For Mach numbers below the shock stall, the aerodynamic characteristics of plain rigid control flaps are not profoundly influenced by the compressibility of the air, in accordance with the indications of the linearized theory (see § 13·2). For definitely supersonic flow the linearized theory leads to results of great simplicity. Thus for two-dimensional flow Ackeret's theory [see equation (13·2, 5)] leads to the following results for a plain flap at the trailing edge

$$a_2 = \frac{4E}{\sqrt{(M^2 - 1)}}, \tag{13·5, 1}$$

$$b_1 = b_2 = \frac{-2}{\sqrt{(M^2 - 1)}}, \tag{13·5, 2}$$

where we have used the notation of § 7·1. On comparison of (13·5, 1) with (7·2, 15) it will be seen that a_2 is much smaller for supersonic flow than for low-speed flow, so control flaps are relatively ineffective in supersonic flight. This is explained by the fact that in supersonic flight deflection of the flap does not

---

* S. Neumark, 'Dynamic Longitudinal Stability in Level Flight, including the Effects of Compressibility and Variations of Atmospheric Characteristics with Height', *R.A.E. Report No. Aero* 2265 (1948). Also 'Longitudinal Stability, Speed and Height', *Aircraft Engineering*, vol. xxii (Nov. 1950), p. 323.

alter the pressure on the aerofoil ahead of the flap, whereas in low-speed flow most of the effect is due to the pressure changes induced on the aerofoil. Another deduction from Ackeret's theory is that a nose flap has the same effectiveness as a trailing edge flap, in strong contrast with the facts for low-speed flow (see § 7·2). A control of rectangular plan form with its span normal to the direction of flight will have the same over-all characteristics as for two-dimensional flow provided that the Mach lines AE and BF springing from the forward corners of the control cut the trailing edge line of the control within the wing area (see Fig. 13·5, 1). Within the area $ABF'E'$ the pressure distribution is the same as for two-dimensional flow,

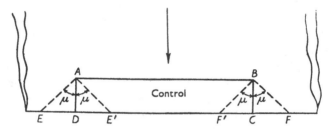

Fig. 13·5, 1. Finite rectangular control flap with span normal to air stream.

but this is not so in ADE' and BCF'. However, the pressure change on the aerofoil within ADE just balances the loss of effectiveness within ADE' and there is a similar compensation at the other end. When E or F lies outside the aerofoil there is some loss of effectiveness.

Little very definite can be said about control characteristics in the transonic region. Experiment shows that controls may become almost wholly ineffective in this region and their characteristics may vary rapidly with Mach number. Such troubles are, broadly speaking, worse with thick aerofoils than with thin ones and are reduced when the trailing edge angle is kept small. According to some experiments by Shaw* a purely aerodynamic reversal of control may occur. In the U.S.A. the phenomenon of 'aileron buzz' has been reported and the

* R. A. Shaw, 'Changes in Control Characteristics with Changes in the Flow Pattern at High Subsonic Speeds, Tests on an EC 1250 Aerofoil with 25 per cent Concave Control Flap', *A.R.C.* 11, 933, (Nov. 1948).

investigations of Smilg* indicate that this is a kind of unstable oscillation involving the single degree of freedom of flap deflection. This is not fully understood but it is clear that the pressure changes bear such a phase relation to the aileron displacements that instability results.

13·6 Trim changes,

Trim changes, sometimes sudden and violent, are often experienced in the transonic region. These are particularly dangerous when they occur in dives and pull-outs. Thus a severe change of trim in the nose-down sense, such as may be caused by a loss of downwash at the tail as the Mach number increases in a dive, may make it impossible for the pilot to pull-out from the dive whereas a change in the nose-up sense developing when the Mach number is reduced in the pull-out may result in excessive normal accelerations. Troubles of this kind can be minimized by good aerodynamic design and special attention must be given to wing roots and nacelles.† Speaking broadly, the thinner the aerofoils are the better.

There is inevitably a large change of trim in flying right through the transonic region. Thus in subsonic flight the centre of pressure on an aerofoil at small incidence lies well forward whereas in supersonic flight it moves back to a position near the centroid of the area, so the pitching moment changes in the nose-down sense.

Trim changes are also associated with changes of the no-lift incidence of cambered aerofoils with Mach number. The constancy of the no-lift incidence at zero for symmetrical aerofoils is a strong reason for their adoption.

13·7 General remarks on stability in high-speed flight

The criterion for static stability of a rigid aircraft is always

$$\text{E} > 0 \qquad (13 \cdot 7, 1)$$

when the dynamical equations are linear with constant co-efficients. The coefficient E is given by equation (5·5, 3) or by

---

* B. Smilg, 'The Prevention of Aileron Oscillations at Transonic Speeds', Sixth International Congress of Applied Mechanics, Paris, 1946.

† The present tendency is to avoid nacelles altogether for high-speed aircraft.

the equivalent expression (5·5, 29); both these are valid when the non-dimensional force and moment coefficients depend on the velocity and they are accordingly applicable when the influence of the compressibility of the air is not negligible. It must be kept in mind that m_u is not in general zero in such circumstances. In the transonic region the non-dimensional force and moment coefficients may change rapidly with speed, so the value of E given by the formulae may only be valid within a very small speed range. It is also to be remarked that the equations of motion will be non-linear unless the deviations from the steady state are small, and the permissible ranges for effective linearity may be very small in the transonic region.

For any given Mach number there will be a certain position of the c.g. on OX for which the static longitudinal stability with fixed elevator will be neutral and this is the 'stick-fixed' neutral point for this Mach number. When the c.g. is at the neutral point E vanishes and for a small range of travel of the c.g. the value of E will be proportional to the c.g. margin $(h_n - h)$ or H_n. Equation (5·5, 29) shows that E is equal to $\left(-\dfrac{dC_m}{dC_R} \right)$ multiplied by a positive quantity, where $\dfrac{dC_m}{dC_R}$ is to be evaluated for fixed elevator and subject to the constancy of $C_R V^2$. We may call $\left(-\dfrac{dC_m}{dC_R} \right)$ the generalized static margin and denote it by K_n. For horizontal or nearly horizontal flight

$$K_n = -\frac{dC_m}{dC_L},$$

and when the non-dimensional force and moment coefficients are independent of speed we have [see equation (10·5, 2)]

$$K_n = H_n.$$

This equation is no longer true when the non-dimensional coefficients are speed dependent. According to the extended theory of static stability due to Gates and Lyon* the general relation is

$$K_n = \psi_1 H_n, \tag{13·7, 2}$$

* S. B. Gates and H. M. Lyon, 'A Continuation of Longitudinal Stability and Control Analysis, Part I, General Theory', *R. & M.* 2027, (Feb. 1944).

where ψ_1 depends on $\mathbf{M}$ and becomes equal to unity for small values of $\mathbf{M}$ (see § 13·8). In the elevator-free case we have the generalized static margin K'_n equal to $\left(-\dfrac{dC_m}{dC_R}\right)$ evaluated for elevator free and subject to the constancy of $C_R V^2$. The corresponding c.g. margin is H'_n and according to Gates and Lyon

$$K'_n = \chi_1 H'_n, \tag{13·7, 3}$$

where again χ_1 reduces to unity for low Mach numbers.

While it is not possible to state simple generalities about stability in the transonic region a simple conclusion can be drawn about the effect on static stability of flying right through this region. Since the aerodynamic centre of an aerofoil lies at about the quarter-chord point for low Mach numbers and moves back to a position near half-chord in supersonic flow there is evidently a tendency for the static stability to become increased.*

13·8 The joint influence of distortion and compressibility on static stability

We have already seen in § 12·6 that the influence of a small amount of structural distortion on static stability may be allowed for by modifying the values of the ordinary aerodynamic derivatives. The modified values are then in general functions of Mach number and of air density. Gates and Lyon† have given what they call 'the extended theory of static stability' in which the derivatives are speed-dependent. This theory is general and thus covers the joint influence of structural distortion, compressibility of the air and the propulsive system. The object of this theory is mainly to show how the stability can be influenced by these agencies and to provide a basis for the interpretation of the results of experiments. We shall now give a brief account of this theory, with some modification of the original notation to bring it into line with that used elsewhere in this book.

---

* It has been frequently asserted that the downwash at the tail is abolished in supersonic flight. If this were true there would be a further strong stabilizing effect (see § 2·6). However, this assertion is not, in general, correct.

† *Loc. cit.*

The basic equations of the theory are

$$C_{L2} = A_1\alpha_0' + A_2\eta_0 + A_3\beta + A_4\delta, \qquad (13\cdot8, 1)$$

and
$$C_H = B_0 + B_1\alpha_0' + B_2\eta_0 + B_3\beta + B_4\delta, \qquad (13\cdot8, 2)$$

where C_{L2} = lift coefficient for the tail plane,

C_H = hinge moment coefficient for the elevator,

α_0' = tailplane incidence to local wind direction at the centre line of the fuselage and with distortion supposed eliminated,

η_0 = angle between the chords of the elevator and tailplane at the fuselage centre line,

β = tab setting relative to elevator at position of operating lever,

$\delta = \epsilon_0 - \epsilon_1$,

ϵ_0 = angle of downwash at the tailplane centre line,

ϵ_1 = angle of downwash at the tailplane tip.

The coefficients A_1 to B_4 are all, in general, functions of **M** and of ρ but these functions are all fixed for a given aircraft. Now when we consider the complete aircraft we obtain

$$C_L = A\alpha + \frac{S'}{S}C_{L2} \qquad (13\cdot8, 3)$$

and
$$C_m = C_{M0} + C_L(h - H_0) - \bar{V}'C_{L2}, \qquad (13\cdot8, 4)$$

where C_m = pitching moment coefficient for the complete aircraft about the c.g.,

C_{M0} = pitching moment coefficient when both C_L and C_{L2} are zero,

h = distance of c.g. aft of datum point, expressed in mean chord lengths,

H_0 = distance of aerodynamic centre of wings and body aft of datum point, expressed in mean chord lengths,

$\bar{V}'$ = modified tail volume ratio [see equation $(10\cdot4, 12)$],

α = angle of incidence of wings measured from the no-lift line,

A = lift slope for wings and body without tail.

It is to be noted that A and C_{M0} are, in general, functions of **M** and of ρ. The value of C_{L2} is given by (13·8, 1) and

$$\alpha_0' = \alpha + \eta_T - \epsilon_0, \tag{13·8, 5}$$

where η_T = setting of tailplane at its centre line relative to no-lift line of wings measured in the undistorted state.

It is further assumed that ϵ_0 and δ are proportional to the lift coefficient of the aircraft less tail.

Thus
$$\epsilon_0 = eA\alpha$$
and
$$\delta = dA\alpha, \tag{13·8, 6}$$

where e and d are constants.

The generalized static margin, elevator fixed, as defined in § 13·7 is

$$K_n = -\frac{dC_m}{dC_R} = -\frac{\partial C_m}{\partial \alpha}\frac{d\alpha}{dC_R} - \frac{\partial C_m}{\partial V}\frac{dV}{dC_R}, \tag{13·8, 7}$$

where C_R and V satisfy the relation

$$C_R V^2 = \text{constant.} \tag{13·8, 8}$$

From (13·8, 1), (13·8, 5) and (13·8, 6) we get

$$C_{L2} = A\alpha\left(\frac{A_1}{A} - A_1 e + A_4 d\right) + A_1\eta_T + A_2\eta_0 + A_3\beta, \tag{13·8, 9}$$

and on substitution of the value of $A\alpha$ from (13·8, 3) we obtain

$$C_{L2}(1 + F) = C_L\left(\frac{A_1}{A} - A_1 e + A_4 d\right) + A_1\eta_T + A_2\eta_0 + A_3\beta, \tag{13·8, 10}$$

where
$$F = \frac{S'}{S}\left(\frac{A_1}{A} - A_1 e + A_4 d\right). \tag{13·8, 11}$$

Equation (13·8, 4) now becomes

$$C_m = C_{M0} + C_L(h - H_0)$$
$$- V_T\left[C_L\left(\frac{A_1}{A} - A_1 e + A_4 d\right) + A_1\eta_T + A_2\eta_0 + A_3\beta\right], \tag{13·8, 12}$$

where
$$V_T = \frac{\bar{V}'}{1+F}.$$
(13·8, 13)

In practice F is small but may reach a value round 0·1.

In the initial state C_m is zero. Hence, when we differentiate (13·8, 12) with respect to C_R we obtain

$$\frac{dC_m}{dC_R} = \frac{dC_{M0}}{dC_R} + (h-H_0)\frac{dC_L}{dC_R} - C_L\frac{dH_0}{dC_R}$$

$$-\frac{dV_T}{dC_R}\frac{C_{M0}+C_L(h-H_0)}{V_T} - V_T G, \quad (13·8, 14)$$

where

$$G = \frac{dC_L}{dC_R}\left(\frac{A_1}{A} - A_1 e + A_4 d\right)$$

$$+C_L\left(\frac{1}{A}\frac{dA_1}{dC_R} - \frac{A_1\dfrac{dA}{dC_R}}{A^2} - e\frac{dA_1}{dC_R} + d\frac{dA_4}{dC_R}\right)$$

$$+\eta_T\frac{dA_1}{dC_R} + \eta_0\frac{dA_2}{dC_R} + A_2\frac{d\eta_0}{dC_R} + \beta\frac{dA_3}{dC_R} + A_3\frac{d\beta}{dC_R}.$$
(13·8, 15)

Equation (13·8, 14) can now be applied to obtain the static margins for the two standard cases.

When the elevator is fixed we have

$$\frac{d\eta_0}{dC_R} = \frac{d\beta}{dC_R} = 0,$$

and η_0 is to be obtained from (13·8, 12) with C_m made zero. From (13·8, 14) and (13·8, 15) we get

$$K_n = (H_0-h)\psi_1 + \psi_0,$$
(13·8, 16)

where
$$\psi_1 = \frac{dC_L}{dC_R} - C_L\left(\frac{\dfrac{dV_T}{dC_R}}{V_T} + \frac{\dfrac{dA_2}{dC_R}}{A_2}\right)$$
(13·8, 17)

and
$$\psi_0 = -\frac{dC_{M0}}{dC_R} + \left(\frac{\dfrac{dV_T}{dC_R}}{V_T} + \frac{\dfrac{dA_2}{dC_R}}{A_2}\right)C_{M0} + C_L\frac{dH_0}{dC_R} + V_T J,$$
(13·8, 18)

with
$$J = \left(\frac{dC_L}{dC_R} - C_L \frac{\frac{dA_2}{dC_R}}{A_2}\right)\left(\frac{A_1}{A} - A_1 e + A_4 d\right)$$

$$+ C_L \left(\frac{\frac{dA_1}{dC_R}}{A} - \frac{A_1 \frac{dA}{dC_R}}{A^2} - e\frac{dA_1}{dC_R} + d\frac{dA_4}{dC_R}\right)$$

$$+ \left(\frac{dA_1}{dC_R} - \frac{A_1 \frac{dA_2}{dC_R}}{A_2}\right)\eta_T + \left(\frac{dA_3}{dC_R} - \frac{A_3 \frac{dA_2}{dC_R}}{A_2}\right)\beta.$$

$$(13\cdot8, 19)$$

It follows from (13·8, 16) that the position of the neutral point is given by

$$h_n = H_0 + \frac{\psi_0}{\psi_1},\qquad (13\cdot8, 20)$$

while the c.g. margin is

$$H_n = h_n - h = \frac{K_n}{\psi_1}.\qquad (13\cdot8, 21)$$

Normally ψ_1 is positive but in extreme cases it may vanish or become negative. When ψ_1 vanishes the static margin K_n remains finite and equal to ψ_0 but the c.g. margin becomes infinite. We may emphasize again that it is the static margin, not the c.g. margin, which is significant for stability.

For the elevator free case we have to bring in the condition that the hinge moment is zero. By equations (13·8, 2), (13·8, 5) and (13·8, 6)

$$C_H = B_0 + A\alpha\left[B_1\left(\frac{1}{A} - e\right) + B_4 d\right] + B_1 \eta_T + B_2 \eta_0 + B_3 \beta.$$

$$(13\cdot8, 22)$$

When we use the last equation to eliminate η_0 from (13·8, 9) it becomes

$$C_{L2} = A\alpha\left(\frac{\bar{A}_1}{A} - \bar{A}_1 e + \bar{A}_4 d\right) + \bar{A}_1 \eta_T + \bar{A}_3 \beta - \bar{B}_0 + \frac{A_2}{B_2}C_H,$$

$$(13\cdot8, 23)$$

where
$$\left.\begin{array}{ll} \bar{A}_1 = A_1 - \dfrac{A_2 B_1}{B_2}, & \bar{A}_3 = A_3 - \dfrac{A_2 B_3}{B_2}, \\[2ex] \bar{A}_4 = A_4 - \dfrac{A_2 B_4}{B_2}, & \text{and}\quad \bar{B}_0 = \dfrac{A_2 B_0}{B_2}. \end{array}\right\} \quad (13\cdot8, 24)$$

The equation for pitching moments, corresponding to (13·8, 12), is

$$C_m = C_{M0} + C_L(h - H_0)$$

$$- \bar{V}_T\left[C_L\left(\frac{\bar{A}_1}{A} - \bar{A}_1 e + \bar{A}_4 d\right) + \bar{A}_1 \eta_T + \bar{A}_3 \beta - \bar{B}_0 + \frac{A_2}{B_2}C_H\right],$$

$$\text{(13·8, 25)}$$

where

$$\bar{V}_T = \frac{\bar{V}'}{1 + \bar{F}} \tag{13·8, 26}$$

and

$$\bar{F} = \frac{S'}{S}\left(\frac{\bar{A}_1}{A} - \bar{A}_1 e + \bar{A}_4 d\right). \tag{13·8, 27}$$

When C_m is zero in the initial state we now derive by differentiation with respect to C_R

$$\frac{dC_m}{dC_R} = \frac{dC_{M0}}{dC_R} + (h - H_0)\frac{dC_L}{dC_R} - C_L\frac{dH_0}{dC_R}$$

$$- \frac{d\bar{V}_T}{dC_R}\frac{C_{M0} + C_L(h - H_0)}{\bar{V}_T} - \bar{V}_T\bar{G}, \tag{13·8, 28}$$

where

$$\bar{G} = \frac{dC_L}{dC_R}\left(\frac{\bar{A}_1}{A} - \bar{A}_1 e + \bar{A}_4 d\right)$$

$$+ C_L\left(\frac{1}{A}\frac{d\bar{A}_1}{dC_R} - \frac{\bar{A}_1\frac{dA}{dC_R}}{A^2} - e\frac{d\bar{A}_1}{dC_R} + d\frac{d\bar{A}_4}{dC_R}\right)$$

$$+ \eta_T\frac{d\bar{A}_1}{dC_R} + \beta\frac{d\bar{A}_3}{dC_R} + \bar{A}_3\frac{d\beta}{dC_R} - \frac{d\bar{B}_0}{dC_R}$$

$$+ \left(\frac{\frac{dA_2}{dC_H}}{B_2} - \frac{A_2\frac{dB_2}{dC_H}}{B_2^2}\right)C_H + \frac{A_2}{B_2}\frac{dC_H}{dC_R}. \tag{13·8, 29}$$

The value of β is given by (13·8, 25) when C_m is zero.

The static margin K_n' for the elevator free case is defined as the value of $-\dfrac{dC_m}{dC_R}$ when the elevator is trimmed and left free, with the tab fixed. In calculating K_n' we must put C_m and C_H zero in (13·8, 25) and C_H, $\dfrac{dC_H}{dC_R}$, $\dfrac{d\beta}{dC_R}$ all zero in (13·8, 28) and (13·8, 29). When β is eliminated from the equations so obtained we find that $\qquad K'_n = -\dfrac{dC_m}{dC_R} = (H_0 - h)\chi_1 + \chi_0,$ $\qquad$ (13·8, 30)

where $\qquad \chi_1 = \dfrac{dC_L}{dC_R} - C_L\left(\dfrac{\frac{d\bar{V}_T}{dC_R}}{\bar{V}_T} + \dfrac{\frac{d\bar{A}_3}{dC_R}}{\bar{A}_3}\right)$ $\qquad$ (13·8, 31)

and $\quad \chi_0 = -\dfrac{dC_{M0}}{dC_R} + \left(\dfrac{\frac{d\bar{V}_T}{dC_R}}{\bar{V}_T} + \dfrac{\frac{d\bar{A}_3}{dC_R}}{\bar{A}_3}\right) C_{M0} + C_L\dfrac{dH_0}{dC_R} + \bar{V}_T\bar{J},$ $\quad$ (13·8, 32)

with $\qquad \bar{J} = \left(\dfrac{dC_L}{dC_R} - C_L\dfrac{\frac{d\bar{A}_3}{dC_R}}{\bar{A}_3}\right)\left(\dfrac{\bar{A}_1}{A} - \bar{A}_1 e + \bar{A}_4 d\right)$

$$+ C_L\left(\dfrac{\frac{d\bar{A}_1}{dC_R}}{A} - \dfrac{\bar{A}_1\frac{dA}{dC_R}}{A^2} - e\dfrac{d\bar{A}_1}{dC_R} + d\dfrac{d\bar{A}_4}{dC_R}\right)$$

$$+ \left(\dfrac{d\bar{A}_1}{dC_R} - \dfrac{\bar{A}_1\frac{d\bar{A}_3}{dC_R}}{\bar{A}_3}\right)\eta_T - \left(\dfrac{d\bar{B}_0}{dC_R} - \dfrac{\bar{B}_0\frac{d\bar{A}_3}{dC_R}}{\bar{A}_3}\right).$$

$$\text{(13·8, 33)}$$

The position of the neutral point (elevator free) is given by $K'_n = 0$, or

$$h'_n = H_0 + \frac{\chi_0}{\chi_1}, \qquad \text{(13·8, 34)}$$

and the corresponding c.g. margin is

$$H'_n = h'_n - h = \frac{K'_n}{\chi_1}. \qquad \text{(13·8, 35)}$$

Thus everything is the same as for the elevator fixed case except that the functions χ_0, χ_1 replace the functions ψ_0, ψ_1 respectively.

The theory of the manœuvre margins given in § 10·7 is based on the assumption that the manœuvres considered occur at constant speed. Hence the theory remains valid when distortion and compressibility are taken into account, but the aerodynamic coefficients must be modified to allow for these effects. Hence these coefficients are, in general, functions of Mach number and of equivalent air speed [see also equation (13·8, 36)].

Lastly, we may consider the application of dimensional analysis to the longitudinal stability of similar aircraft in similar conditions of flight. Let E be a typical angular elastic stiffness of the structure, moment per radian of angular displacement. Then the quantity $\dfrac{\rho l^3 V^2}{E}$ is non-dimensional, where l is a typical linear dimension while ρ and V have their usual meanings. Now the static margins are non-dimensional quantities and they clearly depend on the above non-dimensional parameter and on the Mach number. Hence

$$K_n = f\left(\mathbf{M}, \ \frac{\rho l^3 V^2}{E}\right), \qquad (13\cdot8, 36)$$

and K'_n can be similarly expressed.* Any attempt to use models in the investigation of static stability must be based on the last equation.

---

* Strictly, these static margins also depend on the Reynolds number.

Chapter 14

FLAPS FOR LANDING AND TAKE-OFF

By PROFESSOR A. D. YOUNG

14·1 Introduction

The possibilities of the flap as a device for increasing lift or drag have been apparent since the early days of aeronautics. During the First World War a number of types of aeroplane were equipped with plain flaps and since then various forms of flaps have been developed ranging in complexity from the simple split flap to large chord high lift flaps of two or more components. The practical importance of flaps increased considerably in the nineteen-thirties when, in consequence of the rapid improvements that had by then been made in the aerodynamic efficiency and cleanness of aircraft, accompanied by a rapid increase in wing loadings, gliding angles had decreased, and landing and take-off speeds had increased. Landing and take-off difficulties then threatened to offset the advantages of further aerodynamic improvements and the flap was widely adopted as it provided both lift to reduce the stalling speed and drag to increase the gliding angle. For take-off, the extra drag due to the flap was a disadvantage and this stimulated the development of the slotted flap since it promised a relatively small drag increase at moderate settings for take-off, with a drag increase that could be made comparable with that of a split flap at large settings for landing.

Further increases in wing loading and the stringent landing and take-off requirements of carrier-borne aircraft have stimulated the development of more ambitious flaps giving large lift increments. A variety of forms of such high lift flaps have been produced and they all involve some degree of effective wing area extension. Such flaps have made it possible to attain very high maximum lift coefficients, in some cases approaching 4·0, and when combined with leading edge slots a value of the order of 4·5 has been attained.

In the following the flap chord (c_f) is defined as the length of the projection of the flap on the wing chord line when the flap is retracted. The chord line of the flap is the line fixed relative to the flap which is coincident with the wing chord line when the flap is retracted. The flap deflection (δ_f) is the angle of rotation of the flap chord line relative to the wing chord line. The deflection of a second or auxiliary flap is defined relative to the chord line of the main flap. For conciseness the word 'flap' will be generally understood to mean 'flap for take-off or landing' throughout this chapter.

14·2 Brief description of the main types of simple flaps

Plain flaps

If a portion of the rear of a wing is simply hinged, as illustrated in Fig. 14·2, 1, then that portion is defined as a plain

Fig. 14·2, 1. 0·2c plain flap.

flap. The plain flap is the starting point for all the usual forms of control but is less commonly used as a flap. Its effectiveness derives from the fact that on deflection it changes the effective incidence and camber of the section and so causes a change of circulation and therefore of lift at a given wing incidence (see Chapter 7). The change in chordwise loading produced by a positive deflection is illustrated in Fig. 14·2, 2, having a maximum near the wing leading edge and a secondary peak at the flap hinge and falling to zero at the flap trailing edge.

Split flaps

The term 'split flap' is normally understood to mean a flap that is formed by splitting the wing trailing edge in a roughly chordwise direction forward to a hinge, about which the lower portion can rotate, the upper portion remaining fixed (see Fig. 14·2, 3). The definition has tended to become generalized to cover any flap that is hinged on the wing under-surface and whose operation does not alter the wing upper surface. Split flaps increase the circulation when deflected downwards by increasing the suction behind them and therefore at the wing trailing edge and thus alleviating the adverse pressure gradient

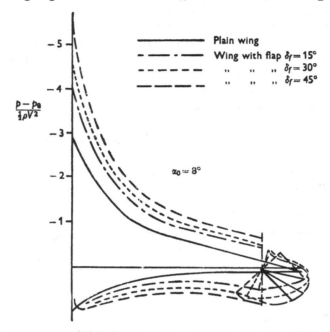

Fig. 14·2, 2. NACA 23012 section and 0·2c plain flap. Change in pressure distribution due to flap deflection.

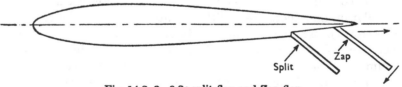

Fig. 14·2, 3. 0·2c split flap and Zap flap.

on the wing upper surface. The resulting change in the chordwise loading distribution is similar in form to that for plain flaps except that there is a discontinuity in pressure across the split flap and a rather greater suction at the trailing edge.

Slotted flaps

When a portion of the rear of a wing can be deflected and displaced to leave a well-defined slot between it and the rest of the wing, then it is referred to as a slotted flap. It may be simply hinged just below its nose, as in the case of the Handley Page type of slotted flap (see Fig. 14·2, 4), or it may be operated with a link or track mechanism, as in the case of the NACA (Fig. 14·2, 5), Blackburn (Fig. 14·2, 6) and Fowler (Fig. 14·2, 7) types of flaps. All slotted flaps have a certain amount of backward movement when operated.

We may note three main factors contributing to the lift increments of slotted flaps. The first is the effective change of camber produced by setting down the flap, as with plain flaps. The second is the flow through the slot, which re-energizes the wing boundary layer when the slot is well designed, and so delays flow separation from the flaps. The flow tends to separate from some point ahead of the flap trailing edge for all flap deflections greater than about 20°, and in consequence the increment in lift due to the flap is limited. This is reflected in the fact that this increment is then considerably less than would be predicted by potential flow theory based on the Kutta-Joukowsky condition being satisfied at the trailing edge. Any device that helps to move the point of separation nearer the flap trailing edge will produce an increase in the lift increment and some reduction of the drag increment. The third factor is the increase of effective lifting surface due to the rearward movement of the flap. In the case of the simple Handley Page type of flap the rearward extension is small and the consequent contribution to the lift increment is comparatively unimportant. In contrast, the rearward extension for the Fowler flap is considerable, being about the length of the flap chord, and the corresponding contribution to the lift increment is very important.

It cannot be too strongly emphasized that the effectiveness of a slotted flap depends critically on the design of the slot. If the slot is not efficient it may be worse than no slot at all, as it may increase rather than reduce the flow breakaway over the flap. Details of efficient slot designs cannot be given here, and reference should be made to original test reports. We may note, however, that flow breakaway in the slot must be avoided and so the slot must converge steadily from the lower to the upper surface. Further, the flow from the slot must merge smoothly into the flow round the wing and flap; hence an

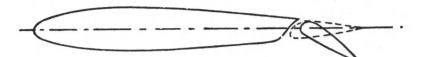

Fig. 14·2, 4. 0·2c slotted flap (Handley Page type).

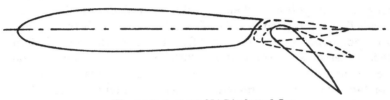

Fig. 14·2, 5. 0·26c NACA slotted flap.

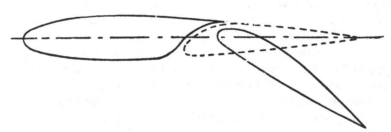

Fig. 14·2, 6. Blackburn slotted flap.

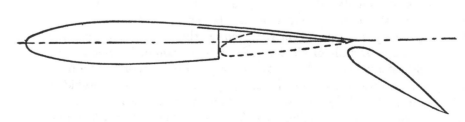

Fig. 14·2, 7. 0·4c Fowler flap (NACA tests).

appreciable length of tip or shroud to the upper surface of the slot is an advantage. In general, it should be possible to fit a smooth curve connecting the wing and flap contours.

The optimum slot shape is a function of the flap angle; the shape for the Handley Page simply hinged flap is usually designed to be at its best for moderate to large flap angles (i.e. 40° to 60°). The NACA slotted flap, however, was developed

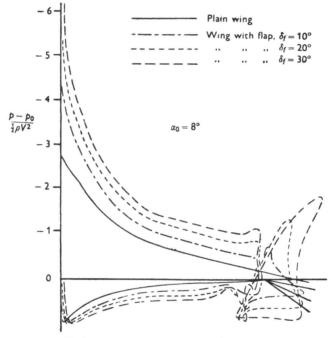

Fig. 14·2, 8. NACA 23012 section and 0·26c slotted flap. Change in pressure distribution due to flap deflection.

on the basis of a series of fairly exhaustive tests, to have, as far as the practical requirements of a track permitted, the optimum shape at all flap settings. The NACA type of flap, therefore, provides rather more lift than the Handley Page type at small to moderate flap angles.

A typical chordwise loading distribution with a slotted flap is shown in Fig. 14·2, 8. It will be seen to be similar to that due to a split or plain flap except for a much greater intensity of loading on the flap itself.

14·3   Lift, drag and pitching moment coefficient increments

Lift coefficient increments

A typical variation of lift coefficient with incidence for a wing with and without a downwardly rotated flap that does not extend the wing chord is illustrated in Fig. 14·3, 1. It will be seen that the effect of the flap movement is to displace the lift curve by a roughly constant amount upwards. In other words, the change in lift coefficient caused by the flap at a

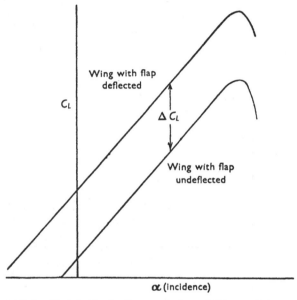

Fig. 14·3, 1. Sketch illustrating variation of lift coefficient with incidence for wing with and without a flap deflected.

given incidence is practically independent of incidence for a large range of incidence. This change is called the *lift coefficient increment* of the flap and is written ΔC_L. By convention it is normally quoted at an incidence of 10° above the no-lift angle of the basic wing, as this incidence is representative of the incidence of a normal aircraft in the process of landing. The lift coefficient increment of a flap varies with the aspect ratio of the wing as does the lift curve slope of the wing alone. It follows that the change in angle of zero lift due to the flap deflection is independent of wing aspect ratio. This fact is linked with the basis of most theoretical work on simple trailing edge flaps, which assumes that the effect of the flap on lift is equivalent to a change of incidence. An interesting feature of the lift coefficient increment of a flap is that it is found to be

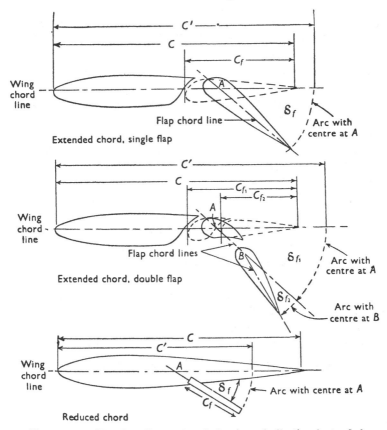

Fig. 14·3, 2. Sketches illustrating derivation of effective (extended or reduced) chord.

almost independent of test conditions, e.g. scale effect, tunnel turbulence, surface finish, etc. In contrast, the increment in maximum lift coefficient due to the flap is very sensitive to test conditions. However, flight tests indicate that under full-scale conditions the increments in $C_{L\text{max}}$ are fairly close to the lift coefficient increments.

When the flaps extend the chord on rotation then at each incidence there is an additional lift coefficient increment proportional to the incidence and the extension in effective lifting surface area. Consequently the curve of lift coefficient against incidence for a given flap setting is no longer parallel to that for the basic wing. For a full span flap the slope is in fact proportional to that for the basic wing multiplied by the ratio of the extended or effective chord (c') to the basic chord (c). The lift coefficient increment then varies with incidence, but for convenience we still define it at an incidence of 10° above the no-lift angle of the basic wing.

The definition of the extended or effective chord (c') for a full span flap is illustrated in Fig. 14·3, 2. If the extended flap is rotated about the point of intersection of the wing and flap chord lines until the two chord lines coincide, then the distance from the leading edge of the wing to the trailing edge of the flap in this position is defined as the extended chord. In the case of a split flap set forward of the usual trailing edge position the same process is applied to determine the effective (in this case reduced) chord. With the aid of this concept of effective chord, results obtained with flaps that extend (or reduce) the chord can readily be correlated with results for flaps that do not. In particular, the lift coefficient increment based on the effective chord ($\Delta C_L'$) for a given value of the ratio of the flap chord (c_f) to the effective chord (c') is independent of the incidence and chord extension. To determine $\Delta C_L'$ and its relation with ΔC_L we note that the lift coefficient of wing plus extended flap based on c' is $C_L' = C_L \dfrac{c}{c'}$, and hence $\Delta C_L'$ is given by

$$\Delta C_L' = C_L \frac{c}{c'} - C_{LW}, \qquad (14\cdot3,\,1)$$

where C_{LW} is the lift coefficient of the basic wing. But $\Delta C_L = C_L - C_{LW}$, and therefore

$$\Delta C_L' = \Delta C_L \frac{c}{c'} - C_{LW}\left(1 - \frac{c}{c'}\right). \qquad (14\cdot3,\,2)$$

Typical variations of the lift coefficient increments of full span split, plain, slotted and Fowler flaps with flap angle on wings of aspect ratio 6 are shown in Fig. 14·3, 3. The marked

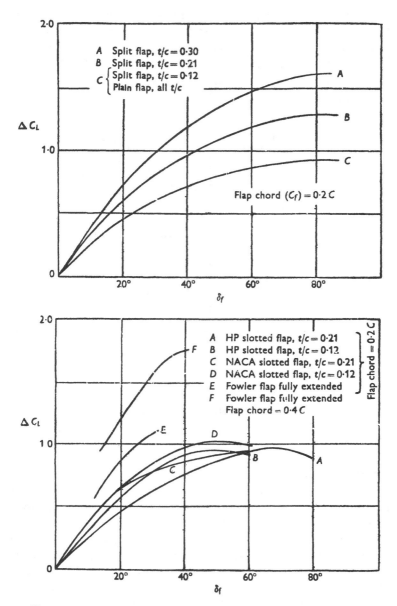

Fig. 14·3, 3. Typical lift coefficient increments for full span split, plain, slotted and fully extended Fowler flaps.

effect of wing thickness/chord ratio on the increment for split flaps will be noted. For a part span flap a rough estimate of the increment can be obtained by applying the factor shown in Fig. 14·3, 4 to the corresponding increment for the full span flap.

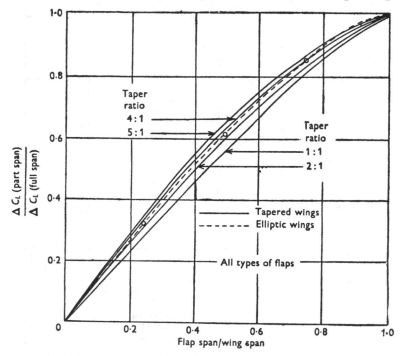

Fig. 14·3, 4. The ratio of lift increment of part span flap to lift increment of full span flap.

Profile drag coefficient increments

The increment in profile drag coefficient due to a flap at a given incidence is rather more influenced by test conditions and incidence than is the lift coefficient increment. Nevertheless, this influence is still sufficiently small over a wide range of incidence for us to accept the increment at a standard incidence as a reliable measure of the profile drag characteristics of a flap. Since our main interest is centred on the effect of the flap drag on take-off the standard incidence has been taken to be 6° above the no-lift angle of the basic wing. This increment is written as ΔC_{DO}. It is assumed to be independent of aspect ratio.

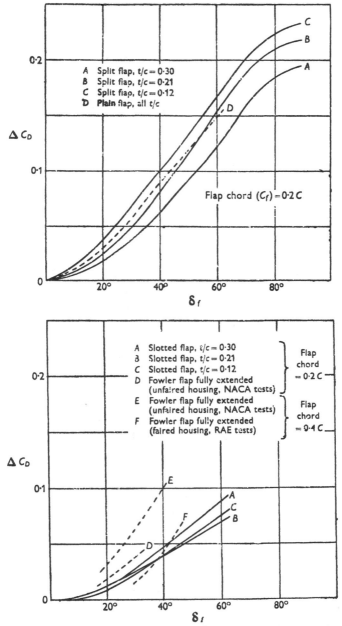

Fig. 14·3, 5. Typical drag coefficient increments for full span split, plain, slotted and fully extended Fowler flaps.

Typical variations of ΔC_{DO} for full span split, plain, slotted and Fowler flaps with flap angle are shown in Fig. 14·3, 5. It is worth noting that for split flaps $\Delta C_{DO} \approx 1\cdot1 \sin^2 \delta_f$, in terms of the area of the flap, where δ_f is the flap angle, whilst for slotted flaps $\Delta C_{DO} \approx 0\cdot5 \sin^2 \delta_f$, in terms of the area of the flap. For a part span flap the drag coefficient increment is found to be fairly closely proportional to the area of the flapped part of the wing.

Pitching moment coefficient increments

The pitching moment coefficient increment of a wing due to a flap at a given incidence is always closely correlated with the lift coefficient increment. Like the latter it is practically independent of test conditions and, if the flap does not extend the chord, it is independent of wing incidence. It is usually quoted at the same standard incidence, viz., 10° above the no-lift angle of the basic wing. In two important respects the pitching moment coefficient increment differs from the lift coefficient increment; firstly, theory shows it to be independent of aspect ratio and, secondly, it is a function of taper ratio. It is written as ΔC_m and, unless otherwise stated, is referred to the wing quarter-chord line.*

Like the lift coefficient increment the results for flaps that extend the chord can be correlated with those for flaps that do not extend the chord if the increments for the former are based on the extended chord. However, we may note that for full span flaps on rectangular wings the pitching moment coefficient increments (written ΔC_{mr}) are roughly a constant fraction of the corresponding lift coefficient increments for a given type of flap. Thus, for split and plain flaps of moderate chords $\dfrac{\Delta C_{mr}}{\Delta C_L} \approx -0\cdot25$, for slotted flaps of the Handley Page and NACA types $\dfrac{\Delta C_{mr}}{\Delta C_L} \approx -0\cdot3$, and for Fowler flaps $\dfrac{\Delta C_{mr}}{\Delta C_L} \approx -0\cdot4$. Here ΔC_L refers to the lift coefficient increment for a wing of aspect ratio 6. These results are consistent with the centre of pressure of the extra lift due to a flap movement being located roughly at the mid-point of the extended chord. For part span or full span flaps on unswept tapered wings ΔC_m can be obtained

---

* Except in § 14·9, the disscusion here refers to wings with unswept quarter-chord lines.

from ΔC_{mr} by means of the factor given in Fig. 14·3, 6. This factor is calculated on the following assumptions. The contribution to the pitching moment increment of each spanwise

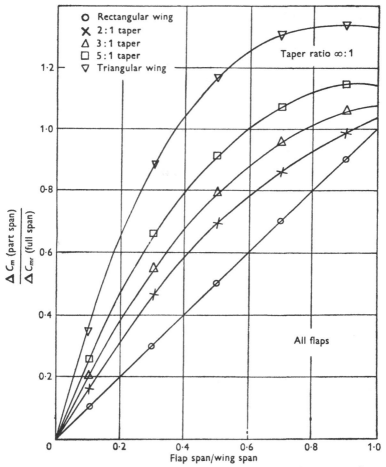

Fig. 14·3, 6. Ratio of pitching moment coefficient increment of part span flap to pitching moment coefficient increment of full span flap on a rectangular wing.

element of the flapped part of the wing is assumed to be the same as for two dimensional flow and is, therefore, 'weighted' as the square of the local chord. On the unflapped part of the wing any change in loading is assumed to be located on the quarter-chord line and therefore does not contribute to the pitching moment coefficient increment. In spite of their crudity, these assumptions are found to lead to results in reasonable agreement with experimental results.

Typical values of the increments for the main types of simple flaps

The following table summarizes some representative measured values of the increments and other relevant data for the main types of simple flaps. The maximum flap settings quoted generally correspond to about the optimum lift coefficient increments.

TABLE 14·3, 1.

Flap type	wing thick- ness t/c	Flap chord c_f/c	Flap angle δ_f	c'/c	ΔC_L $A=6$	ΔC_{DO}	ΔC_{mr}	$\dfrac{-\Delta C_{mr}}{\Delta C_L}$
Split	0·12	0·1	30°	1·0	0·47	0·031	−0·15	0·32
,,	,,	,,	75°	1·0	0·72	0·096	−0·21	0·29
,,	,,	0·2	30°	1·0	0·60	0·066	−0·18	0·30
,,	,,	,,	75°	1·0	0·99	0·205	−0·23	0·23
,,	,,	0·4	30°	1·0	0·87	0·156	−0·19	0·22
,,	,,	,,	60°	1·0	1·23	0·399	−0·25	0·20
,,	0·21	0·1	30°	1·0	0·51	0·018	−0·18	0·35
,,	,,	,,	90°	1·0	0·97	0·091	−0·27	0·28
,,	,,	0·2	30°	1·0	0·79	0·055	−0·24	0·30
,,	,,	,,	90°	1·0	1·31	0·233	−0·33	0·25
,,	,,	0·4	30°	1·0	1·18	0·117	−0·29	0·25
,,	,,	,,	60°	1·0	1·69	0·360	−0·37	0·22
H.P. slotted	0·16	0·2	20°	1·015	0·54	0·015	−0·157	0·29
,,	,,	,,	50°	1·03	0·85	0·073	−0·246	0·29
,,	0·12	0·26	20°	1·02	0·59	0·010	−0·169	0·29
,,	,,	,,	40°	1·04	1·09	0·098	−0·327	0·30
NACA slotted	0·12	0·1	20°	1·018	0·42	0·004	−0·135	0·32
,,	,,	,,	50°	1·030	0·75	0·024	−0·202	0·27
,,	,,	0·26	20°	1·045	0·68	0·007	−0·221	0·33
,,	,,	,,	40°	1·070	1·15	0·57	−0·387	0·34
,,	,,	0·4	20°	1·076	1·01	0·014	−0·304	0·30
,,	,,	,,	30°	1·100	1·38	0·027	−0·416	0·30
,,	,,	,,	40°	1·120	1·30	0·114	−0·367	0·28
,,	0·21	0·26	20°	1·045	0·83	0·014	−0·263	0·32
,,	,,	,,	60°	1·087	1·27	0·119	−0·355	0·33
,,	,,	0·4	20°	1·072	0·92	0·023	−0·282	0·31
,,	,,	,,	50°	1·102	1·28		−0·385	0·30
Fowler	0·12	0·2	15°	1·085	0·45	0·015	−0·169	0·38
,,	,,	,,	30°	1·185	1·06	0·036	−0·443	0·42
,,	,,	0·3	20°	1·181	0·70	0·033	−0·26	0·39
,,	,,	,,	40°	1·280	1·55	0·085	−0·691	0·45
,,	,,	0·4	20°	1·265	0·86	0·039	−0·354	0·41
,,	,,	,,	40°	1·365	1·78	0·099	−0·85	0·48

The data quoted for the other settings of the Fowler flaps refer to partially extended positions following practical tracks but with relatively inefficient slots. The figures refer to full span flaps.

14·4 Multiple flaps

All other types of flaps are basically variants or combinations of the three main types described above. To meet high lift requirements, various multiple types, involving two or more components and appreciable rearward extension of area, have been developed. Examples of such flaps are illustrated in Figs. 14·4, 1 ... 14·4, 7.

Thus, by providing a second slotted flap to the rear of a first a more efficient flap system is obtained, since the effective camber change is attained more smoothly, a further extension of lifting area can be provided, and the beneficial effect of two slots is obtained. An example of this type of flap is the NACA double-slotted flap (Fig. 14·4, 1). The double Fowler flap (Fig. 14·4, 2) is another example, its distinctive feature being the very considerable area increase when fully extended. The case illustrated provided one of the highest lift coefficient increments on record ($\Delta C_L = 2·7$).

A variant of the double-slotted flap scheme involves the use of a flap leading edge slot as is illustrated by the Blackburn arrangement shown in Fig. 14·4, 3. When properly designed this arrangement can be very effective for producing a high lift increment with a relatively low drag increment because of the beneficial effect of the second slot on the flow over the flap and the extra lifting surface area when the flap is extended. Another variant is provided by inset slots in the flaps, also illustrated by a Blackburn arrangement in Fig. 14·4, 4, and with properly designed slots this too can be very effective. However, the simplicity of this device is largely offset by the fact that unless the slot or slots are sealed when the flap is retracted they will cause a serious increase of drag. This difficulty is enhanced by the fact shown by tests that for a slot to function properly it must be cut at a fairly small angle (about 20°) to the flap chord. This implies that the slot opening on the flap upper surface must be fairly well back and the shroud must be very long to seal the slot when the flap is retracted.

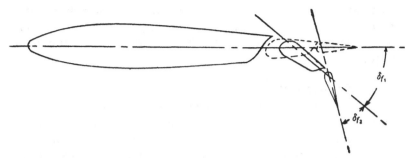

Fig. 14·4, 1. NACA double slotted flap (0·26c, 0·1c).

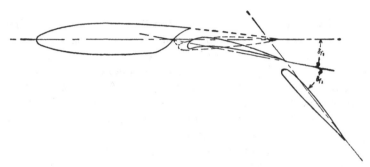

Fig. 14·4, 2. RAE double Fowler flap.

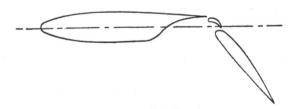

Fig. 14·4, 3. 0·5c Blackburn flap with flap leading edge slat.

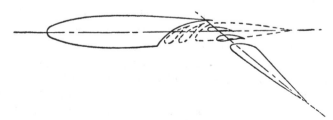

Fig. 14·4, 4. 0·5c Blackburn flap with two inset slots.

Another effective device that has been tested by Blackburn Aircraft Ltd. is a slotted flap combined with a long shroud part of which is hinged and arranged to deflect downwards with the flap (see Fig. 14·4, 5).

Combinations of split and slotted flaps readily suggest themselves and one such arrangement is illustrated in Fig. 14·4, 6.

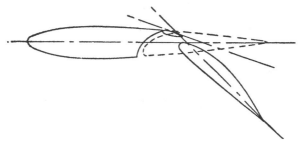

Fig. 14·4, 5. 0·5c Blackburn flap with deflected shroud.

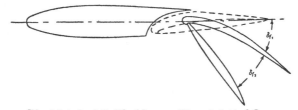

Fig. 14·4, 6. 0·5c Blackburn split and slotted flap.

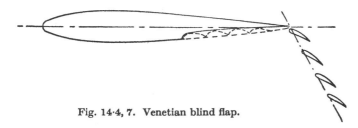

Fig. 14·4, 7. Venetian blind flap.

This arrangement has considerable possibilities where both high lift and variable drag are required.

A logical development of the multi-slotted flap is the so-called venetian blind flap, illustrated in Fig. 14·4, 7. One might anticipate that, if the slots functioned properly, lift coefficient increments approaching the value given by potential flow theory should be attainable. The results of such tests as have been

TABLE 14·4, 1.

Flap type	Wing thickness (t/c)	Flap chords		Flap angles		c'/c	ΔC_L ($A = 6$)	ΔC_{DO}	ΔC_{mr}	$-\dfrac{\Delta C_{mr}}{\Delta C_L}$
		c_{f1}/c	c_{f2}/c	δ_{f1}	δ_{f2}					
NACA double slotted	0·12	0·26	0·1	20°	20°	1·076	1·17	0·015	−0·391	0·33
	,,	0·26	0·1	40°	40°	1·100	1·67	0·099	−0·553	0·33
	,,	0·4	0·26	20°	20°	1·139	1·59	0·052	−0·54	0·34
	,,	0·4	0·26	30°	30°	1·160	2·10	0·114	−0·71	0·34
	0·21	0·4	0·26	20°	20°	1·118	1·50	0·061	−0·56	0·37
	,,	0·4	0·26	40°	30°	1·183	2·13	0·187	−0·78	0·37
Double Fowler	0·16	0·4	0·4	13°	41·7°	1·41	2·65	0·100	−1·08	0·41
Blackburn split and slotted	0·18	0·5	0·5	25°	28°	1·105	2·07	0·104	−0·609	0·29
	,,	0·5	0·5	30°	35°	1·185	2·42	0·189	−0·679	0·28
Blackburn flap with flap and L.E. slat	0·15	0·5, slat chord = 0·086c,		50°		1·134	2·28	0·143	−0·780	0·34
	0·18	0·4, slat chord = 0·19c		40°		1·24	1·631	0·097	−0·569	0·35
Blackburn flap with inset slot	0·15	0·5, 2 inset slots		50°		1·285	2·33	0·117	−0·820	0·35
	0·18	0·4, 1 inset slot		40°		1·09	1·60	0·052	−0·467	0·29
	0·12	0·4, 1 inset slot		40°		1·12	2·05	0·048	−0·602	0·29
Blackburn with deflected shroud	0·15	0·5, shroud = 0·08c		50° flap angle 17° shroud angle		1·27	2·26	0·122	−0·675	0·30
Venetian blind flap	0·12	0·4c, 4 slats of 0·1c each		$\delta_f = 30°$ $\delta_{f1} = \delta_{f2} = \delta_{f3}$ $= \delta_{f4} = 30°$		1·37	1·27	0·021	−0·61	0·48
				$\delta_f = 60°$ $\delta_{f1} = \delta_{f3} = \delta_{f3}$ $= \delta_{f4} = 40°$		1·37	1·88	0·037	−0·96	0·51

made, however, have been rather disappointing. A lift coefficient increment somewhat higher and a drag coefficient somewhat lower than for the corresponding Fowler flap has been attained but the gain hardly justifies the extra mechanical complication and the pitching moment coefficient increments are very high.

Some representative measured values of the lift, drag and pitching moment coefficient increments for full span multiple flaps, together with other relevant data, are summarized in the following table. The maximum flap settings quoted generally correspond to the optimum lift coefficient increments.

For part span flaps, the conversion factors already discussed for the more simple types of flaps apply.

14·5   Flaps and induced drag

In assessing the drag effects of a flap account must be taken of the induced drag changes caused by the flap. The increase of lift will inevitably be accompanied by an increase in induced drag and, in general, the change in the spanwise loading distribution produced by the flap will increase the induced drag still further. For a flap on a wing of elliptic plan form, lifting line theory leads to the result

$$C_{Di} = \frac{C_L^2}{\pi A} + K \frac{(\Delta C_L)^2}{\pi A}, \qquad (14·5, 1)$$

where C_L is the total lift coefficient of the wing plus the flap, ΔC_L is the lift coefficient increment of the flap, and K is a function of the flap span, cut-out, and of the ratio of the aspect ratio (A) to the two-dimensional lift curve slope of the wing (a_0). The first term on the right of equation (14·5, 1) therefore gives the induced drag if the loading due to wing and flap were distributed elliptically, and the second term is the contribution arising from the departure of the loading from the elliptical. The following table lists some representative values of K for flaps without cut-out:

TABLE 14·5, 1

A/a_0 \ Flap span	0·3	0·5	0·7	1·0
0·67	1·45	0·58	0·20	0
1·0	2·03	0·80	0·28	0
2·0	1·80	1·08	0·36	0

If we take a_0 as 6, approximately, these values of A/a_0 correspond to values of A of 4, 6 and 12 approximately. A small cut-out will tend to reduce the factor K slightly.

A few simple calculations will readily indicate the importance of these induced drag effects for typical flap arrangements. It is worth noting that there is a limit to the magnitude of the lift coefficient increment that one can use to advantage during take-off and that this limit is imposed by the accompanying increase of induced drag. For power loadings (lb./b.h.p.) greater than about 8, the minimum take-off distances to a height of 50 ft. are obtained with maximum lift coefficients of the order of 3·0 or less. There is a similar restriction on the maximum lift coefficient that is useful on the climb. For landing no such restriction applies during the approach and touch down. During the ground run, however, a high lift coefficient may be an embarrassment as it reduces the ground load and therefore the effectiveness of the wheel brakes. In some cases it may be necessary to reduce the lift coefficient rapidly once the aeroplane has touched down.

14·6   Wing-body interference

The results quoted in the preceding paragraphs are based on data obtained on wings alone. In the presence of bodies, however, the available evidence indicates that interference effects may modify in some measure the flap characteristics.

The interference effects on the lift coefficient increments are less pronounced than those on the drag coefficient increments and show no systematic trends. The effect on the drag coefficient increments, however, are on the whole favourable for split flaps and unfavourable for slotted flaps. This is presumably due to the fact that split flaps when operated tend to clean up any flow irregularities in wing-fuselage and wing-nacelle junctions, but with slotted flaps there is a definite break at such junctions through which spoiling air may flow. For both split and slotted flaps we may expect the interference effects at the junction to become relatively less important as the flap span is increased.

In spite of these interference effects, however, the over-all picture still remains that slotted flaps cause lower drag coefficient increments than split flaps for a given lift coefficient increment. The difference between the two types of flaps becomes increasingly marked the higher the lift coefficient increment required and the greater the flap span.

14·7   Flaps in combination with wing leading edge slats

Wing leading edge slats, unlike trailing edge flaps, increase the maximum lift coefficient by increasing the stalling incidence. They do not produce any marked increase of lift at a given incidence below the stall, in fact they may in some cases reduce the lift slightly. The most ambitious high lift devices developed so far have combined high lift trailing edge flaps and leading edge slats. When functioning properly, a slat enables the nose of the wing to sustain a much higher lift without subsequent flow breakaway than it would without the slat. In consequence the slat not only increases the stalling incidence but increases the pitching moment and reduces the stability of the wing near the stall.

The following crude relations summarize broadly the main effects of leading edge slats. These effects may be regarded as additive to those produced by a flap if present. The slats are taken to be full span, of chord lengths within the range $0·15c - 0·3c$ and set at about $40°$ to the wing chord line.

(i) $\Delta C_{L\text{max.}} = 3·3 \dfrac{\text{slat chord}}{\text{wing chord}}$, approximately,

(ii) $\Delta C_{m\text{(st)}} = 0·9 \dfrac{\text{slat chord}}{\text{wing chord}}$, approximately,

(iii) $\Delta\alpha_{\text{(st)}} = 10° \ (\pm 3°)$,

(iv) $\Delta\left(\dfrac{\partial C_m}{\partial C_L}\right)_{0·8\,C_{L\text{max.}}} = 0·15 \ (\pm 0·075)$,

where $\Delta C_{m\text{(st)}}$ is the change in pitching moment coefficient at the stall, due to the slat, and $\Delta\alpha_{\text{(st)}}$ is the change in stalling angle.

14·8   Trim and maximum lift coefficient

When the flaps of an aeroplane are lowered they not only alter the pitching moment of the wings, but in addition they alter the downwash at the tail and in consequence alter the pitching moment due to the tail. The latter contribution to the net pitching moment change will be proportional to the tail volume. These two contributions are in general of opposite sign since flaps usually increase the downwash at the tail.

The net change of pitching moment will require some elevator movement to trim if steady flight is to continue, and the amount of elevator movement depends on the conditions that are assumed to govern the steady flight after the flaps are lowered. One can assume that the lift coefficient is unaltered, or one can assume that the ratio of the lift coefficient to the maximum lift coefficient attainable with the corresponding flap setting remains constant. It has been suggested by Duddy* that the latter assumption is rather more in accordance with practice. The change of elevator angle required may, in either case, be of positive or negative sign, but the larger the tail volume the greater becomes the downwash effect and the more likely is the required elevator movement to be positive. Duddy estimates that even with a full span Fowler flap fully deflected the elevator movement required to trim will be positive for tail volume ratios greater than about 0·75. We may also note that an increase in tail volume ratio permits a rearward movement of the c.g. relative to the wing for the same degree of static stability. This will tend to reduce the pitching moment on the wing about the c.g. due to the flap deflection.

This latter fact is important in assessing the effect of the trimming load required on the tailplane on the maximum attainable lift coefficient of the whole aeroplane. If the trimming load required near the stall on the tailplane is downwards with the flap deflected and is greater in magnitude when expressed as a coefficient than the corresponding load required with the flap retracted, then the increment in $C_{L\text{max}}$ for the whole aeroplane due to the flap will be less than that obtained on the wing alone. The trimming load must balance the change of pitching moment of the wing about the c.g. due to the flap. As noted above, an increase of tail volume permits a rearward movement of the c.g. for a given degree of static stability and hence permits this change of pitching moment to become more positive and the consequent trimming load on the tailplane to decrease (if measured as positive in the downward direction). Duddy† estimates in a specimen example that with a 50 per

* R. R. Duddy, 'High Lift Devices and their Uses', *J. Roy. Aero. Soc.*, vol. **53** (1949), p. 859 *et seq.*   † *Loc. cit.*

cent span split flap the change in C_{Lmax} due to trimming out the pitching moment caused by the flap is zero for a tail volume ratio of about 0·75. With a 70 per cent span slotted flap the corresponding tail volume ratio required is 0·85, and for a full span Fowler flap this increases to 1·15. It will be clear then that even with relatively ambitious flap arrangements it is possible, by adopting a large enough but by no means unacceptable tail volume, to ensure little change of trim or reduction of available C_{Lmax}.

14·9   Flaps on sweptback wings

The discussion has so far been confined to flaps on unswept wings. The effect of sweepback on flap characteristics is a subject that is by no means fully explored as yet, although it is of increasing importance, and the available data are scattered and unsystematic. A few tentative generalizations are possible and they will be discussed here, but they await further systematic and comprehensive experiments for confirmation.

The most noteworthy feature of the characteristics of a flap on a sweptback wing is the fact that the lift coefficient increment at an incidence below the stall is reduced relatively little by the sweepback, but the stalling incidence is reduced markedly below that of the wing with flap retracted. In consequence the increment in maximum lift coefficient falls with increase of sweepback. These remarks are illustrated in Fig. 14·9, 1.

An analysis of the available data suggests the rough rule that

$$\frac{\Delta C_{Lmax.}}{\Delta C_{Lmax.} (\gamma = 0)} \approx \cos^3 \gamma,$$

where γ = angle of sweep, and $\Delta C_{Lmax. (\gamma=0)}$ is the increment in maximum lift coefficient due to the flap on a wing of zero sweep. The angle of sweep is here defined by the angle between the pitching axis and the projection of the quarter-chord line on the plane containing the direction of flight at, say, zero incidence and the pitching axis. Thus, on a wing with a sweepback angle of 45° the increment in maximum lift coefficient will be about one-third of that for a wing of zero sweepback angle. In contrast, the data for the lift coefficient increment at incidences below the stall indicate a slight increase for sweepback angles up to about 20° and then a slow fall with further increase of sweepback angle. For 45° of sweep-back the lift coefficient increment is reduced about 25 per cent. It should be noted that the data analysed refer to wings of aspect ratio within a range from about 5 to 6·5. For wings of small aspect ratio the results may be different.

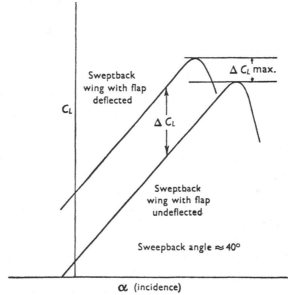

Fig. 14·9, 1. Sketch illustrating effect of sweepback on stalling incidence, lift cofficient increment and in max. lift coefficient due to flap.

The above discussion applies to wings with flaps that do not extend the chord. One or two isolated tests on sweptback wings with flaps that extend the chord indicate that then the maximum lift coefficient increments are somewhat larger than would be predieted by the above formula. An analysis of the resulting data suggests that the increment in maximum lift coefficient can be estimated on the assumption that the part of the increment due to the area extension is not subject to the $\cos^3 \gamma$ factor, whilst the remaining part of the increment is subject to this factor. The problem as a whole is one that will undoubtedly attract considerable interest, as any device for raising the maximum lift coefficients of sweptback wings will be of vital importance in the design of tailless and high-speed aircraft. A similar analysis of data on the profile drag increments of flaps on sweptback wings leads to the rough rule

$$\frac{\Delta C_{DO}}{(\Delta C_{DO})_{\gamma=0}} \approx \cos \gamma.$$

The pitching moment increment due to a flap on a sweptback wing can be regarded as made up of two contributions. The first derives from the change of pitching moment due to the flap about the local aerodynamic centre of each spanwise element of the flapped part of the wing. The second derives from the change in spanwise lift distribution produced by the flap. With the usual c.g. positions these two contributions are in general of opposite sign, the first being negative and the second being positive. Thus it is possible to arrange for the net change of pitching moment with flap deflection to be small. An accurate and quick method for estimating this net change has yet to be developed. For details of an approximate method, which has been found to be fairly reliable up to moderate sweepback angles of the order of 30°, see a paper by Dent and Curtis.*

* M. M. Dent and M. F. Curtis, 'A Method of Estimating the Effect of Flaps on the Pitching Moment and Lift of Tailless Aircraft', *R.A.E. Report No. Aero.* 1861 (1942), A.R.C. 7270.

GENERAL REFERENCES FOR CHAPTER 14

The literature of the subject of flaps is too extensive for an adequate reference list to be given in the space here available. However, the following general reviews provide comprehensive bibliographies for the reader who wishes to follow in more detail a particular aspect of the subject. The subject matter of this chapter has been based mainly on the first reference.

A. D. Young, 'The Aerodynamic Characteristics of Flaps', *R.A.E. Report No.* 2185, *R. & M.* 2622 (1947).
A. R. Weyl, 'High Lift Devices and Tailless Aeroplanes', *Aircraft Engineering*, vol. 17 (Oct. 1945), p. 292 *et seq.*
H. B. Irving, 'Wing Brake Flaps, a Review of their Properties and the Means for their Operation', *Aircraft Engineering*, vol. 7 (Aug. 1935), p. 189 *et seq.*
R. P. Alston, 'Wing Flaps and Other Devices as Aids to Landing', *J. Roy. Aero. Soc.*, vol. 39 (1935), p. 637 *et seq.*

Useful information on flaps will be found in the Data Sheets issued by the Royal Aeronautical Society.

Chapter 15

SUNDRY TOPICS

15·1 Motion with a control free

We have already considered static stability with elevator free (see §§ 10·4, 10·5 and 10·6) but have not treated the more general cases of dynamic stability with a free control or controls. It is to be understood that the word 'free' merely indicates that the control surface is not rigidly constrained and that the theory covers the case of constraint by a spring.

It is clear that motion with a control surface free is just a particular case of motion with distortion and the theoretical approach is exactly the same for the two problems. The methods for dealing with motion involving distortion have been discussed in Chapter 12. The 'method of distortional co-ordinates' (see §§ 12·2 and 12·4) is the most appropriate here and there will be one new displacement co-ordinate for each control flap or tab which is independently free in the motion. The co-ordinates will most conveniently be the angular deflections measured relative to the surface to which the flap or tab is attached and at the section where the controlling moment is applied. For example, for motion with free rudder the additional degree of freedom will be the angle of deflection of the rudder relative to the fin measured in the section where the rudder controls are attached.

The method of constructing the dynamical equations has been explained in § 12·4 and is based on the Lagrangian method of virtual work. For any rigid control flap or tab the new dynamical equation will, with the co-ordinates chosen as above, express the balance of the hinge moments (including those of inertial origin) for the surface. The other dynamical equations will contain additional terms representing the influence of the angular displacement, velocity and acceleration of the surface.

As an example we may consider the dynamical equations for longitudinal-symmetric motion with free elevator. These are formally identical with the equations given in § 12·5 for a flexible fuselage with a single distortional co-ordinate Φ. It is only necessary to substitute the elevator angle η for Φ and to interpret the other symbols accordingly. The function $f(x)$ which represented the bending mode is now zero for all points not on the elevator and equal to distance from the hinge for points on the elevator.

The 'method of the modification of derivatives' (see § 12·6) can be used whenever the motions are so slow that the terms in the angular velocity and acceleration in the hinge moment equation are negligible in comparison with the displacement term. For example, when investigating the influence of freeing the elevator on the phugoid motion we could (provided that the elevator is aerodynamically underbalanced) disregard the angular velocity and acceleration of the elevator and use the equation of hinge moments to eliminate the elevator co-ordinate η from the remaining dynamical equations.

The full dynamical equations with even one free control are rather complicated and lead to a determinantal equation of the sixth degree. It is sometimes possible to disregard one or more of the components of the motion of deviation and to simplify the treatment accordingly. For example, when investigating 'snaking', which is a rather rapid oscillation of the lateral-antisymmetric type occurring with free rudder, it may sometimes be a legitimate approximation to disregard the lateral movement of the C.G. and the rolling and to treat the motion as having just two degrees of freedom, namely, rudder angle and angle of yaw.

While it is possible for stability to be improved by freeing a control, the chief concern of the designer is to ensure that stability is not worsened. Now freeing a control will have no influence on the stability when the motion of the flap is uncoupled to the other motions occurring. There are two measures which secure this uncoupling, completely or nearly so:

(a) Mass balance of the flap (see § 7·7).

(b) Making zero the aerodynamic hinge moment coefficient b_1 (see § 7·1).

These measures should not be adopted blindly but only after careful consideration of the circumstances of the case.

It is beyond our scope to enter into a detailed investigation of 'snaking', but we shall record the following conclusions which emerge from the investigations of Bryant and Gandy,[*] Green-berg and Sternfield,[†] Neumark[‡] and others:

(i) When the rudder is mass-balanced and b_1 is zero, freeing the rudder has no influence on the stability (see above).

(ii) Divergence may occur when b_1 is large and negative.

(iii) Oscillatory instability may occur when a numerically small b_2 (close aerodynamic balance) is associated with positive values of b_1.

(iv) Mass-underbalance of the rudder (c.g. aft of hinge) is destabilizing.

(v) Frictional resistance to motion of the rudder may give rise to an oscillation of limited amplitude.

With reference to (v), it has been shown[§] that a constant frictional resistance (such as may arise from 'solid' friction) can have a destabilizing effect only when the corresponding linear velocity damping also has a destabilizing effect.

An oscillation which has been called 'longitudinal snaking' may occur when the elevator is free. Instability in this mode can be avoided by measures similar to those which are effective for lateral snaking.

15·2   Automatic stabilization

The general scheme of automatic stabilization consists in arranging a mechanism which moves the control surfaces in such a manner that desirable dynamical characteristics of the complete system—aircraft cum stabilizer—are attained. In order that the pilot may still be able to control the aircraft it is arranged by a suitable linkage that the movement of the control is compounded from components contributed by the pilot and by the stabilizer. Thus, when the pilot holds his control lever fixed the aircraft is controlled by the stabilizer. On the other

[*] L. W. Bryant and R. W. Gandy, 'An Investigation of the Lateral Stability of Aeroplanes with Rudder Free', *R. & M.* 2247 (Dec. 1939).

[†] H. Greenberg and L. Sternfield, 'A Theoretical Investigation of the Lateral Oscillations of an Airplane with Free Rudder, with Special Reference to the Effect of Friction', *NACA Advance Restricted Report* (March 1943), A.R.C. 6987.

[‡] S. Neumark, 'A Simplified Theory of the Lateral Oscillations of an Air-craft with Rudder Free, including the Effect of Friction in the Control System', *R. & M.* 2259 (May 1945).

[§] R. A. Frazer, W. J. Duncan and A. R. Collar, *Elementary Matrices*, Cambridge University Press, 1938.

hand, quick control movements by the pilot will be transmitted almost unmodified to the control surface. The mechanism is called a stabilizer because it is commonly used to overcome some tendency towards instability and because satisfactory dynamical behaviour implies stability of the system.

A stabilizer may in general be regarded as built from the following elements:

(a) A detecting unit which detects a deviation from the desired steady state of flight and provides a suitable physical output or signal which is some definite function of the deviation. Usually the signal will be proportional to the deviation. There may be several such detectors responding to distinct deviations.

(b) A mixing unit in which the signals from the detectors are compounded in suitable proportions. In general, one of the compounded signals will be proportional to the displacement of the control surface and have a negative coefficient (negative feed-back). The output from the mixer will be proportional to the required further movement of the control surface but will usually be at a low power level.

(c) An amplifier or servo motor which gives an output proportional to the signal from the mixer and at a power level sufficient to operate the control surface.

The negative feed-back or 'follow-up' mentioned under (b) may alternatively be provided in the servo motor.

Ideally the control surface would respond instantly to the deviations from the steady state. However, on account of the inertia and aerodynamic damping of the control surface itself and of inevitable lags in the automatic stabilizer there is always a lag in the response of the control flap. In addition, the aerodynamic forces and moments do not instantly reach the steady values corresponding to the flap deflection (Wagner effect). Any lag has, in general, a destabilizing effect, so it is very important to reduce the lags in the stabilizer system as far as possible. The destabilizing effect of a given lag is greater the higher the frequency of the oscillation dealt with so the avoidance of lag is of particular importance when oscillations of high frequency occur. There are various electrical and other devices for providing an artificial lead which can be arranged to cancel the lags to a large extent.

A stabilizer which has good qualities theoretically may be objectionable in practice because in certain circumstances it 'handles' the controls roughly and imposes unduly heavy loads on the structure. This trouble can be avoided by a careful selection of the deviations to which the stabilizer responds and of the gear ratios and other characteristics.

A stabilizer would be said to provide the maximum of stability when the least damped of all the free modes of the complete system comprising aircraft and stabilizer had a maximum value; this would necessarily occur when two modes had the same damping or at a point of bifurcation (see § 4·12). However, detailed studies of the response of stabilized aircraft to disturbances have shown that the greatest steadiness is *not* attained when the stability is at a maximum. The optimum stabilization is difficult to define in a manner which is both useful and precise on account of the great variety of disturbances which can occur. Guidance in selecting the best arrangement can be got from theoretical studies but extended experience in flight must be the final criterion. It is obvious that the stabilizer must be arranged to suit the aircraft. This need can usually be met satisfactorily by adjusting the parameters (mixing and gear ratios, etc.) of a stabilizer of given type.

Let x_n now denote some particular deviation and let η be the displacement of the control flap. Then η may be made to depend on x_n, $\dot{x}_n$, $\ddot{x}_n$ (if x_n is a displacement) and on $\int x_n dt$ or on a combination of these for two or more deviations x_n. This does not exhaust the possibilities. For instance, in what is called a rate/rate control $\dot{\eta}$ would be made proportional to $\dot{x}_n$ or to a linear combination of such quantities. Only a few of the extremely numerous possible combinations have been tried and it is possible that still better ones may yet be found.

As a simple example we may consider the stabilization of longitudinal-symmetric motion by the application without lag of an elevator angle η proportional to the angle of pitch θ. We shall suppose that the displacement of the elevator gives rise merely to a pitching moment in proportion to the displacement. The effect is to add a constant to the bottom right-hand element of the determinant in equation (5·2, 14). The co-factor of this element is a quadratic in λ so the equation for λ remains a quartic and the coefficients of λ^4 and λ^3 are unaltered. Hence the sum of the roots of the equation is also unaltered. Let the roots corresponding to the phugoid and short period oscillation be $\mu_1 \pm i\omega_1$ and $\mu_2 \pm i\omega_2$ respectively. Then by equation (5·2, 15) we have

$$2(\mu_1 + \mu_2) = -\mathbf{B} \qquad (15·2, 1)$$

where $\mathbf{B}$ is independent of the action of the stabilizer. Consequently the stabilizer can only increase the stability of one of the oscillations at the expense of reducing the stability of the other. However, the short period oscillation is usually heavily damped and an over-all improvement in stabilization is secured by transferring some damping from this to the phugoid oscillation which, at the best, is only slightly damped. The transfer of damping can be increased by increasing the gearing of the elevator control until the phugoid oscillation degenerates into a pair of subsidences (point of bifurcation). However, the best stabilization is obtained with a less powerful control.

The foregoing example shows that when the deflection of a flap is made proportional to an angular deviation the result is a redistribution of the dampings among the modes without any increase of the sum of the dampings. However, matters can be so arranged that this sum is increased when the flap deflection depends on the time rate of change of the angular deviation. Some degree of 'anticipation' is secured by making the flap respond to the second derivative or acceleration of the deviation.

A stabilizer may be employed to secure immunity from instability in some particular condition of flight, for instance in transonic flight. Clearly the stabilizer can only be effective so long as the control surface through which it operates itself remains effective. A stabilizer can also be used to make good a deficiency in some particular aerodynamic derivative. For instance, an automatic stabilizer operating on the rudder may be used to make good a deficiency in N_v.

15·3 Automatic pilots

The essential difference between automatic stabilization and automatic control is that with the latter the control by the human pilot is altogether eliminated. In principle a unit providing complete stabilization for both longitudinal and lateral motions is an automatic pilot but it might not be a satisfactory one. One essential feature in a satisfactory automatic pilot is freedom from slow wander of the datum whereas such a wander may be of no importance in a stabilizer. As an instance, an automatic stabilizer may render the lateral motion of an aircraft amply stable for control by a human pilot and yet allow a slow wander in azimuth if left entirely alone by the pilot. Evidently it would be necessary to provide the stabilizer here with some additional device to prevent the long-term change in the direction of flight. A device of this kind which stabilizes the datum is called a *monitor* and a magnetic compass could be used as a monitor for direction. We recognize then that automatic pilots require to be adequately monitored. As emphasized in § 1·2 the automatic pilot must render the aircraft completely stable in all the conditions of flight which arise.

Nearly all automatic pilots incorporate one or more gyroscopes. The gyroscope consists of a rotor in the form of a solid of revolution which can rotate freely about its axis of figure. The rotor is carried by the inner gimbal ring on bearings which are as nearly as possible frictionless and is provided with a pneumatic or electric drive to maintain a steady and high rate of rotation. The inner gimbal ring is freely pivoted to the outer gimbal ring which in turn is freely pivoted on a frame attached to the aircraft. The three axes—of rotor, inner gimbal and outer gimbal—are intersecting and are mutually perpendicular in the standard state of the instrument. Thus the axis of the rotor is free to take up any direction and, in the absence of friction and applied loads, the axis of a perfectly balanced rotor would maintain a fixed direction in space. This remains true when the instrument is placed in a uniform gravitational and accelerational field, provided that the c.g. of the rotor and of each gimbal ring lies at the point of intersection of the axes. However, since friction in the pivots cannot be altogether avoided, even the best gyroscope is subject to slow precession (wander of the direction of the rotor axis). Hence a monitor must be provided. There is, moreover, a second reason for providing a monitor and to make this clear let us consider a gyroscope whose rotor axis is horizontal and intended to maintain a constant geographical bearing. If the instrument were perfect the axis would maintain a fixed direction in space, but as the aircraft carrying it changed its position relative to the earth the bearing would, in general, change and the axis cease to be horizontal; the rotation of the earth also in general causes the apparent direction of the axis to change.

In order to provide a complete monitor for a gyroscope we must define *two* directions, which in practice are the vertical and a horizontal datum in azimuth. The latter can, as already mentioned, be provided by a compass and gravity defines the vertical. It must be remembered, however, that any device carried by the aircraft responds to the apparent gravity, i.e. to the vector resultant of gravity and the reversed acceleration. This difficulty is overcome in practice by making the monitoring couple extremely weak so that short-term deviations of apparent gravity from the true vertical have a negligible effect.

In order to control the direction of the gyroscope rotor axis the monitor must apply a couple about such an axis that the resulting precession reduces any directional error. This implies that the axis of the applied couple is perpendicular to the rotor axis and to the axis of the required precession.

As an illustrative example of a monitored gyroscopic stabilizer we may take the British Mark IA automatic elevator control. In this instrument the rotor axis lies fore-and-aft horizontally, the axis of the inner ring is horizontal and transverse while that of the outer ring is vertical, the pivot bearings being carried by a frame attached to the aircraft. The tilt of the inner ring relative to the frame is made to control the elevator by means of a servo motor. Gravitational monitoring is provided by the following device. A small weight is attached to the outer gimbal at the starboard side. So long as the axis of the outer ring is vertical the weight gives no couple about that axis but any tilt in the fore-and-aft plane gives rise to a couple about the axis and thus causes precession of the gyroscope about the transverse axis.* Hence the elevator is moved until the precession ceases when the axis of the outer ring is again vertical. The equilibrium tilt of the axis can be varied at will by applying a couple with a vertical axis to the outer ring by means of a spring whose tension can be adjusted.

* Note that this is perpendicular to the axes of the rotor and of the applied couple.

The steady setting is then such that the couples due to the weight and to the spring are in balance. In this manner the equilibrium attitude of the aircraft is under control.

A gyroscope constrained by a spring is a convenient means for obtaining a signal proportional to some component of angular velocity of the aircraft, since the couple is proportional to the rate of precession and may be measured by the strain of the spring.

As mentioned in § 15·2 there is an enormous variety of possible systems of automatic stabilization. However, many of the automatic pilots already used depend on the simple system of detecting the three angular deviations θ, ϕ and ψ in pitch, roll and yaw respectively. In some instruments one or more of the angular velocities or accelerations are used in addition. The British Mark VII automatic pilot dispensed with the θ control and depended on detecting u and $\dot{u}$. This was found to be unsatisfactory in service as the instrument tended to apply an excessive elevator angle in response to a horizontal gust.

15·4 Response calculations

We have very briefly considered response calculations in § 4·7 but we shall now give a list of integrals which greatly facilitate the application of impulsive admittances to such calculations.

It has been shown in § 4·6 that the response in any degree of freedom to an applied force $F(t)$ applied in the same or some other degree of freedom is given by the integral

$$\int_0^t F(\tau)\alpha(t-\tau)d\tau, \qquad (15\cdot4, 1)$$

where $\alpha(t)$ is the appropriate impulsive admittance and the system is in its datum steady state at the instant $t = 0$. Provided that the roots of the determinantal equation are all distinct, $\alpha(t)$ will be the sum of terms $a(t)$ where, corresponding to a real root μ,

$$a(t) = Ae^{\mu t}, \qquad (15\cdot4, 2)$$

and, corresponding to the complex pair of roots $\mu \pm i\omega$,

$$a(t) = Ae^{\mu t}\sin{(\omega t + \epsilon)}. \qquad (15\cdot4, 3)$$

Since
$$\sin\left(\omega t + \epsilon + \frac{\pi}{2}\right) = \cos\left(\omega t + \epsilon\right), \qquad (15\cdot4, 4)$$

it follows that the results for
$$a(t) = Ae^{\mu t}\cos\left(\omega t + \epsilon\right)$$

can be obtained from those corresponding to (15·4, 3) by substituting $\left(\epsilon + \frac{\pi}{2}\right)$ for ϵ throughout the formulae. The function

$$R(t) = \int_0^t F(\tau)a(t-\tau)d\tau, \qquad (15\cdot4, 5)$$

is worked out and listed for the functions $a(t)$ given by (15·4, 2) and (15·4, 3) and for a number of useful functions $F(t)$.

Note that for stable systems μ is always negative.

(I) *Real exponential term in the admittance.* The function $a(t)$ is given by (15·4, 2).

Case IA. $F(t) = f$, a constant, when t is positive.
$$R(t) = \frac{Af}{\mu}(e^{\mu t} - 1). \qquad (15\cdot4, 6)$$

Case IB. $F(t) = ft$ (linearly applied force).
$$R(t) = \frac{Af}{\mu^2}[e^{\mu t} - (1 + \mu t)]. \qquad (15\cdot4, 7)$$

Case IC. $F(t) = ft^m$ with m a positive integer.
$$R(t) = \frac{m!\,Af}{\mu^{m+1}}\left\{e^{\mu t} - \left[1 + \mu t + \frac{(\mu t)^2}{2!} + \dots + \frac{(\mu t)^m}{m!}\right]\right\} \quad (15\cdot4, 8)$$
$$= \frac{m!\,Af}{\mu^{m+1}}\left[\frac{(\mu t)^{m+1}}{(m+1)!} + \frac{(\mu t)^{m+2}}{(m+2)!} + \dots ad\ inf\right]. \quad (15\cdot4, 9)$$

Case ID. $F(t) = fe^{nt}$ (n real, $n \neq \mu$).
$$R(t) = \frac{Af}{(n-\mu)}(e^{nt} - e^{\mu t}). \qquad (15\cdot4, 10)$$

Case IE. $F(t) = f(1 - e^{-kt})$ (k real, $k \neq -\mu$).
$$R(t) = \frac{Af}{\mu(k+\mu)}[k(e^{\mu t} - 1) - \mu(1 - e^{-kt})]. \qquad (15\cdot4, 11)$$

Case I$_F$. $F(t) = f\sin(pt + \eta)$.

$$R(t) = \frac{Af}{\mu^2 + p^2}\{e^{\mu t}(\mu\sin\eta + p\cos\eta)$$
$$- [\mu\sin(pt+\eta) + p\cos(pt+\eta)]\}. \quad (15\cdot4, 12)$$

Case I$_G$. $F(t) = f\cos(pt + \eta)$.

$$R(t) = \frac{Af}{\mu^2 + p^2}\{e^{\mu t}(\mu\cos\eta - p\sin\eta)$$
$$- [\mu\cos(pt+\eta) - p\sin(pt+\eta)]\}. \quad (15\cdot4, 13)$$

(II) *Exponential sinusoidal term in the admittance.* The function $a(t)$ is given by equation (15·4, 3).

Case II$_A$. $F(t) = f$, a constant.

$$R(t) = \frac{Af}{\mu^2 + \omega^2}\{e^{\mu t}[\mu\sin(\omega t + \epsilon) - \omega\cos(\omega t + \epsilon)]$$
$$+ \omega\cos\epsilon - \mu\sin\epsilon\}. \quad (15\cdot4, 14)$$

Case II$_B$. $F(t) = ft$.

$$R(t) = \frac{Aft}{\mu^2 + \omega^2}(\omega\cos\epsilon - \mu\sin\epsilon)$$
$$+ \frac{Af}{(\mu^2 + \omega^2)^2}[2\mu\omega\cos\epsilon + (\omega^2 - \mu^2)\sin\epsilon]$$
$$- \frac{Afe^{\mu t}}{(\mu^2 + \omega^2)^2}[2\mu\omega\cos(\omega t + \epsilon) + (\omega^2 - \mu^2)\sin(\omega t + \epsilon)].$$
$$(15\cdot4, 15)$$

Case II$_C$. $F(t) = ft^m$. The function $R(t)$ can be worked out from the results given in § 103 of Volume I of *Treatise on the Integral Calculus*, by J. Edwards. See, however, the remarks at the end of this section.

Case II$_D$. $F(t) = fe^{nt}$ (n real).

$$R(t) = \frac{Af}{(\mu - n)^2 + \omega^2}[e^{nt}\{\omega\cos\epsilon - (\mu - n)\sin\epsilon\}$$
$$- e^{\mu t}\{\omega\cos(\omega t + \epsilon) - (\mu - n)\sin(\omega t + \epsilon)\}]. \quad (15\cdot4, 16)$$

Case IIE. $F(t) = f(1 - e^{-kt})$ (k real).

$$R(t) = \frac{Af}{\mu^2 + \omega^2}\{e^{\mu t}[\mu \sin(\omega t + \epsilon) - \omega \cos(\omega t + \epsilon)]$$

$$+ \omega \cos \epsilon - \mu \sin \epsilon\}$$

$$+ \frac{Af}{(\mu+k)^2 + \omega^2}\{e^{\mu t}[\omega \cos(\omega t + \epsilon) - (\mu + k)\sin(\omega t + \epsilon)]$$

$$- e^{-kt}[\omega \cos \epsilon - (\mu + k)\sin \epsilon]\}. \tag{15·4, 17}$$

Case IIF. $F(t) = f\sin(pt + \eta)$.

$$R(t) = \frac{1}{2}\frac{Af}{(\omega+p)^2 + \mu^2}\{e^{\mu t}[(\omega + p)\sin(\omega t + \epsilon - \eta)$$

$$+ \mu \cos(\omega t + \epsilon - \eta)]$$

$$+ (\omega + p)\sin(pt + \eta - \epsilon) - \mu \cos(pt + \eta - \epsilon)\}$$

$$- \frac{1}{2}\frac{Af}{(\omega-p)^2 + \mu^2}\{e^{\mu t}[(\omega - p)\sin(\omega t + \epsilon + \eta)$$

$$+ \mu \cos(\omega t + \epsilon + \eta)]$$

$$- (\omega - p)\sin(pt + \eta + \epsilon) - \mu \cos(pt + \eta + \epsilon)\}. \tag{15·4, 18}$$

The expression for $R(t)$ when $F(t) = f\cos(pt + \eta)$ can be obtained from (15·4, 18) on substituting $\frac{\pi}{2} + \eta$ for η.

When either or both of the functions $F(t)$ and $a(t)$ contain a sine or cosine function the expression for $R(t)$ can be written in a more concise form by the use of auxiliary symbols.

Take *Case* IIA by way of example and put

$$\mu = r\cos\phi, \tag{15·4, 19}$$

$$\omega = r\sin\phi, \tag{15·4, 20}$$

where r is to be positive, i.e.

$$r = + \sqrt{(\omega^2 + \mu^2)}. \tag{15·4, 21}$$

The angle ϕ is such that

$$\tan\phi = \frac{\omega}{\mu}, \tag{15·4, 22}$$

but this has two solutions lying between 0 and 2π, say ϕ_1 and $\phi_1 + \pi$. The particular solution must be chosen which gives μ

and ω in (15·4, 19) and (15·4, 20) respectively the correct signs, with r positive. We can now write equation (15·4, 14) as

$$R(t) = \frac{Af}{r}[e^{\mu t}\sin(\omega t + \epsilon - \phi) - \sin(\epsilon - \phi)]. \quad (15·4, 23)$$

With this same notation the solution (15·4, 15) for *Case* IIB can be written

$$R(t) = \frac{Af}{r^2}[e^{\mu t}\sin(\omega t + \epsilon - 2\phi) - \sin(\epsilon - 2\phi) - rt\sin(\epsilon - \phi).]$$

$$(15·4, 24)$$

These expressions also have the advantage that when the function $a(t)$ is $e^{\mu t}\cos(\omega t + \epsilon)$ we have merely to substitute cosines for sines throughout the formulae (15·4, 23) and (15·4, 24).

A similar treatment can be used in *Cases* IF, IG, IIC, IID, IIE and IIF.

15·5 Influence of curvature of the airstream: camber derivatives

The forces and moments on an aerofoil or hinged flap are modified by the curvature of the stream in which it is placed. This matter is of some importance since the flow induced by the vortex system of an aircraft is not in general rectilinear; it appears that the curvature of the induced flow may be of special consequence near the wing tips [see § 11·3(B)]. The effect is also of importance in relation to the calculation of wind tunnel corrections.* As a first approximation we may regard the curvature of the streamlines as uniform in the neighbourhood of any given section of the aerofoil. When the stream is curved the apparent angle of incidence varies along the chord and it is necessary to adopt a convention as to the position in the chord where the incidence is measured. We shall here adopt the usual convention of measuring the incidence at mid-chord and relative to the chord line. Then, for a stream of given velocity, density and temperature, the aerodynamic forces and moments per unit of span will be functions of the incidence α and the curvature σ of the streamlines.

* R. C. Pankhurst and H. H. Pearcey, 'Camber Derivatives and Two-dimensional Tunnel Interference at Maximum Lift', A.R.C. Current Papers No. 28 (1950).

So long as we are dealing with very thin aerofoils of small camber the effect of the curvature σ is the same as that of giving the aerofoil, supposed to be in a rectilinear stream, a contrary camber of the same numerical curvature. Clearly if we give the aerofoil mean line the same curvature as the stream the net effect will be zero and the influence of any departure from this neutral state will be the same as for rectilinear flow when the camber is small. Hence the influence of the curvature σ can be calculated by the theory of thin aerofoils. Suppose that the undisturbed streamlines are convex towards the upper surface of the aerofoil. Then the camber line will be given by the parabola

$$y = \tfrac{1}{2}\sigma x(c-x), \qquad (15\cdot5, 1)$$

which has constant curvature $-\sigma$ for $0 \leqslant x \leqslant c$ when σ is small.* It is convenient to measure the camber of the parabola by the non-dimensional quantity

$$\gamma = \frac{\text{maximum ordinate}}{\text{chord}} \qquad (15\cdot5, 2)$$

$$= \tfrac{1}{8}\sigma c. \qquad (15\cdot5, 3)$$

We now define the camber derivatives by the equations

$$a^1 = \frac{\partial C_L}{\partial \gamma}, \qquad (15\cdot5, 4)$$

$$m^1 = \frac{\partial C_m}{\partial \gamma}, \qquad (15\cdot5, 5)$$

$$b^1 = \frac{\partial C_H}{\partial \gamma}, \qquad (15\cdot5, 6)$$

where C_m refers to the quarter-chord axis. The theory of thin aerofoils (see § 7·2) leads to the values for infinite aspect ratio

$$a^1 = 4\pi, \qquad (15\cdot5, 7)$$

$$m^1 = -\pi, \qquad (15\cdot5, 8)$$

$$b^1 = -\frac{1}{E^2}\left[2(\pi-\phi)\cos\phi + \frac{3}{2}\sin\phi + \frac{1}{6}\sin 3\phi\right], \qquad (15\cdot5, 9)$$

* The difference between the ordinates of this parabola and of a circle of curvature $-\sigma$ passing through the leading and trailing edges is of second order when σ is taken as of first order.

where the last refers to a flap without nose balance and with sealed hinge. The derivatives are somewhat influenced by the thickness of the aerofoil and, more particularly, by the boundary layer but the further discussion of this is beyond our scope.

15·6 Buffeting

Buffeting is defined as the more or less irregular oscillation of part of an aircraft caused by the wake from some other part. It is also possible for a surface to be set in oscillation by its own wake and the term auto-buffeting has been applied to this phenomenon. The classical and most important case of buffeting is tail buffeting caused by the wake from the wings. In general buffeting is a nuisance and may limit the usable lift coefficient of an aircraft. Its only positive value lies in its serving as a warning to the pilot of the approach to the stall (see § 11·2).

The wake from a fully stalled wing is violently turbulent and any aerofoil surface immersed in it will be strongly buffeted. In the wake proper there is a depression of total head but it is found that severe buffeting may occur outside the region of reduced total head. However, it is true that the violence of the buffeting falls away both above and below the wake and ultimately becomes negligible. Since the wings of an aircraft in normal service never become fully stalled the kind of buffeting which is of most practical importance is pre-stall buffeting. If this is severe it certainly indicates bad aerodynamic design or faulty construction. Experience shows that pre-stall buffeting usually has its origin in a breakaway of flow at a wing root or at a wing-nacelle junction. The following are features which tend to promote buffeting:

(a) The occurrence of air channels at roots and junctions which flare outwards too steeply towards the rear. This can be avoided by suitable basic design and by the provision of suitable fillets. Surfaces intersecting at angles appreciably less than a right angle are to be avoided.

(b) The leakage into a wing root or junction of stagnant air or air whose velocity relative to the aircraft is small.

The writer knows of an instance where a particular example of a type of fighter aircraft, which is normally quite free from buffeting, suffered from severe pre-stall buffeting. This was ultimately traced to the bad fit of some of the cowling panels which permitted 'dead' air to leak into the wing-fuselage junction.

(c) Excrescences or roughnesses in the junctions.

Item (b) is of particular importance and does not always receive the attention it deserves.

Experiments have shown that, with the possible exception of the wake from fully stalled wings, there is no clearly predominant frequency of blow in buffeting. In other words, we have a continuous spectrum of frequency in the applied aerodynamic force. For this reason it is not possible for the structure to resonate with the aerodynamic forces, except possibly when the wings are fully stalled.

It has been found that the usable lift coefficient of aircraft flying at high Mach numbers is sometimes limited by buffeting to values considerably below those corresponding to the true stall. There is thus a 'buffeting boundary' in the diagram in which C_L is plotted against M. The only remedy for this would appear to be first-class aerodynamic design with special reference to wing roots, general aerodynamic cleanness and the avoidance of leaky surfaces.

15.7   Ground effect

The proximity of the ground to an aircraft, as in landing and take-off, influences the lift and drag and may have a very important effect on the pitching moment. These effects may conveniently be regarded as caused by the image of the aircraft in the ground plane. Behind the wings of the aircraft itself there is a region of downwash which becomes progressively less intense as the distance below the plane of the wings increases. The image is inverted and lies as much below the ground plane as the aircraft is above it. Hence the velocity at the aircraft induced there by the image is an upwash which, however, is comparatively weak on account of the remoteness of the image. The net result is that the downwash is reduced by an amount which increases as the aircraft approaches the ground, on account of the approach of the image to the aircraft. The reduction of the downwash has the following effects:

(a) For a given apparent angle of incidence the lift is increased. This is sometimes called the cushioning effect.

(b) For a given value of the lift the induced drag is reduced.

(c) The lift on the tail is increased, giving a nose-down pitching moment which must be corrected by an upward movement of the elevator.

(d) The wing wake and slipstream (if any) are raised relative to the tail and this may give rise to important changes of pitching moment.

The quantitative exploration of ground effect is most conveniently done by tests with models in a wind tunnel provided with a movable floor representing the ground. This does not provide a perfect representation of the full-scale conditions because of scale effect and because the boundary layer at the fixed floor does not correspond to the full-scale state of affairs.* When the particulars of the downwash field of the aircraft are known (or have been estimated) the modified field which exists in the presence of the ground can be obtained immediately by vector summation of the induced velocities for the aircraft itself and its inverted image. If results are required for a given value of the lift coefficient it must be kept in mind that this is influenced by the ground effect.

Ground effects are increased when flaps are put down and are at their maximum with the flaps in the setting for final landing, unless there is some unusual slipstream effect (see item (d) above).

15.8 Accelerated flight and the Froude number

The importance of acceleration in a set of similar systems in similar circumstances depends on the Froude number. Let l be a typical length of a member of the family of similar systems, V a typical velocity and a a typical acceleration. Then the Froude number is

$$F = \frac{V^2}{la} \qquad (15 \cdot 8, 1)$$

and it is easy to verify that this is non-dimensional. The classical instance of the Froude number is where a is identified with g, the acceleration in free fall under gravity. This arises inevitably in the theory of aircraft manœuvres as will be illustrated by some examples.

* This could be remedied by having a flexible floor stretched between rollers and run with the axial speed equal to the tunnel speed.

(a) *Circling flight* (see p. 19)

Here we identify l with the radius r of the horizontal circular path, so

$$F = \frac{V^2}{rg}.$$ (15·8, 2)

Accordingly the angle of bank ϕ (see equation (2·4, 3)) is given by

$$\tan \phi = F$$ (15·8, 3)

and the lift is given by

$$\frac{L}{W} = \sqrt{(1 + F^2)}.$$ (15·8, 4)

(b) *Phugoid oscillation of small amplitude* (see p. 31)

We now identify l with the wave length λ of the undulatory path and find that the Froude number is constant for a family of dynamically similar aircraft.

(c) *Flight in a non-uniform atmosphere* (see p. 17)

Here the effective Froude number is

$$F = \frac{V_e^2}{2g} \frac{d}{dh}\left(\frac{1}{\sigma}\right),$$ (15·8, 5)

where $\frac{d}{dh}\left(\frac{1}{\sigma}\right)$ plays the part of the reciprocal of the typical length. Then the equation (2·3, 4) for the thrust becomes

$$D + W(1 + F) \sin \Theta = T.$$ (15·8, 6)

When the influence of rate of change of flight speed on some phenomenon is under investigation it is convenient to invert the Froude number and use the non-dimensional acceleration number

$$A = \frac{al}{V^2}.$$ (15·8, 7)

This is significant for aeroelastic phenomena, such as flutter; for example, the critical flutter speed will depend on A. A valuable method for treating the stability of systems in accelerated motion has been developed by Collar.* While not completely general, this method is adequate for many practical applications.

* A. R. Collar, 'On the Stability of Accelerated Motion: Some Thoughts on Linear Differential Equations with Variable Coefficients', *Aero. Quarterly*, vol. VIII (1957), p. 309.

15.9   Instability in high speed rolling

In the argument given in § 3·9 demonstrating the independence of motions of the longitudinal-symmetric and lateral-antisymmetric types it is postulated that the initial motion of the
aircraft is one of steady rectilinear symmetric flight in equilibrium and without rotation. In general, when any of the postulated conditions are violated the two types of motion cease to
be independent. An important example is provided by the case
where the initial motion involves rapid rolling for, when the
angular velocity p is large, the equation of pitching moments
(3·4, 18) contains the gyroscopic term $(A - C)\,rp$ which couples
the motions in yaw and pitch and this is not of second order
when the initial value of p is large. Similarly the yawing
moment is influenced by the angular velocity in pitch through
the term $(B - A)\,pq$ in equation (3·4, 19). It has been shown by
Phillips* that these coupling effects give rise to instability in
certain circumstances. For simplicity neglect gravity and
damping actions and assume that the initial motion is rectilinear with constant angular velocity in roll p_0. Then there will
be a divergence, predominantly in pitch, if

$$\omega_\theta < p_0 < \omega_\psi \sqrt{\frac{C}{B-A}}, \qquad (15\text{·}9,\ 1)$$

or a divergence, predominantly in yaw, if

$$\omega_\theta > p_0 > \omega_\psi \sqrt{\frac{C}{B-A}}, \qquad (15\text{·}9,\ 2)$$

where ω_θ and ω_ψ are the angular frequencies in motions of pure
pitch and pure yaw, respectively, and it is assumed that
$C = A + B$.

The moment of inertia A about the longitudinal axis is much
smaller than B, the moment of inertia about the transverse
axis, for many modern high speed aircraft of small aspect ratio
and the gyroscopic moments are correspondingly large. The
inequalities (15·9, 1) and (15·9, 2) show that the critical rates of
roll become smaller as $(B - A)$ increases and as ω_θ and ω_ψ are

---

* W. H. Phillips, 'Effect of Steady Rolling on Longitudinal and Directional
Stability', *N.A.C.A. Tech. Note* 1627, 1948. W. J. G. Pinsker, 'Critical Flight
Conditions and Loads Resulting from Inertia Cross-coupling and Aero-
dynamic Stability Deficiencies', *A.R.C. Current Papers*, No. 404, 1958.

reduced, as happens when the longitudinal and lateral static stability margins become small.

15.10 Treatment of distortion by the superposition method

If we are given the distribution of the angle of incidence along a wing we can apply aerodynamic theory to obtain the lift per unit span and location of the centre of pressure at all points of the span. Then the twisting and bending moments can be computed and from these the contributions to the angle of incidence caused by distortion can be found. In this manner we work backwards from the resultant incidence (i.e. the sum of the initial incidence and that caused by distortion) to the initial incidence. On account of the linearity of the relationships, any number of solutions obtained in this manner may be superposed and the initial incidence at any wing section will be the sum of those for the individual cases. Suppose that α_{1r} is the total or resultant incidence in the rth solution, α_{0r} the corresponding initial incidence and F_r the corresponding value of some force, moment or stress. Then by the principle of superposition we have the following more general system of corresponding incidences and forces:

$$\left. \begin{aligned} \alpha_0 &= \Sigma k_r \alpha_{0r}, \\ \alpha_1 &= \Sigma k_r \alpha_{1r}, \\ F &= \Sigma k_r F_r. \end{aligned} \right\} \qquad (15\cdot10,\ 1)$$

If we can choose the arbitrary coefficients k so that α_0 is a good approximation to the actual initial distribution of incidence, then a satisfactory solution to the whole problem will be attained.* The simplest procedure is to use the method of *collocation* in which α_0 as given by (15·10, 1) is made to agree exactly with the true initial incidence at a number n of selected stations, where n is the number of basic solutions used (say 3). Other mathematical techniques of fitting, such as least squares, may also be adopted. This method has the advantage that the basic solutions can be obtained with the most refined

* R. B. Brown, K. F. Holtby and H. C. Martin, 'A Superposition Method for Calculating the Aeroelastic Behaviour of Swept Wings', *Jour. Aero. Sci.* vol. XVIII (1951), p. 531.

aerodynamic methods available and the final result will have an accuracy of the same order, provided of course that the elastic specification of the structure is sufficiently detailed and accurate.

15.11   Artificial 'feel'

When the controls of an aircraft are operated by the pilot, either manually or by foot, and without assistance by a servo-motor, the resistances to the movements of the controls and their general reaction to handling, which are collectively known as the 'feel' of the controls, are of great value to the pilot since they help to provide him with information about the behaviour of the aircraft. This information is sometimes fully appreciated by the pilot at the conscious level but very often unconsciously or with only partial consciousness. However, when the controls are power-operated, the reactions on the stick and pedals are independent of the aerodynamic reactions on the control surfaces, or nearly so, and the pilot is deprived of a valuable source of information. Attempts are made to remedy this deficiency by providing the pilot with artificial or 'synthetic' feel. Ideally, the artificial feel would reproduce the natural feel in all circumstances, with an appropriately chosen level of intensity. An approximation to this could be attained by using strain gauges on the gear directly operating the control flaps together with suitable electronic and electrical devices. However, less elaborate devices are commonly used. One fairly simple system provides a stick force proportional jointly to flap deflection and to $q = \frac{1}{2}\rho V^2$; this is called 'q feel'.

15.12   Importance of the derivative $M_{\dot{w}}$

It may appear somewhat arbitrary that, in dealing with longitudinal-symmetric motion on p. 91, only the derivative $M_{\dot{w}}$ among all the acceleration derivatives was explicitly retained in the equations of motion, although it was explained that many of these derivatives could be absorbed into the inertial coefficients of the aircraft. It is therefore not superfluous to show why $M_{\dot{w}}$ is a derivative of very considerable importance.

In the first place, it follows from the argument given on p. 214 that the effective damping coefficient for a pure pitching oscillation about the c.g. is $-(M_q + VM_{\dot{w}})$ and $M_{\dot{w}}$ may contribute importantly to the total. Again, it appears from equation (5·2, 16) on p. 129 that $M_{\dot{w}}$ is associated linearly with M_q in contributing to the coefficient B in the determinantal equation (5·2, 15), and this particular coefficient is the effective sum of all the damping coefficients.* In the second place, it can be shown that the presence of $M_{\dot{w}}$ is, in certain circumstances and from a certain viewpoint, equivalent to a shift of the C.G. of the aircraft. Thus in 'longitudinal' motion the normal acceleration at the C.G. is $(\dot{w} - qV)$ and when the angular velocity q is zero this reduces to $\dot{w}$. Now, on account of the properties of the centre of mass (identical with the C.G.), a uniform normal acceleration of the whole body does not give rise to any moment about the centre of mass. But when $M_{\dot{w}}$ is not zero there is a moment proportional to the acceleration $\dot{w}$ which would correspond to a certain fore-and-aft shift of the C.G.

GENERAL REFERENCES FOR CHAPTER 15

J. Whatham and H. M. Lyon, 'A Theoretical Investigation of Dynamic Stability with Free Elevators', *R. & M.* 1980 (1943).

W. S. Brown, 'The Longitudinal Stability of an Aircraft with Free Elevator', *R. & M.* 1981 (1943).

S. Neumark, 'The Disturbed Longitudinal Motion of an Uncontrolled Aircraft and of an Aircraft with Automatic Control', *R. & M.* 2078 (1943).

R. W. Gandy, 'The Response of an Aeroplane to the Application of Ailerons and Rudders, Part I, Response in Roll', *R. & M.* 1915 (1943).

J. Whatham and E. Priestley, 'Longitudinal Response Theory by the Method of the Laplace Transform', *R.A.E. Report No. Aero.* 2160, A.R.C. 10375 (1946).

---

* It is to be noted that the non-dimensional forms of the derivatives appear in (5·2, 16).

INDEX

Printed in the United States
By Bookmasters